The Rust Fungi

The Rust Fungi

Edited by

K. J. Scott and A. K. Chakravorty

Department of Biochemistry
University of Queensland
St. Lucia
Queensland
Australia

1982

ACADEMIC PRESS

A Subsidiary of Harcourt Brace Jovanovich, Publisher

London   New York
Paris   San Diego   San Francisco   São Paulo
Sydney   Tokyo   Toronto

ACADEMIC PRESS INC. (LONDON) LTD.
24–28 Oval Road
London NW1 7DX

U.S. Edition published by
ACADEMIC PRESS INC.
111 Fifth Avenue
New York, New York 10003

British Library Cataloguing in Publication Data

The Rust Fungi.
  1. Rusts (Fungi)
  I. Scott, K. J.   II. Chakravorty, A. K.
  589.2'25   QK627.A1

ISBN 0-12-633520-6

LCCCN 81-71584

Printed in Northern Ireland by
W. & G. Baird Ltd., Greystone Press, Antrim.

Contributors

Dr H. Buchenauer, *Institut für Pflanzenkrankheiten, Bonn, Fed. Rep. of Germany.*

Dr A. K. Chakravorty, *Department of Biochemistry, University of Queensland, St. Lucia, Queensland, Australia, 4067.*

Dr M. Heath, *Botany Department, University of Toronto, Toronto, Ontario, Canada.*

Dr Y. Hiratsuka, *Canadian Forest Service, Northern Forest Research Centre, Edmonton, Alberta, Canada.*

Dr R. A. McIntosh, *Plant Breeding Institute, The University of Sydney, Castle Hill, Australia.*

Dr D. J. Maclean, *Department of Biochemistry, University of Queensland, St. Lucia, Queensland, St. Lucia, Queensland, Australia, 4067.*

Prof. S. Sato, *Faculty of Agriculture and Forestry, Tsukuba University, Ibaraki-ken, Japan.*

Prof. K. J. Scott, *Department of Biochemistry, University of Queensland, St. Lucia, Queensland, Australia, 4067.*

Prof. I. A. Watson, *Plant Breeding Institute, The University of Sydney, Castle Hill, Australia.*

Preface

The rust fungi, comprising about 4600 species (and which constitute the whole order uredinales), are probably the cause of more crop losses than any other group of plant pathogens. These fungi are of great interest not only for the economic problems that they cause, but also for their highly specialized relationships with host plants. This specialization is seen in interesting and complex life cycles, a mode of nutrition which makes it very difficult to culture them away from their hosts, and an extreme form of host specificity. A complete understanding of this specificity must await identification of the specific gene products of host and parasite which are responsible for compatability and incompatibility, and critical studies on the synthesis of and interaction between these molecules. Until the last decade or so these fungi were considered classical examples of obligate fungal parasites. However, in the intervening time some species have been cultured on laboratory media, which in turn has allowed us the opportunity to observe the metabolism and somatic genetics of these organisms. The challenge to culture other, more recalcitrant members of this group of fungi still remains.

This book represents the first comprehensive account of the rusts since 1946, and it attempts to bring together current information on these fungi from diverse fields of study, and to present possible directions future investigations may take. Aspects covered include taxonomy and morphology as well as culture and metabolism. The genetics and biochemistry of host/parasite relationships are discussed along with host defense mechanisms, and biological and chemical control. Because of the wide range of topics covered, it is hoped that the book will prove a useful reference and stimulus for future research to a diverse spectrum of readers, from students in plant pathology and mycology, through to academics and research workers in these and allied fields.

The contributors to this book have been selected from a wide variety of disciplines, and are all internationally recognized in their own fields. No attempt has been made to standardize the style or to alter the interpretation of data. Although a few authors were a little slow in presenting their chapter and this necessitated some updating by other contributors, this is now in the past, and the editors express their sincere appreciation to the authors for their time and effort. Finally, we would like to acknowledge our apprecia-

tion for the technical and clerical assistance provided by the Department of
Biochemistry, University of Queensland.

May, 1982 K. J. Scott and A. K. Chakravorty

Contents

1. Morphology and Taxonomy of Rust Fungi

Y. HIRATSUKA[1] and S. SATO[2]

[1]Northern Forest Research Centre, Canadian Forestry Service, Edmonton, Alberta, Canada and
[2]Institute of Agriculture and Forestry, University of Tsukuba, Ibaraki-ken, Japan

CONTENTS

I. Introduction

Rust fungi, or fungi belonging to the order Uredinales, constitute one of the largest groups in Basidiomycetes. About 5000–6000 species have been recognized, and about 300 generic names have been proposed (Laundon, 1965) of which between 100–125 genera are recognized as "good" genera (Cummins, 1959; Thirumalachar and Mundkur, 1949, 1950; Hiratsuka, 1955). The majority of species in temperate regions of the northern hemisphere, Australia and New Zealand have been well catalogued, but many new genera and species are still expected to be found in tropical and subtropical regions such as South America, Africa and Southeastern Asia.

Many species of rust fungi are the cause of internationally important plant diseases. Some examples are wheat stem rust (*Puccinia graminis* Pers.), wheat leaf rust (*Puccinia recondita* Rob. ex Desm.), white pine blister rust (*Cronartium ribicola* J. C. Fischer ex Rahb.), corn leaf rusts (*Puccinia sorghi* Schw. and *Puccinia polysora* Underw.), flax rust (*Melampsora lini* (Ehrenb. Lév.), and coffee leaf rust (*Hemileia vastatrix* Berk. and Br.).

One of the unique features of rust fungi is that they have up to six functionally and morphologically different spore states in their life cycles. This is further complicated because, in addition to the different numbers of spore states, they often need two unrelated groups of host plants to complete their life cycles (heteroecious life cycle). Some can complete their life cycles on only one kind of host plant (autoecious life cycle).

Hosts of rust fungi range from ferns and gymnosperms to highly evolved families of dicotyledons such as Leguminosae, Euphorbiaceae and Compositae, and families of monocotyledons such as Liliaceae and Orchidaceae. Several genera of fungi such as *Jola*, *Herpobasidium* and *Eocronartium* that parasitize mosses and ferns previously were classified in Tremellales but are also considered rust fungi by some authors (Jackson, 1935; Hennen, personal communication). Also, one species of rust fungus has been reported for the first time on a plant of the palm family (Palmae) (*Cerradoa palmaea* Hennen and Ono on *Attalea ceraensis* Barb.-Rodr.) from Brazil (Hennen and Ono, 1978). There are definite close relationships between rusts and their host plants. Often identification of the host species is the most important first step in identification of a rust. Moreover, host–rust relationships often give useful information for the phylogeny and classification of higher plants (Savile, 1979).

In recent years many species of rusts have been cultured successfully on artificial media (Williams *et al.*, 1966, 1967; Harvey and Grasham, 1970; Turel, 1969; Coffey and Shaw, 1972; Coffey *et al.*, 1970). In natural ecosystems, however, rust fungi appear to survive only as obligate parasites of living plants.

It is very important, therefore, to have a good understanding of morphological features, life cycles, and host–rust relationships in the study of rust fungi as important plant pathogens or interesting biological agents.

II. Morphology

A. *Definition and Terminology of Spore States*

The basic terminology for the spore states of rust fungi was proposed by de Bary, Tulasne, and others in the middle of the last century and modified by many uredinologists since then (Arthur, 1905, 1925; Azbukina, 1970; Cunningham, 1930; Cummins, 1959; Hiratsuka, 1973b, 1975; Laundon, 1967, 1972; Savile, 1968). Several names have been proposed for each spore state. Major variations for each of the spore states are as follows: spermatia (in spermogonia), pycniospores (in pycnia), pycnidiospores (in pycnidia); aeciospores (in aecia), aecidiospores (in aecidia); urediniospores (in uredinia), urediospores (in uredia or uredosori); teliospores (in telia), teleutospores (in teleutosori); basidiospores (on basidia), and sporidia (on promycelia). The first names listed are the most accepted and are designated often in Roman numerals as O, I, II, III and IV, respectively.

Not only are several different terms used for one spore state, but two different bases have been applied in the definitions and terminology of spore states. In certain cases, therefore, the same spore state has been interpreted differently and called by different names. The two systems are the morphologic and the ontogenic. They have been used separately or sometimes mixed either consciously or unconsciously. Using illogical criteria or mixing the two different systems in defining and naming spore states creates confusion and therefore is undesirable.

The morphologic system emphasizes the morphology of spores as the basis for defining spore states. This concept has been followed by early European mycologists and recently was defended by Laundon (1967, 1972) and Holm (1973). In this system, for example, aeciospores are defined as "spores produced in chains with characteristic 'verrucose' spore ornamentation" (Laundon, 1972), and urediniospores are defined as "pedicellate spores with characteristic 'echinulate' ornamentation" (Laundon, 1967, 1972). On the other hand, the ontogenic system that was proposed by Cummins (1959) and defended by Hiratsuka (1973b, 1975) emphasizes positions of the spore states in the life cycle rather than clearly recognizable morphological entities. Accordingly, "a sorus may look like an aecial cup and spores may have typical 'verrucose' ornamentaiton but actually be an uredium (uredinium) or a telium" (Cummins, 1959). Definitions of spore states according to the ontogenic system proposed by Hiratsuka (1973b) are as follows:

Spermatia are haploid gametes.
Aeciospores are nonrepeating vegetative spores produced usually from the

result of dikaryotization and germinate to initiate dikaryotic mycelium, thus
usually associated with spermogonia.
Urediniospores are repeating vegetative spores produced usually on dikaryotic
mycelium.
Teliospores are basidia producing spores.
Basidiospores are haploid spores produced on basidia usually as the result of
meiosis.

Terminologies in either system are the same for many typical species such
as *Puccinia graminis*, *Cronartium ribicola* and *Melampsora lini* but differ for
a considerable number of genera and species. For example, urediniospores
(by the ontogenic system) of *Chrysomyxa* and *Coleosporium*, which are
produced on nonconiferous alternate hosts, are produced in chains and have
verrucose spore ornamentation and so have to be called aeciospores by the
morphologic system. Also, so much variation exists in the spore morphology
of teliospores (see B. 4. below) that one cannot define them morphologically. The only reasonable definition of teliospore is that "the spore gives
rise to basidia upon germination", which is an ontogenic definition. We
think that the ontogenic system is more logical and creates less confusion
than the morphologic system. We therefore will use terminology and definitions of spore states based on the ontogenic system.

B. *Morphology of Spore States*

1. *Spermatia*

Spermatia are small, hyaline, single-celled spores produced in spermogonia
(singular—spermogonium). They are contained in sugar-rich nectar that
oozes out as droplets from spermogonia. The infectious or sexual function of
spermatia was not known until Craigie (1927) conclusively demonstrated
with *Puccinia graminia* and *P. helianthi* that spermatia are functional as
gametes in heterothalic rusts. Nectar containing spermatia was found to
attract many kinds of insects and is considered to be important in cross-
fertilization of haploid pustules of the opposite mating type (Buller, 1950).
Hiratsuka and Cummins (1963) studied the morphology of spermogonia of
68 genera and recognized 11 morphological types. They emphasized the
significance of spermogonial morphology to rust taxonomy. Recently,
Hiratsuka and Hiratsuka (1980) expanded the work, added a new type (type
12), and classified morphological types into six groups (Fig. 1, Table I).
Major types are shown in Figs 2 to 11.
  Spermatia are produced in basipetal succession at the apex of sperma-
tiophores and are considered to be phialospores (Olive, 1944; Hughes,

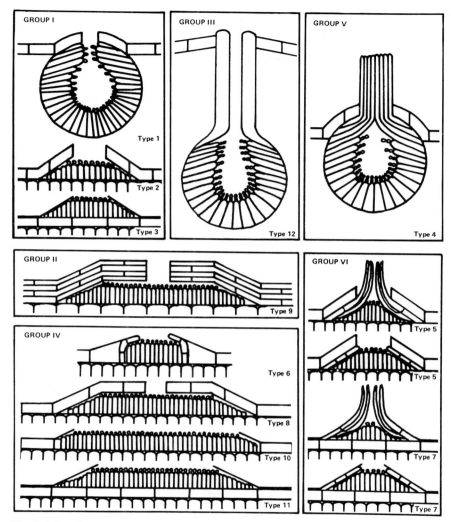

Fig. 1. Morphological types of spermogonia and six groups. (From Hiratsuka and Hiratsuka, 1980.)

1970). This was confirmed by an electron microscope study of spermogonia of *Gymnosporangium juniperi-virginianae* by Mims *et al.* (1976). Rijikenberg and Truter (1974b) and Harder and Chong (1978), however, concluded that spermatia of *Puccinia sorghi* and *P. coronata*, respectively, are annelospores rather than phialospores, although figures in these papers suggest that

Table I. Summary of six groups of spermogonia types (Hiratsuka and Hiratsuka, 1980).

Groups	Types of spermogonia	Representative genera	Type of aecia	Life cycle	Remarks
I	1, 2, 3	Milesina, Uredinopsis, Hyalopsora, Chrysomyxa, Melampsora, Coleosporium, Melampsoridium, Puccinastrum, Melampsorella	peridermioid (caeomoid)*	heteroecious (autoecious)	*Melampsora only 0·I on conifers except Melampsora (p.p.)
II	9	Cronartium	peridermioid	heteroecious	0·I on Pinus
III	12	Mikronegeria, Caeoma (p.p.)	caeomoid	heteroecious	0·I on tropical conifers
IV	6, 8, 10, 11	Gerwasia, Gymnoconia, Hamaspora, Frommea, Phragmidium, Xenodochus, Kuehneola, Triphragmium, Kunkelia, Teleconia	uredinoid caeomoid	autoecious	All on Rosaceae except few species of Kuehneola
V	4	Puccinia, Uromyces, Gymnosporangium, Miyagia, Cumminsiella, Baeodromus, Cionothrix, Chrysopsora, Chrysocelis, Chardoniella, Maravalia, Didymopsora, Polioma, Zaghouania, Gambleola, Pucciniosira, Endophyllum	aecidioid (uredinoid)	heteroecious autoecious	
VI	5, 7	Achrotelium, Arthuria, Cerotelium, Chaconia, Crossopsora, Dasturella, Diabole, Dicheirinia, Diorchidium, Hapalophragmium, Lipocystis, Masseeella, Ochropsora, Olivea, Phakopsora, Phragmidiella, Physopella, Pileolaria, Pucciniostele, Ravenelia, Scopella, Sorataea, Spumula, Tegillum, Tranzschelia, Uropyxis, Cystomyces, Dasyspora, Didymosporella, Poliotelium, Skierka, Uromycladium	uredinoid (caeomoid) (aecidioid)	autoecious (heteroecious)*	*Some species of Cerotelium, Ochropsora and Tranzschelia are heteroecious

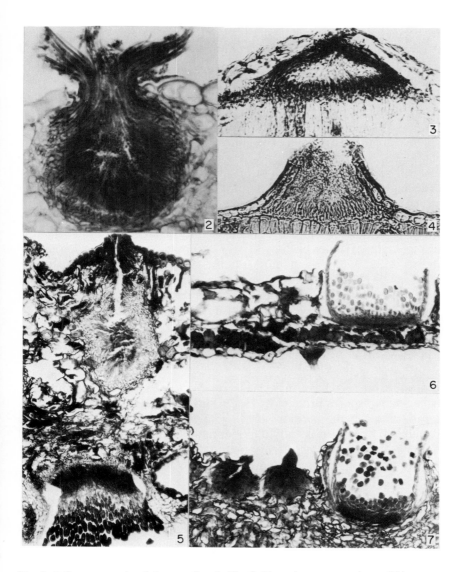

Figs 2–7. Spermogonia of the rust fungi. *Fig. 2:* Type 4 spermogonium of *Uromyces euphorbiae* Cke. and Pk. (× 300). *Fig. 3:* Type 5 spermogonium of *Dasyspora gregaria* (Kunge) P. Henn. (× 300). *Fig. 4:* Type 5 spermogonium of *Cystomyces costaricensis* Syd. (× 300). *Fig. 5:* Type 12 spermogonia and caeomoid aecium of *Micronegeria alba* Oehrens and Peterson (× 65). *Fig. 6:* Type 7 spermogonium and aecidioid aecium of *Phakopsora ampelopsidis* Diet. et Syd. (× 65). *Fig. 7:* Type 4 spermogonia and aecidioid aecium of *Puccinia coronata* Rob. ex Desm. (× 65). (Figs 2–4 from Hiratsuka and Cummins, 1963.)

the spermatia are also phialospores. It is interesting to note that, apart from in the rusts, phialospores are known only in the Ascomycetes, as was pointed out by Hughes (1970).

Spermogonia of many species are known to have "flexuous hyphae", and the flexuous hyphae are considered to function as receptive hyphae or trichogynes (Buller, 1950). Savile (1939) questioned the distinction between stiff and tapered ostiolar periphysis and the thin-walled flexuous hyphae recognized by Buller. The presence of the receptive hyphae with the spermogonia is the main reason why some uredinologists prefer not to use the term spermogonia, which suggests the structures are male gamete-producing organs. Savile (1976) considered the organ analogous to the entomophylous perfect flower of higher plants because of the presence of receptive hyphae and the production of nectar. Others think, however, that hyphae protruding from stomata in the vicinity of aecia primordia (the real receptive organs) may be the primary receptive structure (Allen, 1934; Andrus, 1931, 1933; Wang and Martens, 1939; Hiratsuka and Hiratsuka, 1980). Savile (1939) also recognized these hyphae often, especially in moist greenhouse conditions, but concluded that they could seldom function in nature because they were pinched off by closing of stomata.

2. Aeciospores

Aeciospores are produced in aecia (singular—aecium). They are nonrepeating vegetative spores produced as the result of dikaryotization and usually are associated with spermogonia (Figs 5, 6, 7). Aecia traditionally are divided into five types: aecidioid, peridermioid, roesterlioid, caeomoid and unredinoid. The five types correspond to the morphology of the imperfect genera *Aecidium, Peridermium, Roestelium, Caeoma* and *Uredo*, respectively. The first three types of aecia (aecidioid, peridermioid and roestelioid) have surrounding wall structures called peridia, which are made up of specialized spores (Fig. 12). Although these aecia are generally reasonably distinct in morphology, they are difficult to define and separate purely on morphological bases. To avoid confusion, by "gentleman's agreement" (Cummins, 1959) the three types have been defined as follows: peridermioid aecia are peridiate aecia produced on coniferous hosts; roestelioid aecia are aecia of the genus *Gymnosporangium*; and the rest of aecia having peridia are called aecidioid aecia.

Caeomoid aecia have catenulate aeciospores without peridia (Figs 5, 13) but sometimes a clear distinction from peridiate aecia is difficult. For example, aecia of *Melampsora* species on conifers (*M. epitea, M. medusae, M. larici-populina,* etc.) have rudimentary but definite peridial cells. Also,

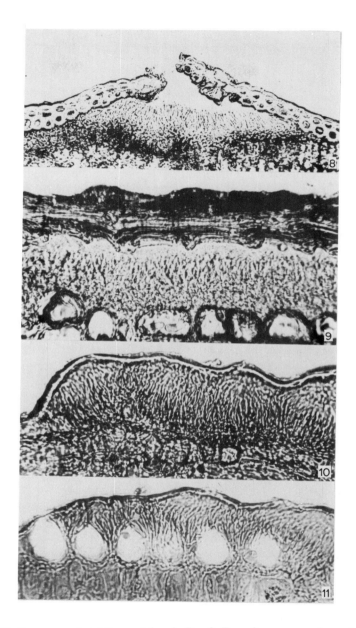

Figs 8–11. Spermogonia of the rust fungi. *Fig. 8:* Type 2 spermogonium of *Coleo-sporium jonesii. Fig. 9:* Type 9 spermogonium of *Cronartium ribicola. Fig. 10:* Type 10 spermogonium of *Phragmidium barnardi. Fig. 11:* Type 11 spermogonium of *Phragmidium sanguisorbae* (all × 430). (From Hiratsuka and Cummins, 1963.)

caeomoid aecia of such genera as *Phragmidium* and *Xenodocus* have peripheral paraphysis (Fig. 14). Uredinoid aecia (Fig. 16) have pedicellate spores similar to many urediniospores. They are often called primary uredinia or aecial uredinia.

In addition to the above-mentioned types of aecia, most of the major reference works describe *Dasyspora gregaria* (Kunze) P. Henn. as having unique hyphoid aecia (Fig. 15); however, Hennen (personal communication, 1980) recently concluded that the structures do not belong to the rust but to the sporangia of an alga present with the rust.

Most aeciospores, except uredinoid aeciospores and few unique types, are catenulate (Figs 12–14) and are characterized by verrucose spore ornamentation (Figs 17–20). Hughes (1970) considered common aeciospores to be catenulate meristem arthrospores. It is interesting that meristem arthrospore production is found only in the imperfect states of several Ascomycetes besides rusts (Hughes, 1970).

Morphogenesis of spores and spore surface ornamentation of typical

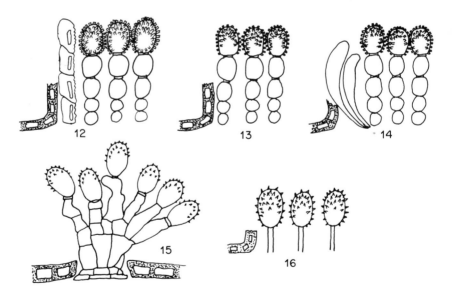

Figs 12–16. Morphological types of aecia (schematic). *Fig. 12:* Aecidium with catenulate aeciospores with peridium. (*Aecidium, Peridermium, Roestelium*). *Fig. 13:* Aecidium with catenulate aeciospores, without bounding structure (*Caeoma*). *Fig. 14:* Aecidium with peripheral paraphyses. *Fig. 15:* Hyphoid aecium of *Dasyspora gregaria* (redrawn from Cummins, 1959), but may represent a structure of an alga (Hennen, personal communication, 1980). *Fig. 16:* Uredinoid aecium (*Uredo* like aecium) with pedicellate spores.

Figs 17–20. Aeciospores of the rust fungi. SEM. *Fig. 17:* Aeciospore of *Puccinia graminis* Pers. on *Berberis thunbergii* D. C. (× 4800). *Fig. 18:* Aeciospore of *Puccinia coronata* Corda on *Berchemia ramosa* Sieb. et Zucc. (× 4800). *Fig. 19:* Aeciospore of *Cronartium comandrae* Pk. on *Pinus contorta* Dougl. (× 2100). *Fig. 20:* Aeciospore of *Cronartium coleosporioides* Arth. on *Pinus contorta* Dougl. (× 2300).

aeciospores are illustrated well by Holm and Tibell (1974) (*Puccinia graminis*) and Rijikenberg and Truter (1974a) (*Puccinia sorghi*). According to their observations, electron transparent plates that are to become warts are laid next to the plasma membrane and under the thin primary walls when the spores are young. The space between the plates is filled with electron dense material, which a few authors call the second primary wall (PW-2). After the warts are fully developed, the secondary walls develop under the warts as primary walls disappear to uncover the warts. The matrix between the warts disappears, exposing the warts to the surface. The warts were often found to have a layered configuration, with up to seven or eight layers as in the species of *Cronartium* (Hiratsuka, 1971) (Fig. 20). Sometimes each layer has a characteristic shape and size, as in *Chrysomyxa ledicola*, while other species have no obvious layers, as on the warts of *Cronartium comandrae* (Fig. 19). Smooth spots on the aeciospores of some species such as *Cronartium coleosporiodes* (Fig. 20) are made up of undivided wart layers. Besides regular warts, many aeciospores have larger ornamentation called pore plugs or refractive granules, which were described and named by Dodge (1924). They are found on aeciospores of many *Puccinia* species such as *Puccinia graminis* (Fig. 17), *P. caricina* and *P. mitriformis* but not on species such as *P. recondita*, *P. coronata* (Fig. 18) and *P. andropogonis* or species of many other genera (Figs 19, 20). These structures have been variously called refractive granules (Holm, 1963) and platelets (Plattchen) (Klebahn, 1912–14). Savile (1973) recognized five types of aeciospores among *Puccinia* and *Uromyces* species attacking Cyperaceae, Juncaceae and Poaceae based on the presence and absence of pore plugs and their distribution on aeciospores. He considered them to be valuable taxonomic characters.

3. Urediniospores

Urediniospores are produced in uredinia (singular—uredinium). Urediniospores are defined as repeating vegetative spores produced usually on dikaryotic mycelium and are repeatedly produced on a host several times during the growing season of the host. Because of this, the uredinial state is the most destructive spore state of such rusts as wheat stem rust (*Puccinia graminis*), coffee leaf rust (*Hemileia vastatrix*) and poplar leaf rust (*Melampsora medusae*).

Kenny (1970) recognized 14 different morphological types of uredinia among rust fungi based on such morphological characteristics as the presence or absence of bounding structures, position of the hymenium in the host tissue, and growth pattern of the hymenium. Several representative types are illustrated in Figs 21–29.

Uredinia of *Melampsoridium* are surrounded by peridia with pointed ostiolar cells (Fig. 21). Uredinia of such genera as *Pucciniastrum*, *Milesina*, *Uredinopsis*, *Hyalopsora*, *Melampsorella*, *Uropyxis* and *Achrotelium* also have peridia; their cells are simple or sometimes are composed of specialized cells (Fig. 22). Uredinia of *Phakopsora*, *Dicheirinia*, *Prospodium*, *Hamaspora*, *Crossopsora*, *Cerotelium*, *Dasturella* and several other genera have basally united and septated paraphyses (Fig. 23). Uredinia of many genera

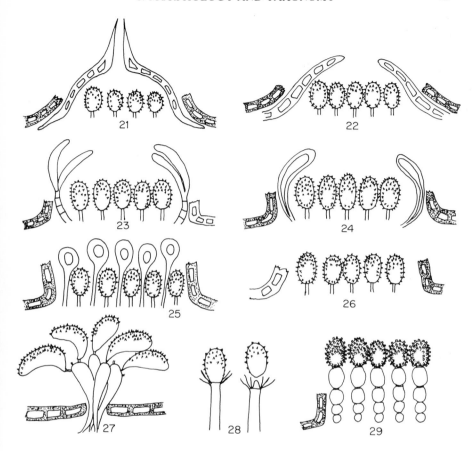

Figs 21–29. Morphological types of uredinia (schematic). *Fig. 21:* Uredinium with peridium, pointed ostiolar cells, and pedicellate urediniospores. *Fig. 22:* Uredinium with peridium, less developed ostiolar cells and pedicellate urediniospores. *Fig. 23:* Uredinium with basally united and septate peripheral paraphyses, and pedicellate urediniospores. *Fig. 24:* Uredinium with peripheral paraphyses and pedicellate urediniospores. *Fig. 25:* Uredinium with intrasoral paraphyses and pedicellate urediniospores. *Fig. 26:* Uredinium without bounding structure and with pedicellate urediniospores. *Fig. 27:* Superstomatal uredinium of *Hemileia* sp. *Fig. 28:* Urediniospores produced by successive proliferation at the tips of sporophores (*Intrapes paliformis*). *Fig. 29:* Uredinium with calenulate urediniospores.

have peripheral paraphyses (Fig. 24) either with uniformly thin paraphysal walls (*Phragmidium, Hamaspora, Frommea, Kuhneola* and *Triphragmium*) or with dorsally thickened paraphysal walls (*Uropyxis, Chaconia, Phragmopyxis* and *Diorchidium*).

Many species of *Ravenelia* and *Melampsora* have capitate paraphyses, which are scattered among the uredinia (intrasoral paraphyses) (Fig. 25). Uredinia of many species of *Puccinia* and species of several other genera (*Pileolaria, Scopella, Zaghouania, Cumminsiella* and *Polioma*) have no accessory structure (Fig. 26). Uredinia of *Hemileia, Desmella* and *Blastospora* are produced on sporophores that protrude from stomata (Fig. 27). Urediniospores of *Chrysomyxa* and *Coleosporium* are produced in chains and have verrucose spore ornamentation similar to their aeciospores (Fig. 29). Several species of *Puccinia* and other genera also have uredinia similar to the aecial states of the same species (aecial uredinia or repeating aecia).

Urediniospores of *Kernkampella breyniae-patensis* (Mund. and Thirum.) Rajendren (Rajendren, 1970) and the newly described imperfect rust *Intrapes paliformis* Hennen and Figueiredo (Hennen and Figueiredo, 1979) are produced in a unique manner. Their spores are produced by successive proliferation at the tips of sporophores, and the lower parts of the pedicels of previously produced spores form collars around the bases of the newly formed spores (Fig. 28).

The majority of the urediniospores are produced singly on pedicels and have characteristic echinulate spore surface ornamentation (Figs 30, 32). Urediniospores of such genera as *Chrysomyxa* and *Coleosporium*, however, are produced in chains and have verrucose surface ornamentation (Fig. 33) similar to aeciospores of the same species and of many other genera. Many variations of surface ornamentation are known (Fig. 31). Also, occasional two-celled urediniospores have been reported in *Gymnosporangium gaumannii* (Hiratsuka, 1973a).

Some species such as *Uredinopsis* spp., *Hyalopsora polypodii*, *Puccinia vexans* and *Puccinia atrofusca* produce modified urediniospores called amphispores in addition to regular urediniospores. Amphispores have thicker cell walls than ordinary urediniospores and usually have darker pigmentation in the cell walls. They probably are capable of enduring an extended period of unfavorable climatic conditions.

Morphogenesis of typical echinulate urediniospores has been well illustrated and documented with electron micrographs (Littlefield and Bracker, 1971—*Melampsora lini*; Harder, 1976c—*Puccinia coronata* and *Puccinia graminis*; Ehrlich and Ehrlich, 1969—*Puccinia graminis*). These studies indicate that small conical electron transparent spines are first initiated below the primary wall next to the plasma membrance in invaginated pockets. The spines are pushed upward as they increase in size and gradually are exposed to the surface by the progressive thickening of an inner secondary wall beneath the spines and the disintegration of the primary wall.

Germ pores of urediniospores are of good taxonomic character and can be divided into three groups depending on distribution: scattered, bizonate

Figs 30–33. Urediniospores of the rust fungi. SEM. *Fig. 30:* Urediniospore of *Puccinia coronata* Corda on *Avena sativa* L. (× 3300). *Fig. 31:* Urediniospore of *Pileolaria shiraiana* (Diet. et Syd.) S. Ito on *Rhus trichocarpa* Miquel. (× 1600). *Fig. 32:* Urediniospore of *Uromyces polygoni-aviculariae* (Pers.) Karst. on *Polygonum aviculare* L. (× 3300). *Fig. 33:* Urediniospore of *Coleosporium eupatorii* Arth. ex *Cumm.* on *Euputorium fortunei* Turcz. var. *simplicifolium* Nakai. (× 3700).

and equatorial (Cummins, 1936). Although the three types of germ pore distribution are seen in various groups of rusts and no clear subdivision of the rusts can be made on the basis of this character, this character is consistent at a specific level. Cummins (1936) concluded that the presence of numerous

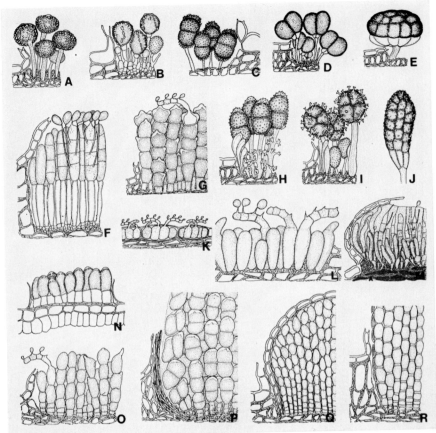

Fig. 34. Morphological variations of telial states of Uredinales. A. *Pileolaria brevipes*
Berk. and Rav. B *Trachyspora intrusa* (Grev.) Arth. C. *Dasyspora gregaria* (Kunze)
P. Henn. D. *Diorchidiella australe* (Speg.) Lindq. E. *Ravenelia mera* Cumm. F.
Chrysella mikaniae Syd. G. *Kuehneola uredinis* (Lk.) Arth. H. *Prospodium appendi-
culatum* (Wint.) Arth. I. *Sphaerophragmium acaciae* (Cke.) Magn. J. *Cumminsina
clavispora* Petr. K. *Melampsorella symphyti* Bubak. L. *Chrysocelis lupini* Lagh. and
Diet. M. *Goplana dioscoreae* Cumm. N. *Lipocystis caesalpiniae* (Arth.) Cumm. O.
Phragmidiella markhamiae P. Henn. P. *Didymopsora africana* Cumm. Q. *Dietelia
verruciformis* (P. Henn.) P. Henn. R. *Endophylloides portoricensis* Whet. and
Olive. (Reproduced with permission from Cummins, 1959, from Hiratsuka, 1973.)

scattered pores was a sign of a primitive condition in the rust fungi that was
found in rusts on more primitive host plants. The equatorial distribution of
pores represents an advanced condition that is found more on highly evolved
plant groups.

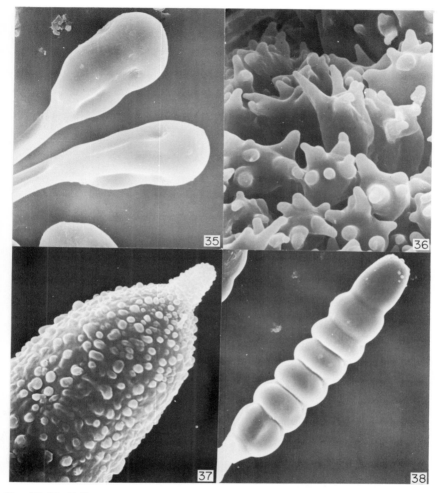

Figs 35–38. Teliospores of the rust fungi. SEM. *Fig. 35:* Teliospore of *Puccinia graminis* Pers. on *Agrostis clavata* Trin. (× 1500). *Fig. 36:* Teliospores of *Puccinia coronata* Corda on *Calamagrostis arundinacea* Roth. var. *genuina* Hack. (× 1500). *Fig. 37:* Teliospore of *Phragmidium rosae-multiflorae* Diet. on *Rosa polyantha* Sieb. et Zucc. var. *genuina* Nakai. (× 3300). *Fig. 38:* Teliospore of *Xenodocus carbonarius* Schlech. on *Sanguisorba officinalis* L. (× 800).

4. Teliospores

Teliospores are basidia-producing spores, and sori that produce teliospores are called telia (singular—telium). Many morphological variations exist in

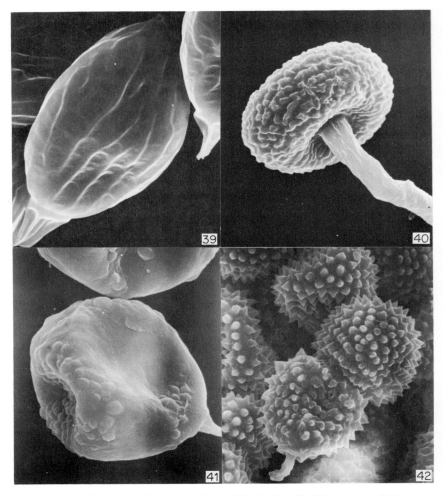

Figs 39–42. Teliospores of the rust fungi. SEM. *Fig. 39:* Teliospore of *Uromyces erythronii* Passer. on *Erythronium japonium* Mak. (× 2300). *Fig. 40:* Teliospore of *Pileolaria kluqkistiana* Diet on *Rhus javanica* L. (× 1750). *Fig. 41:* Teliospore of *Triphragmidium ulmariae* Link on *Filipendula kamtshatica* Maxim. (× 1500). *Fig. 42:* Teliospore of *Tranzschelia* sp. on *Prunus* sp. (× 1750).

teliospores, and some of them are illustrated in Figs 34–42. Teliospores may be one—(Figs 34 A, B; 39, 40, 41), two—(Figs 34 C, D, H; 35, 42) or multicelled spores (Figs 34 E, F, G, I, J; 38) produced on pedicels. They may be a single layer of cells (Figs 34 K, L) or a multilayer of cells (Figs 34 M, N) without pedicels. They may be produced in chains (Figs 34 O, P, Q, R) with

(Figs 34 Q, R) or without peridial cells (Figs 34 O, P). The surface of teliospores may be smooth (Figs 34 D, E; 35, 36, 38), echinulate (Figs 34 H, 42) verrucose (Figs 34 A, 37, 40, 41), or reticulate or striate (Figs 34 B, 39). They may be thick walled and heavily pigmented (Figs 34, A, B, H, I, J) and germinate after dormancy or thin walled and hyaline (Figs 34 L, O) and germinate without dormancy.

Hughes (1970) identified three methods of teliospore morphogenesis: (a) melampsoraceous, a simple budding process by sporogenous cells producing teliospore within the host epidermis (*Pucciniastrum, Melampsora, Coleosporium* p.p.); (b) cronartiaceous, meristem arthrospore production of teliospores in chains (*Cronartium, Chrysomyxa, Didymopsora, Coleosporium* p.p.); (c) pucciniaceous, sympodioconidial production of pedicellate teliospores (*Puccinia, Gymnosporangium, Phragmidium, Cystopsora, Chaconia, Scopella*). The careful examination of teliospore ontogeny in this manner should give much-needed information to help classify rusts into more natural groups.

5. Basidiospores

Basidiospores are produced on basidia. Basidia are usually four-celled and produce four basidiospores. Two-celled basidia are often produced in species of various genera. The majority of rust genera produce so-called external basidia, but several genera (*Coleosporium, Chrysella, Chrysopsora, Goplana, Achrotelium* and *Ochropsora*) are known to produce internal basidia. Still others (*Zaghouania* and *Blastospora*) are known to produce basidia that can be called semi-internal basidia. In external basidia, basidia emerge through germ pores, or the tips of the spores elongate and produce basidia completely outside the teliospores (Fig. 43). Internal basidia are produced by division of the teliospore contents into four cells; each cell germinates and produces basidiospores outside of the teliospores (Figs 44, 45). The name semi-internal basidia is used when only parts of the original teliospores become parts of the basidia and parts of the basidia are extended to the outside of the teliospores (Figs 46, 47, 48). Some species such as *Uromyces aloes* (Cke.) Magn. (Thirumalachar, 1946) and *Endocronartium harknessii* (J. P. Moore) Y. Hirat. (Hiratsuka *et al.*, 1966) produce infection hyphae directly on the basidia without forming basidiospores (Fig. 49). Also, two-celled basidia are known to be formed in a few species (see discussions under "Nuclear cycle").

According to the classification and terminology of basidia by Donk (1954) and Talbot (1973), teliospores of rusts can be called probasidia because nuclear fusions take place in the structures, and basidia are metabasidia

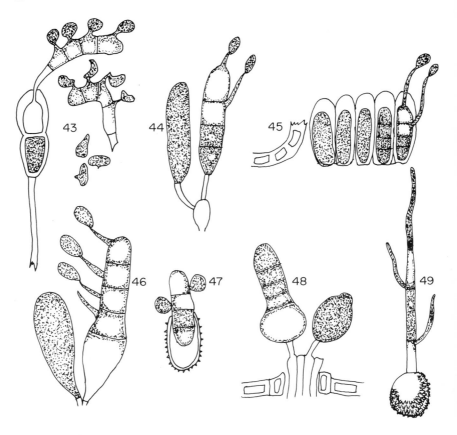

Figs 43–49. Variations in basidia formation of the rust fungi (schematic). *Fig. 43:* External basidium of *Puccinia* sp. (redrawn from Hiratsuka, 1955). *Fig. 44:* Internal basidia of *Achrotelium ichnocarpi* Syd. *Fig. 45:* Internal basidia of *Coleosporium* sp. *Fig. 46:* Semi-internal basidium of *Mikronegeria fagi* Diet. et. Neg. *Fig. 47:* Semi-internal basidium and sessile basidiospores of *Zaghouania phillyreae* Pat. *Fig. 48:* Semi-internal basidium of *Blastospora itoana* Togashi et Onuma. *Fig. 49:* Basidium of *Endocronartium harknessii* (J. P. Moore) Y. Hiratsuka without basidiospores.

because meiosis occurs in them. Basidia are sometimes also called promycelia. Most of the basidia produce sterigmata (Figs 43, 44, 45, 46) and basidiospores are produced at the tips of sterigmata; species of the genus *Zaghouania*, however, are sessil, and basidiospores are produced directly on the basidia (Fig. 47).

6. Hyphae and haustoria

Hyphae are septate and hyaline. Hyphae of gametophytic mycelia are mostly haploid and monokaryotic, and hyphae of sporophytic mycelia are mostly dikaryotic. The presence of clamp connections has been reported in a few species, but dikaryotic hyphae of rust fungi are generally known to lack clamp connection (Berkson, 1970). Hyphae of both kinds grow mostly between host cells (intercellular), obtaining nutrients from the host cells by means of specialized structures called haustoria.

Moss (1926) compared haustoria of the genera of Pucciniastreae and concluded that morphological characters of haustoria had some taxonomic value. He reported simple elongated haustoria in the genera of *Pucciniastrum* and *Hyalopsora* and lobed complex haustoria in *Uredinopsis, Milesina* and *Melampsorella*.

The septal structures of the hyphae of rust fungi are considered to be much simpler than septa found in other basidiomycetous fungi. They lack the dolipore apparatus and parenthesomes (Littlefield and Bracker, 1971; Jones, 1973; Harder, 1976b) but are more complex than simple perforated septa of ascomycetes. Littlefield and Heath (1979) considered, however, that the septal apparatus of rusts and the higher Basidiomycetes were equally complex. Khan and Kimbrough (1980) concluded that septal pore structures of rusts are essentially identical to those of Septobasidiales (Dykstra, 1974) and *Eocronartium muscicola* (Fr.) Fitz. (the parasitic Auriculariales is considered a rust by some authors).

C. Life Cycles

1. Variations of life cycles

Three basic types of life cycles in the rust fungi, are known, depending on the kinds of spores they have. The life cycles are macrocyclic, demicyclic and microcyclic. The macrocyclic life cycle has all spore states, the demicyclic life cycle lacks the uredinial state, and the microcyclic life cycle lacks the aecial state as well as the uredinial state and has only spermogonial, telial and basidial states. In all three types of life cycles, the spermogonial state may be absent. Combined with the host alternating (heteroecious) and nonhost alternating (autoecious) nature of the life cycles, the following five basic variations are recognized.

(a) (0), I–II, III, IV—heteromacrocyclic;
(b) (0), I, II, III, IV—automacrocyclic;
(c) (0), I–III, IV—heterodemicyclic;

(d) (0), I, III, IV—autodemicyclic;

(e) (0), III, IV—microcyclic.

The teliospores of certain microcyclic species are obviously derived from the aecial states of related macrocyclic or demicyclic species. In those species the teliospores have morphological features of aeciospores of parental species but germinate to produce basidia. The life cycles of such microcyclic rusts are called endocyclic. The genera proposed for endocyclic rusts are *Endophyllum*, *Kunkelia* and *Endocronartium*.

The complete life cycle of many rusts is not known. Species of rusts known only in the spermogonial and aecial states are called by form genus names such as *Aecidium*, *Peridermium*, *Roestelium* and *Caeoma*. Species of rusts known only by the uredinial state are classified under the genus *Uredo* or a few other form genera established for uredinial states.

2. Concept of correlated species and Tranzschel's Law

According to Jackson (1931), a heteromacrocyclic species often shortens its life cycle in various ways and produces fungi having different life cycles with various numbers of spore states that are recognized as different species. Such species, which are considered to be related to each other but have different life cycles, are called correlated species (Cummins, 1959). An example of a group of correlated species cited by Cummins (1959) is:

Puccinia intervenience Bethel—heterodemicyclic;

P. graminiella Diet. and Holw.—autodemicyclic;

P. sherardiana Koern.—microcyclic;

Endophyllum tuberculatum (Ell. and Kell.) Arth. and Fromme—microcyclic (endocyclic).

When a rust species with a longer life cycle has been reduced to a microcyclic species, there is a definite pattern or rule as to the habit of its telial state. The pattern or rule is called Tranzschel's Law because the concept was first pointed out by Tranzschel (1904). Tranzschel's Law has been defined by Cummins (1959) as follows: "Telia of microcyclic species simulate the habit of the aecia of the parental macrocyclic species and occur on the aecial host plants of the latter". An example of Tranzschel's Law cited by Cummins (1959) is the relationship between *Puccinia monoica* Arth. (heteromacrocyclic) and *P. thalaspeos* Schub. (microcyclic). The uredinia and telia of *P. monoica* occur on *Koeleria cristata* (L.) Pers. (June grass), and the systemic aecial state is produced on species of *Arabis*. Teliospores of *P. thalaspeos* are morphologically similar to those of *P. monoica* but occur on species of *Arabis* in a systemic infection similar to the aecial state of *P. monoica*. Also, in endocyclic species the telia are derived from aecia of the parental species and therefore follow Tranzschel's Law.

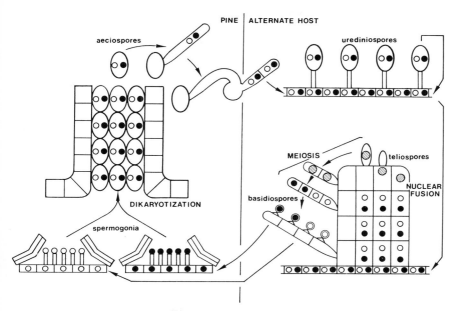

Fig. 50. Nuclear cycle of a heteromacrocyclic rust (*Cronartium* sp.) (schematic) (from Hiratsuka, 1973).

3. Nuclear cycle

The basic nuclear cycle associated with the life cycle of a macrocyclic rust is illustrated in Fig. 50. Spermogonia are produced on haploid mycelia and produce haploid spermatia. Spermatia are male gametes and are responsible for the dikaryotization (plasmogamy) of aecia primordia of the opposite mating type. In addition to dikaryotization by spermogonia, several other ways of dikaryotization are proposed (Buller, 1950), such as (1) fusion of haploid mycelia of the opposite sex in host plant tissue in the case of a compound infection, (2) fusion of a dikaryotic mycelium with a haploid mycelium in autoecious macrocylic rusts such as *Puccinia helianthi*, (3) spontaneous dikaryotization in the case of homothallic species such as *Puccinia malvacearum*, *P. xanthii*, *P. adoxae* and *P. buxi.* (homothallic species usually lack spermogonia), and (4) self-dikaryotization of perennial systemic rusts such as species of *Cronartium* spp.

After dikaryotization, aecia and dikaryotic aeciospores are formed. Aeciospores germinate to initiate dikaryotic mycelium on which uredinia and telia are formed. Urediniospores occur in a dikaryotic repeating stage (conidia) and initiate dikaryotic mycelia. Teliospores are first dikaryotic,

B

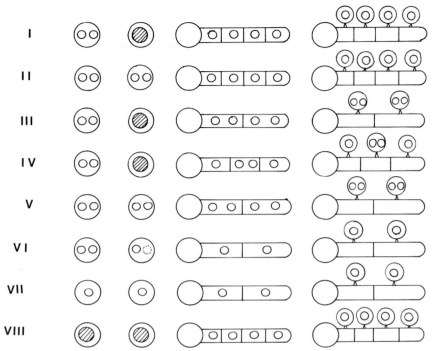

Fig. 51. Variations of nuclear conditions in teliospores and basidia (⬚ diploid nucleus, ○ haploid nucleus). I: Nuclear fusion in teliospores, the fusion nuclei divided twice (meiosis), and produce four celled basidia with four uninucleate basidiospores. Occur in most macrocyclic and demicyclic species and a few microcyclic species [*Puccinia malvacearum* Bert. (Allen, 1933), *Endophyllum sempervivi* (Alb. et Schw.) dBy. II: No nuclear fusion in binucleate teliospores, the two nuclei divide once in the basidia and produce four celled basidia with four uninucleate basidiospores. e.g. *Endophyllum euphorbiae-sylvaticae* (DC) Wint. III: Nuclear fusion in teliospores and meiosis in basidia produce four nuclei, produce two celled basidia each of which produce binucleate basidiospores. e.g. *Puccinia arenariae* (Schum.) Wint., *P. horiana* P. Henn. (Kohno *et al.*, 1975), *Cystopsora oleae* Butl. pp. (Thirumalachar, 1945). IV: Variation of I. with three celled basidia with two uninucleate and one binucleate basidiospores. e.g. *Sphenospora kevorkianii* Lind. (Olive, 1953). V: No nuclear fusion in binucleate teliospores and the two nuclei divide once in basidia, producing two celled basidia with binucleate basidiospores. e.g. *Endophyllum spilanthes* Thirum. and Govindu (Thirumalachar and Govindu, 1954). VI: One of the nuclei in binucleate teliospores disappear and the remaining nucleus divide once producing two celled basidia with two uninucleate basidiospores. e.g. *Endophyllum valerianae-tuberosae* (Jackson, 1931). VII: Uninucleate (haploid) teliospores produce two celled basidia with uninucleate basidiospores. e.g. *Kunkelia nitens* (Schw.) Arth. pp. (Olive, 1953), *Uromyces rudbeckiae* Diet and Holw. (Jackson, 1931). VIII: Uninucleate (diploid) teliospores produce four celled basidia with uninucleate basidiospores, e.g. *Endophyllum euphorbiae* Plowr. (Olive, 1953).

but nuclear fusions (karyogamy) occur in the spore to produce diploid nuclei. Upon germination, teliospores produce basidia. Meiosis takes place in the basidia to produce four haploid cells, each of which germinate to produce haploid basidiospores. The basidiospores germinate and initiate haploid mycelia-producing spermogonia and aecia primordia.

The nuclear condition of teliospores and basidia is basically as explained above for most of the macrocyclic and demicyclic rusts, including *Puccinia graminis*, *P. helianthi*, *Cronartium ribicola* and *Melampsora lini* and several microscopic species such as *P. malvacearum* and *Endophyllum sempervivi*. Many variations in the nuclear condition of teliospores and basidia have been reported among microcyclic species and especially in endocyclic species (Jackson, 1935; Olive, 1953). Some of the variations are illustrated in Fig. 51. Also, additional nuclear divisions have been observed often in uninucleate basidiospores producing binucleate or sometimes tetranucleate basidiospores (Allen, 1933; Olive, 1949; Berkson and Britton, 1969; Kohno *et al.*, 1977).

III. Taxonomy

A. Phylogeny

All major taxonomic schemes of fungi include rusts in the class Basidiomycetes and subclass Heterobasidiomycetidae. Under the subclass, rusts are often classified together with smuts (Ustilaginales) and jelly fungi (Dacryomycetales. Tremellales and Auriculariales) (Martin, 1950; Gäumann, 1964; Wolf and Wolf, 1949) or are placed with smuts under Teliomycetidae (Teliosporae) (Bessey, 1950; Ainsworth, 1973; Alexopoulos and Mims, 1979). Von Arx (1967, 1970) however, did not agree with this common practice of putting Ustilaginales (smuts) close to rusts and recognized a separate class, Endomycetes, to include smuts and other basidiomycetous and ascomycetous yeasts (Endomycetales, Torulopsidales) together with Taphrinales. He kept Uredinales in Basidiomycetes. The results of recent ultrastructural and other studies indicate that rusts may not be as closely related to smuts as has been considered (Donk, 1972, 1973). Also, in all cases of successful axenic cultures of rusts, mycelial-type colonies were produced, and there was no sign of the yeast-like growth that is characteristic of smuts in artificial cultures. This further indicates that rusts and smuts may have been derived from different ancestral lines and probably are not closely related.

Several theories have been proposed about the probable origin of rusts and their relationship to other groups of fungi. Dietel (1928), Gäumann

(1964), Leppik (1961), Donk (1972, 1973) and others speculated that the rusts have evolved from parasitic Auriculariales. Rogers (1934) and Olive (1957) considered that the rusts were derived from tremellaceous ancestral forms, which in turn were derived from Ascomycetes. Petersen (1975) speculated that most of the basidiomycetous fungi, including rusts, were derived from a pool of ascomycetous fungi by the production of "exogenous" ascospores. Hughes (1970), examining the spore ontogeny of rusts, also suggested there was a close relationship between rusts and ascomycetous fungi. Savile (1955, 1968, 1976) speculated that rusts were derived from a simple ascomycete much like the genus *Taphrina* (Proto-*Taphrina*) parasitizing ancient ferns or fern ancestors. Bessey (1950) considered Discomycetes to be ancestors of Basidiomycetes, including rusts, and he speculated that Ascomycetes were derived from some floridean algae.

Most of the rusts have specific host species or a group of related host species. Because of the close relationship of rusts with their hosts, they are considered to have evolved hand in hand. Rust–host relationships, therefore, are very important in considering the phylogeny and origin of rust fungi. Not only can we depend on the host–rust relationship in classifying and tracing the evolutional history of rusts, but rusts can be helpful in the taxonomy and classification of higher plants and in speculation on their phylogenetical relations (Nannfeldt, 1968; Savile, 1971, 1979; Anikister and Wahl, 1979).

Most of the rusts that occur on ferns in the temperate regions (*Uredinopsis, Milesina* and *Hyalopsora*) have heteromacrocyclic life cycles and are considered to be primitive groups of rusts by many authors. Their spermogonial and aecial states occur on northern conifers, especially *Abies*. Not only do they have all five spore states, but many species of such genera as *Uredinopsis* and *Hyalopsora* have an extra kind of urediniospore called an amphispore. On the other hand, groups of rusts on more advanced hosts have proportionally more autoecious and microcylic species. For these reasons it generally is considered that the heteromacrocyclic life cycle is the primitive type and that variously reduced life cycles are derived from them (Jackson, 1931).

It has been a vexing problem to imagine that the most complicated life cycles are the most primitive forms. Spore states of *Milesina, Hyalopsora* and *Uredinopsis* are by no means simple or show primitive characteristics, except that their teliospores are simple in form and are produced inside the host tissues and some spore states lack pigmentation. Several authors (Leppik, 1967; Savile, 1955; Hennen and Buritica, 1980) came up with more plausible explanations of this apparent contradiction. They speculated that simple autoecious auriculariaceous rusts on mosses, ferns and some angiosperms (*Eocronartium, Jola, Platycarpa, Herpobasidium* and *Xenogloea*) in tropical and subtropical regions can be considered to be close to the original primitive rusts. They further suggested that those simple rusts with only gelatinous telia and basidia expanded their life cycles to include other states (aecial or uredinial states) becoming rusts with "partially expanded life cycles", and finally they achieved fully expanded life cycles (macrocyclic life cycles). From fully expanded rusts many kinds of reduced life cycles have been derived, as was explained by Jackson (1931). According to this theory, rusts having simple life cycles can be either primitive unexpanded forms or much more recent kinds that emerged as the result of reductions in life cycles of species with fully expanded life cycles (Hennen and Buritica, 1980).

Leppik (1953) introduced the idea of a "hologenetic ladder" to explain how rusts evolved with host plant groups by shifting their spore states to newly emerged, ecologically prosperous groups of plants. As an example he analysed the existing species of *Melampsora* and speculated the sequence of possible shifts of hosts in the group. He suggested that teliosporous rusts on

Table II. Subdivisions of *Melampsora* according to Leppik (1953) and examples of species.

Subdivision of Melampsora Hosts	Fern	Conifer	Salicales	Liliaceae Saxifragaceae Euphorbiaceae etc.
Proto-*Melampsora*	III			
?	?			
Pino-*Melampsora*	III	0, I		
?	?	?		
Salico-*Melampsora*		0, I	II, III	
M. *larici-epitea* Kleb.		Larix	Salix	
M. *medusae* Thuem.		Larix	Populus	
Neo-*Melampsora*			II, III	0, I
M. *arctica* Rostr.			Salix	Saxifraga
M. *magnusiana* Wagner			Populus	Chelidonium
M. *orchidi-repentis* (Plowr.) Kleb.			Salix	Orchis
M. *allii-salicis albae* Kleb.			Salix	Allium
M. *allii-populina* Kleb.			Populus	Allium
M. *ribesii-viminalis* Kleb.			Salix	Ribes
Auto-*Melampsora*				0, I, II, III
M. *euphorbiae* (Schub.) Cast.				Euphorbia
M. *vernalis* Nissl.				Saxifraga
M. *kusanio* Dietel				Hypericum
M. *hirculi* Lind				Saxifraga

ferns (proto-*Melampsora*) existed first, and then the aecial state was produced on conifers (pino-*Melampsora*), after which telial and uredinial states
were shifted to Salicales (salicio-*Melampsora*). In the next stage the rusts
shifted their aecial states from coniferous hosts to more modern angiosperms such as Saxifragaceae, Euphorbiaceae and Liliaceae (neo-
Melampsora) and finally produced many autoecious species on these hosts
(auto-*Melampsora*). In Table II some existing species of *Melampsora* are
arranged according to Leppik's subdivisions of the genus. He also explained
the evolution of the genus *Gymnosporangium* (and *Coleopuccinia*) (1956),
grass rusts (1961) and legume rusts (1967) in similar manner. Leppik (1953,
1967) introduced the term "biogenic radiation" to speculate on the sequence
of the hologenetic ladder. According to him, in a group of heteroecious rusts
often one generation is restricted to a host or a group of hosts, while another
generation seems to be radiated out having a wider range of hosts. This
suggests that the generation having the wider host range was established
later in the hologenetic ladder. For example, he speculated that for present-
day fir-fern rusts (*Uredinopsis*, *Milesina* and *Hyalopsora*) *Abies* is the primary host and many ferns are secondary hosts established later in geological
time. The rusts radiated out from *Abies* because of a wider range of fern
hosts. Leppik (1973) further proposed that an even older group of fern
(tropical or subtropical) must have been the primary hosts of conifer rusts
before uredinial and telial states were shifted to modern ferns (temperate).
He discussed the origin and distribution of conifers and their rusts in the light
of recent knowledge about continental drift.

B. Taxonomic Criteria

Because of the plaeomorphic nature and variable life cycles of rusts together
with their host specificity, it is important but often difficult to put relative
values on certain morphological or other characters for taxonomic and
classification purposes.

   One of the difficult problems in rust classification is that some morphological characters are very similar in supposedly widely separated rusts, but at
the same time, other characters are very different in obviously closely
related rusts. Petersen (1975) recognized this phenomenon and called it the
"reticulate nature of taxonomic characters". For example, the morphology
of urediniospores of species from such distinct genera as *Puccinia*, *Cronar-*
tium and *Ravenelia* is very similar, but their teliospore morphology is very
different. On the other hand, if one puts emphasis on a character such as the
number of cells in the teliospores, even closely related species can be
classified in different genera. Hiratsuka and Hiratsuka (1966) showed such

examples in rusts of the genus *Allium*. They found that *Puccinia allii* (DC) Rud. and *Uromyces durus* Diet. are very closely related, except for the number of cells in the teliospores. They even found several transitional forms having variable proportions of one- and two-cell teliospores. These rusts that are in different genera are in fact much closer to each other than most species of the same genus.

Also, different characters are valuable at different levels of classification. Some characters such as the morphological types of spermogonia, mode of teliospore germination, and morphological types of uredinia or aecia are valuable in generic and supergeneric classification or in discussing the phylogeny of rusts, but they are of little value in distinguishing rusts at the species level. On the other hand, fine differences in spore surface ornamentation or the number and arrangement of germ pores of urediniospores are good characters by which to distinguish rusts at the species level but are of no significant value in generic or supergeneric classification.

C. Subdivisions of Rusts

Although rust fungi comprise one of the best-described major groups of fungi, suprageneric subdivisions are by no means settled in a satisfactory manner acceptable to the majority of uredinologists.

Traditionally rusts are divided into two families, namely Melampsoraceae and Pucciniaceae (Arthur, 1934; Hiratsuka, 1955; Dietel, 1928; Clements and Shear, 1931). Often several other families such as Coleosporiaceae and Cronartiaceae are added to them (Klebahn, 1913; Wilson and Henderson, 1966). Those divisions are based on teliospore characters (sessil for Melampsoraceae, pedicellate for Pucciniaceae, internal basidia for Coleosporaceae and catenulate for Cronartiaceae). Many subfamilies have been proposed in each family. This heavy dependence on teliospore characters as the basis for major subdivisions of rusts has been questioned often (Hiratsuka and Cummins, 1963; Leppik, 1972; Petersen, 1975; Laundon, 1974; Savile, 1976). Because of this, Cummins (1959) did not even mention the names of families in his "Illustrated Genera of Rust Fungi," and Laundon (1974) avoided presenting any particular system in his summary of rusts.

Hiratsuka and Cummins (1963) examined spermogonia of many genera and concluded that morphological types of spermogonia are dependable characters in rust taxonomy. Recently, Hiratsuka and Hiratsuka (1980) expanded the work and categorized morphological types of spermogonia into six groups (Fig. 1, Table I). They concluded that those groups may represent rather homogeneous subdivisions of rusts. Considering the spermogonial types as important criteria, Savile (1976) recognized

Raveneliaceae and Phragmidiaceae in addition to Pucciniaceae in an attempt to separate obviously distinct groups that are formerly classified as Pucciniaceae. Several other proposals for subdivisions have been presented by Leppik (1972) and Azbukina (1974), who both took morphological types of spermogonia into consideration.

A disproportionately large number of species belongs to about one dozen genera: *Puccinia* (about 3000–4000 species), *Uromyces* (about 600–700 species), *Ravenelia* (more than 150 species), *Melampsora* (about 100 species), *Coleosporium* (about 80 species), *Phragmidium* (about 70 species), *Gymnosporangium* (about 50 species), and *Pucciniastrum, Prospodium, Hemileia, Milesina, Cronartium* and *Uredinopsis* (each about 30–40 species). Schemes of subdivisions of rusts so far proposed tend to emphasize these larger genera and put other smaller genera (many of them monotypic) into subdivisions based on the larger genera. Although small, many of these genera singly or in groups may represent distinct enough entities to warrant family status. Careful examination and comparison of the many small genera, especially those from tropical and subtropical regions, are essential to arrive at satisfactory subdivisions of the rust fungi. Besides teliospore characters and morphological types of spermogonia, other characters such as spore ontogeny (Hughes, 1970), morphological types of uredinia (Kenny, 1970), host–rust associations, and types of life cycles should be considered carefully to arrive at a more natural classification of rusts.

Acknowledgement

We would like to thank Dr M. Kakishima of Tsukuba University, Japan for letting us use his photographs for Figs 17, 18, 30–33, 35–42 inclusive.

References

Ainsworth, G. C. (1973). Introduction and keys to higher taxa. *In* "The fungi—an advanced treatise". (G. C. Ainsworth, F. K. Sparrow and A. S. Sussman, eds), Vol. IVA. Academic Press, New York and London.

Alexopoulos, C. J. and Mims, C. W. (1979). "Introductory Mycology." 3rd edn. John Wiley, New York and London.

Allen, R. E. (1933). A cytological study of the teliospores, promycelia, and sporidia in *Puccinia malvacearum*. *Phytopathology* **23**, 572–586.

Allen, R. E. (1934). A cytological study of heterothallism in flax rust. *J. Agric. Res.* **49**, 765–791.

Andrus, C. F. (1931). The mechanism of sex in *Uromyces appendiculatus* and *U. vignae*. *J. Agric. Res.* **42**, 559–587.

Andrus, C. F. (1933). Sex and accessory cell fusions in the Uredineae. *J. Wash. Acad. Sci.* **23**, 544–557.

Anikister, Y. and Wahl, I. (1979). Coevolution of the rust fungi on Gramineae and Liliaceae and their hosts. *Ann. Rev. Phytopathol.* **17**, 367–403.

Arthur, J. C. (1905). Terminology of the spore structures in the Uredinales. *Bot. Gaz.* **39**, 219–222.

Arthur, J. C. (1925). Terminology of Uredinales. *Bot. Gaz.* **80**, 219–223.

Arthur, J. C. (1934). "Manual of the rusts in United States and Canada." Purdue Research Foundation, Lafayette.

Azbukina, Z. M. (1970). On terminology of rust fungi and its connection with the position of some taxa in the system. (In Russian). *Mikologiyae Fitopathologiyn* **4**, 340–345.

Azbukina, Z. M. (1974). "Rust Fungi of Far East." (In Russian). Hayka, Moscow.

Berkson, B. M. (1970). Cytological studies of the telial stage of *Cerotelium dicentrae*. *Amer. J. Bot.* **57**, 899–903.

Berkson, B. M. and Britton, M. P. (1969). Cytological studies on the teliospore germination in *Puccinia lobata*. *Mycologia* **61**, 981–986.

Bessey, E. A. (1950). "Morphology and Taxonomy of the Fungi." Blakiston Co.

Buller, A. H. R. (1950). "Researches of Fungi." Vol. VII. University of Toronto Press, Toronto.

Cain, R. F. (1972). Evolution of the fungi. *Mycologia* **64**, 1–14.

Clements, F. E. and Shear, C. L. (1931). "The Genera of Fungi." H. W. Wilson, New York.

Coffey, M. D. and Shaw, M. (1972). Nutritional studies with axenic cultures of the flax rust *Melampsora lini. Physiol. Pl. Path.* **2**, 37–46.

Coffey, M. D., Bose, A. and Shaw, M. (1970). *In vitro* culture of the flax rust, *Melampsora lini. Can. J. Bot.* **48**, 773–776.

Craigie, J. H. (1927). Discovery of the function of the pycnia of the rust fungi. *Nature* **120**, 765–767.

Cummins, G. B. (1936). Phylogenetic significance of the pores in urediospores. *Mycologia* **28**, 103–132.

Cummins, G. B. (1959). "Illustrated Genera of Rust Fungi." Burgess, Minneapolis.

Cunningham, G. H. (1930). Terminology of the spore forms and associated structures of the rust fungi. *New Zealand J. Sci. Technol.* **12**, 123–128.

Dietel, P. (1928). Uredinales. *In* "Engler u. Prantl. Natürl. Pflanzenfam. II". **4**, 24–98.

Donk, M. A. (1954). A note on sterigmata in general. *Bothalia* **6**, 301–302.

Donk, M. A. (1972). The heterobasidiomycetes: a reconnaissance II. Some problems connected with the restricted emendation. *Koninkl. Nederl. Akad. Wetensch.* **C75**, 376–390.

Donk, M. A. (1973). The heterobasidiomycetes: a reconnaissance III. How to recognize a basidiomycete? *Koninkl. Nederl. Akad. Wetensch.* **C76**, 1–22.

Dykstra, M. J. (1974). Some ultrastructural features in the genus *Septobasidium. Can. J. Bot.* **52**, 971–972.

Ehrlich, M. A. and Ehrlich, H. G. (1969). Uredospore development in *Puccinia graminis. Can. J. Bot.* **47**, 2061–2064.

Gäumann, E. (1964). "Die Pilze". Birkhauser Verlag, Basel.

Harder, D. E. (1976a). Mitosis and cell division in some cereal rust fungi. I. Fine structure of the interphase and premitotic nuclei. *Can. J. Bot.* **54**, 981–994.

Harder, D. E. (1976b). Mitosis and cell division in some cereal rust fungi. II. The process of mitosis and cytokinesis. *Can. J. Bot.* **54**, 995–1009.

Harder, D. E. (1976c). Electron microscopy of urediospore formation of *Puccinia coronata avenae* and *P. graminis avenae*. *Can. J. Bot.* **54**, 1010–1019.

Harder, D. E. and Chong, J. (1978). Ultrastructure of spermatium ontogeny in *Puccinia coronata avenae*. *Can. J. Bot.* **56**, 395–403.

Harvey, A. E. and Grasham, J. L. (1970). Growth of *Cronartium ribicola* in the absence of physical contact with host. *Can. J. Bot.* **48**, 71–73.

Hennen, J. F. and Buritica, P. (1980). A brief summary of modern rust taxonomy and evolutionary theory. *Rep. Tottori Mycol. Inst.* **18**, 243–256.

Hennen, J. F. and Figueiredo, M. B. (1979). *Intrapes*, a new genus of fungi imperfecti (Uredinales) from Brazilian cerrado. *Mycologia* **71**, 836–840.

Hennen, J. F. and Ono, Y. (1978). *Cerradoa palmaea*: the first rust fungus on Palmae. *Mycologia* **70**, 569–576.

Hiratsuka, N. (1955). "Uredinological Studies". Kasai, Tokyo.

Hiratsuka, Y. (1971). Spore surface morphology of pine stem rusts of Canada as observed under a scanning electron microscope. *Can. J. Bot.* **49**, 371–372.

Hiratsuka, Y. (1973a). Sorus development, spore morphology and nuclear condition of *Gymnosporangium gaumannii* ssp. *albertense*. *Mycologia* **65**, 137–144.

Hiratsuka, Y. (1973b). The nuclear cycle and the terminology of spore states in Uredinales. *Mycologia* **65**, 432–443.

Hiratsuka, Y. (1975). Recent controversies on the terminology of rust fungi. *Rept. Tottori Mycol. Inst.* **12**, 99–104.

Hiratsuka, Y. and Cummins, G. B. (1963). Morphology of spermogonia of the rust fungi. *Mycologia* **55**, 487–507.

Hiratsuka, N. and Hiratsuka, T. (1966). Studies on *Uromyces durus* Dietal and its related species parasitic on *Allium grayi* in Japan. *Trans. Mycol. Soc. Jap.* **7**, 160–173.

Hiratsuka, Y. and Hiratsuka, N. (1980) Morphology of spermogonia and taxonomy of rust fungi. *Rept. Tottori Mycol. Inst.* **18**, 257–268.

Hiratsuka, Y., Morf, W. and Powell, J. M. (1966). Cytology of the aeciospores and aeciospore germ tubes of *Peridermium harknessii* and *P. stalactiforme* of the *Cronartium coleosporioides* complex. *Can. J. Bot.* **44**, 1639–1643.

Holm, L. (1963). Études urédinologiques. 1. Sur les écidies de Oenothéracées. *Svensk Bot. Tidskr.* **57**, 129–144.

Holm, L. (1973). Some notes on rust terminology. *Rep. Tottori Mycol. Inst. (Japan)*. **10**, 183–187.

Holm, L. and Tibell, L. (1974). Studies on the fine structure of aeciospores. III. Aeciospore ontogeny in *Puccinia graminnis*. *Svensk Bot. Tidskr.* **68**, 136–152.

Hughes, S. J. (1970). Ontogeny of spore forms in Uredinales. *Can. J. Bot.* **48**, 2147–2157.

Jackson, H. S. (1931). Present evolutionary tendencies and the origin of life cycle in the Uredinales. *Mem. Torrey Bot. Club.* **18**, 1–108.

Jackson, H. S. (1935). The nuclear cycle in *Herpobasidium filicinum* with a discussion of the significance of homothallism in Basidiomycetes. *Mycologia* **27**, 553–572.

Jones, D. R. (1973). Ultrastructure of septal pore in *Uromyces dianthi*. *Trans. Br. Mycol. Soc.* **61**, 227–235.

Kenny, M. J. (1970). Comparative morphology of the uredia of the rust fungi. Ph.D. Thesis, Purdue University, Indiana.

Khan, S. R. and Kimbrough, J. W. (1980). Ultrastructure and the taxonomy of *Eocronartium*. *Can. J. Bot.* **58**, 642–647.

Klebahn, H. (1912–1914). Uredineen. *In* "Kryptogamen flora der Mark Bran-
debrug." Leipzig.

Kohno, M., Nishimuia, T., Ishizaki, H. and Kunoh, H. (1975). Cytological studies
on rust fungi (III) Nuclear behaviors during the process from teliospore stage
through sporidial stage in two short-cycled rusts, *Kuehneola japonica* and *Puccinia
horiana*. *Bull. Fac. Agric. Mie Univ.* **49**, 21–29.

Kohno, M., Nishimura, T., Noda, M., Ishizaki, H. and Kunoh, H. (1977). Cytological studies on rust fungi (VII). The nuclear behavior of *Gymnosporangium
asiaticum* Miyabe et Yamada during the stages from teliospore germination
through sporidium germination. *Trans. Mycol. Soc. Japan.* **18**, 211–219.

Laundon, G. F. (1965). The generic names of Uredinales. *Mycol. Pap.* **99**,
1–22.

Laundon, G. F. (1967). Terminology of the rust fungi. *Trans. Brit. Mycol. Soc.* **50**,
189–194.

Laundon, G. F. (1972). Deliniation of aecial from uredial states. *Trans. Brit. Mycol.
Soc.* **58**, 344–346.

Laundon, G. F. (1974). Uredinales. *In* "The fungi—an advanced treatise." (G. C.
Ainsworth, F. K. Sparrow and A. S. Sussman, eds), Vol. IV-B. Academic Press,
New York and London.

Leppik, E. (1953). Some viewpoints on the phylogeny of rust fungi. I. Coniferous
rusts. *Mycologia* **45**, 46–74.

Leppik, E. (1956). Some viewpoints on the phylogeny of rust fungi. II. Gymnosporangium. *Mycologia* **48**, 637–654.

Leppik, E. (1961). Some viewpoints on the phylogeny of rust fungi. III. Origin of
grass rusts. *Mycologia* **51**, 512–528.

Leppik, E. (1967). Some viewpoints on the phylogeny of rust fungi. VI. Biogenic
radiation. *Mycologia* **59**, 568–579.

Leppik, E. (1972). Evolutionary specialization of rust fungi (Uredinales) on the
Leguminosae. *Ann. Bot. Fenn.* **9**, 135–148.

Leppik, E. (1973). Origin and evolution of conifer rusts in the light of continental
drift. *Mycopathol. et Mycol. Appl.* **49**, 121–136.

Littlefield, L. J. and Bracker, C. E. (1971). Ultrastructure and development of
urediniospore ornamentation in *Melampsora lini*. *Can. J. Bot.* **49**, 2067–2073.

Littlefield, L. J. and Heath, M. (1979). "Ultrastructure of rust fungi." Academic
Press, London and New York.

Martin, G. W. (1950). "Outline of fungi." Wm. C. Brown, Dubuque.

Mimms, C. W., Seabury, F. and Thurston, E. L. (1976). An ultrastructural study of
spermatium formation in the rust fungus *Gymnosporangium juniperi-virginianae*.
Amer. J. Bot. **63**, 997–1002.

Moss, E. H. (1926). The uredo stage of *Pucciniastreae Ann. Bot.* **40**, 813–847.

Nannfeldt, J. A. (1968). Fungi as plant taxonomist. Festskrift till Torgny Segerstedt
85–95. Acta Univ. Uppsala.

Olive, L. S. (1944). Spermatial formation in *Gymnosporangium clavipes*. *Mycologia*
36, 211–214.

Olive, L. S. (1949). A cytological study of typical and atypical basidial development
in *Gymnosporangium clavipes*. *Mycologia* **41**, 420–426.

Olive, L. S. (1953). The structure and behavior of fungus nuclei. *Bot. Rev.* **19**,
439–586.

Olive, L. S. (1957). Two new genera of the Ceratobasidiaceae and their phylogenetic
significance. *Am. J. Bot.* **44**, 429–435.

Petersen, R. H. (1975). The rust fungus life cycle. *Bot. Rev.* **40**, 453–513.

Rajendren, R. B. (1970). Cytology and developmental morphology of *Kernkampella breyniae-patentis* and *Ravenelia hobsoni*. *Mycologia* **62**, 1112–1121.
Rijkenberg, F. H. J. and Truter, S. J. (1974a). The ultrastructure of *Puccinia sorghi* aecial stage. *Protoplasma* **81**, 231–245.
Rijkenberg, F. H. J. and Truter, S. J. (1974b). The ultrastructure of sporogenesis in the pycnial stage of *Puccinia sorghi*. *Mycologia* **66**, 319–326.
Rogers, D. P. (1934). The basidium. *Stud. Nat. Hist. Univ. Iowa* **16**, 160–183.
Savile, D. B. O. (1939). Nuclear structure and behavior in species of the Uredinales. *Am. J. Bot.* **26**, 585–609.
Savile, D. B. O. (1955). A phylogeny of the Basidiomycetes. *Can. J. Bot.* **33**, 60–104.
Savile, D. B. O. (1968). The case against "Uredinium". *Mycologia* **60**, 459–464.
Savile, D. B. O. (1971). Generic disposition and pycnium type in the Uredinales. *Mycologia* **63**, 1089–1091.
Savile, D. B. O. (1973). Aeciospore types in *Puccinia* and *Uromyces* attacking Cyperaceae, Juncaceae and Poaceae. *Rep. Tottori Mycol. Inst.* **10**, 225–241.
Savile, D. B. O. (1976). Evolution of the rust fungi (Uredinales) as reflected by their ecological problems. *Evolutionary Biol.* **9**, 137–207.
Savile, D. B. O. (1979). Fungi as aids in higher plant classification. *Bot. Rev.* **45**, 377–503.
Talbot, P. H. B. (1973). Towards unification in basidial terminology. *Trans. Brit. Mycol. Soc.* **61**, 497–512.
Thirumalachar, M. J. (1946). A cytological study of *Uromyces aloes*. *Bot. Gaz.* **108**, 245–254.
Thirumalachar, M. J. and Govindu, H. C. (1954). Morphological and cytological studies of a bisporidial species of *Endophyllum*. *Bot. Gaz.* **115**, 391.
Thirumalachar, M. J. and Mundkur, B. B. (1949). Genera of rusts. *Indian Phytopath.* **2**, 65–101, 193–244.
Thirumalachar, M. J. and Mundkur, B. B. (1950). Genera of rusts. *Indian Phytopath.* **3**, 4–42, 203–204.
Tranzschel, W. (1904). Über die Möglichkeit, die Biologie wirtswechselnder Rostpilze auf Grund morphologisher Merkmale vorauszusehen. *Trav. Soc. Imp. Nat. St. Petersburg Compt. Rend.* **35**, 311–312.
Turel, F. L. (1969). Saprophytic development of flax rust, *Melampsora lini*, race No. 3. *Can. J. Bot.* **47**, 821–823.
von Arx, J. A. (1967). "Pilzkunde." J. Cramer, Lehre.
von Arx, J. A. (1970). "The genera of fungi sporulating in pure culture." J. Cramer, Lehre.
Wang, Yung-Chang and Martens, P. (1939). Sur l'origine de la dicaryophase chez quelques Urédinées. *La Cellule* **48**, 215–245.
Williams, P. G., Scott, K. J. and Kuhl, J. L. (1966). Vegetative growth of *Puccinia graminis* f. sp. *tritici in vitro*. *Phytopathology* **56**, 1418–1419.
Williams, P. G., Scott, K. J. and Kuhl, J. L. (1967). Sporulation and pathogenicity of *Puccinia graminis* f. sp. *tritici* grown on an artificial medium. *Phytopathology* **57**, 326–327.
Wilson, M. and Henderson, D. M. (1966). "British Rust Fungi." Cambridge University Press, London and New York.
Wolf, F. A. and Wolf, F. T. (1949). "The Fungi." Volume I. John Wiley, London.

Monographs, Regional Floras and other
Important Literature

Ahmad, S. (1956). *Biologia* **2**, 26–101. (Uredinales of West Pakistan.)

Arthur, J. C. (1929). "The Plant Rusts (Uredinales)." John Wiley, New York.

Arthur, J. C. (1934). "Manual of Rusts in United States and Canada" with supplement by G. B. Cummins. 1962. Hafner, New York.

Azbukina, L. M. (1974). "Rust Fungi of Far East." (In Russian). Hayka, Moscow.

Cummins, G. B. (1959). "Illustrated Genera of Rust Fungi." Burgess, Minneapolis.

Cummins, G. B. (1971). "The Rust Fungi of Cereals, Grasses and Bamboos." Springer-Verlag, Berlin and New York.

Cummins, G. B. (1978). "Rust Fungi on Legumes and Composites in North America." University of Arizona Press, Tucson.

Cunningham, G. H. (1931). "The Rust Fungi of New Zealand." Palmerston North, New Zealand.

Fischer, E. (1904). "Die Uredineen der Schweiz." Bern.

Gäumann, E. (1959). "Die Rostpilze Mitteleuropas." Büchler, Bern.

Gjaerum, H. (1974). "Nordens Rustsopper." Fungiflora, Oslo, Norway.

Gonzales-Fragoso, R. (1924, 1925). Flora Ibérica. Uredinales. Vol. I and Vol. II. Madrid.

Hiratsuka, N. (1936). "A monograph of the Pucciniastreae." Tottori.

Hiratsuka, N. (1955). "Uredinological Studies." Kasi, Tokyo.

Hiratsuka, N. (1958). "Revision of Taxonomy of the Pucciniastreae." Kasai, Tokyo.

Hiratsuka, Y. and Powell, J. M. (1976). "Pine Stem Rusts of Canada." Forestry Technical Report 4. Canadian Forestry Service, Department of the Environment, Ottawa.

Hylander, N. I., Jørstad, I. and Nannfeldt, J. A. (1951). *Opera Bot.* **1**, 1–102. (Enumeratio Uredinearum Scandinavicarum).

Jørstad, I. (1951). Skr. Norske Vid.-Akad. Oslo I. Mat.-Nat. Kl. 1951, 1–87. (The Uredinales of Iceland).

Kern, F. D. (1972). "A Revised Taxonomic Account of Gymnosporangium." Pennsylvania State University Press, University Park and London.

Kuprevicz, V. T. and Tranzschel, V. H. (1957). Uredinales. Fasc. 1. Fam. Melampsoraceae. Fl. Cryptog. URSS. Vol. IV. Moscow.

Laundon, G. F. (1965). *Mycol. Pap.* **99**, 1–24. (The generic names of Uredinales).

Liro, J. I. (1908). "Uredineae Fennicae." Helsinki.

Littlefield, L. J. and Heath, M. C. (1979). "Ultrastructure of Rust Fungi." Academic Press, London and New York.

McAlpine, D. (1906). "The Rusts of Australia." R. S. Brain, Melbourne.

Parmelee, J. A. (1965). *Can. J. Bot.* **43**, 239–267. (The genus *Gymnosporangium* in Eastern Canada).

Parmelee, J. A. (1971). *Can. J. Bot.* **49**, 903–926. (The genus *Gymnosporangium* in Western Canada).

Peterson, R. S. (1967). *Bull. Torrey Bot. Club.* **94**, 511–542. (The *Peridermium* species on the pine stems).

Peterson, R. S. (1973). *Rept. Tottori Mycol. Inst.* **10**, 203–223. [Studies of *Cronartium* (Uredinales)].

Roure, L. A. (1963). "The Rusts of Puerto Rico." Technical Papers 35, Agricultural Experiment Station, University of Puerto Rico.

Săvlescu, T. R. (1953). "Monografia Uredinalelor din Republica Popalară Română. Vol. II, pp. 337–1166. Bucurest.

Sydow, P. and Sydow, H. (1902–1924). "Monographia Uredinearum Vols. 1–4." Borntraeger, Leipzig.

Thirumalachar, M. J. and Mundkur, B. B. (1949). *Indian Phytopathol.* **2**, 65–101, 193–244. (Genera of rusts I and II).

Thirumalachar, M. J. and Mundkur, B. B. (1950). *Indian Phytopathol.* **3**, 4–42, 203–204. (Genera of rusts III and Appendix).

Tranzschel, V. H. (1939). "Conspectus Uredinalium URSS."

Wilson, M. and Henderson, D. M. (1966). "British Rust Fungi." Cambridge University Press, London and New York.

Ziller, W. G. (1974). "The Tree Rusts of Western Canada." Publication No. 1329, Canadian Forestry Service, Department of the Environment, Victoria.

2. Axenic Culture and Metabolism of Rust Fungi

D. J. MACLEAN

Department of Biochemistry, University of Queensland, Australia

Part A: Isolation and Characteristics of Axenic Cultures

CONTENTS OF PART A

I. Introduction

Axenic culture is defined as the growth of organism(s) of a single species in the absence of living organisms or cells of any other species (Dougherty, (1953), (for example a "pure culture" of an organism growing on nutrient agar). Rust fungi are highly specialized plant parasites which obtain their nutrients from the living tissues of their host; if the host dies, the fungus dies also. These fungi resisted all attempts to grow them apart from their hosts and in axenic culture until some controversial experiments on the culture of *Gymnosporangium juniperi-virginianae* and two other rusts which were first reported by Hotson and Cutter in 1951. Subsequently, Williams *et al.* (1966) demonstrated culturability of the wheat stem rust fungus (*Puccinia graminis* f.sp. *tritici*) in a sustained series of experiments starting from 1966. Even at the present time however, no standard technique has been developed which can guarantee the successful axenic culture of wild isolates of rust species. A major aim of this chapter therefore, will be to examine carefully our present knowledge in an attempt to find unifying principles of practical value for future culture work.

The axenic culture of rust fungi is no longer an end in itself however. It is now necessary to consider ways in which these cultures can be used as experimental tools both to increase our general knowledge of the biology and physiology of these organisms, and to assist in understanding how they interact with host tissues when growing parasitically. Therefore, it is hoped that this chapter will help create a background perspective which will enable future research to use axenic cultures, rather than merely to provide the stimulus to culture new species of rusts as objects of scientific curiousity.

The early work on attempts to grow rust fungi in axenic culture has been covered in reviews by Scott and Maclean (1969), Brian (1967), Staples and Wynn (1965), and Yarwood (1956), while other authors who have dealt with some aspects of this subject include Mains (1917), Arthur (1929), Allen (1954, 1959, 1965), Cochrane (1960), Shaw (1963, 1964, 1967) and Thrower (1965). Other reviews which have discussed recent developments in the axenic culture of rust fungi include those of Scott (1972, 1976), Williams (1975a), Wolf (1974) and Coffey (1975).

Success in the axenic culture of rust fungi has made it necessary to reconsider some concepts of fungal nutrition (cf. Brian, 1967; Scott and Maclean, 1969; Lewis, 1973). Inability to grow an organism in axenic culture (cf. Yarwood, 1956) can no longer be regarded as a necessary characteristic of an obligate parasite, and this term should now revert to its original definition, i.e. organisms which in natural habitats grow only in association with a living host. However, it is more useful to use the terms biotroph, necrotroph (Link, 1933) and saprotroph (Lewis, 1973) to refer to how and where an organism obtains its nutrients in a particular defined situation. Thus, rust fungi grow as biotrophs when they extract nutrients from the living cells of their host, and grow as saprotrophs on laboratory media.

Sections II and III of this chapter will deal with the methods used and problems encountered in inducing rust fungi to grow as saprotrophs in axenic culture, while Section IV will discuss the characteristics of established axenic cultures.

II. Axenic Culture from Spores

A. Type of Spore

Undoubtedly spores are the most easily handled starting material from which to initiate axenic cultures. Uredospores have been used almost exclusively because they can be used to perpetuate genetically homogeneous clones by repeated re-infection of susceptible plants of the same host species (the primary host), and can thus be obtained in great profusion. Because more experiments have been carried out on the culture of *Puccinia graminis* f.sp. *tritici* than any other rust species, this organism will be used as a case study to illustrate the techniques applied and problems encountered when using uredospores as inoculum.

B. Media and Methods for Puccinia graminis f.sp. tritici

Perhaps the most puzzling aspect of the initiation of saprotrophic growth of *P. graminis* or any other rust species from uredospores is the variability observed within and between experiments. The key component in the medium is the organic nitrogen source: Difco yeast extract in the first experiments (Williams *et al.*, 1966), later supplemented with or replaced by Evans peptone, a proteolytic digest of whale or calf muscles, as a more effective nitrogen source (Williams *et al.*, 1967). However, despite careful attempts to standardize techniques in experiments designed to determine optimum concentrations of Evans peptone and other nutrients, growth was unreliable, prompting Kuhl *et al.* (1971) to comment that "The erratic growth between replicates in many of our experiments makes it difficult to draw a general conclusion about the value of a particular medium or given set of conditions. Negative results must therefore be treated with caution". This variability and its possible cause may become clearer if each of the parameters involved in initiating axenic cultures from uredospores is examined.

1. Organic nitrogen source

Early experiments used rather dilute (0·1%) concentrations of Difco yeast extract or Evans peptone, either singly or in combination (Williams *et al.*, 1966, 1967; Bushnell, 1968; Coffey *et al.*, 1969; Wong and Willetts, 1970; Bose and Shaw, 1971; Bushnell and Stewart, 1971). Subsequent workers have omitted the yeast extract and increased the peptone concentration to 0·4–0·6% to obtain both more profuse and more consistent growth (Williams, 1971; Bushnell, 1976; Foudin and Wynn, 1972; Green, 1976). Tests of alternative nitrogen sources showed that other commercially-available peptones were inferior to Evans peptone, some preparations failing to support the initiation of any saprotrophic growth at all (Bushnell and Rajendren, 1970; Wong and Willetts, 1970; Foudin and Wynn, 1972; Kuhl *et al.*, 1971). However, acid hydrolysates of casein (Bushnell, 1976; Foudin and Wynn, 1972) appeared to be equal or superior to Evans peptone. Enriched media are recommended for future work, e.g. 0·5% Evans peptone (Williams, 1971); 0·4% Evans peptone plus 0·6% Difco Casamino acids (Bushnell, 1976), 0·6% Acidicase, B.B.L. (Foudin and Wynn, 1972), or 0.5% acid hydrolysed casein, Merck No. 2238 (Grambow and Riedel, 1977).

Defined nitrogen sources have been formulated in a number of laboratories, first by Scott and Maclean (1969) and Kuhl *et al.* (1971) who observed growth on liquid or agar media containing 22 mM aspartic acid with 1·65 mM cysteine or cystine or glutathione (but not methionine). Green (1976) subsequently reported growth of North American isolates of *P. graminis* on a cysteine-glutamine medium. In contrast, Bose and Shaw (1974a, 1974b) found that North American isolates grew very poorly on similar cysteine-based media containing added calcium, although these media supported very good growth of an Australian isolate, race 126-ANZ-6, 7.

The importance of a correct balance of amino acids in a complex mixture was demonstrated by Foudin and Wynn (1972) who devised agar-solidified media containing 16 amino acids in concentrations based on analyses of acid hydrolysed casein. When mixed with trace elements this amino acid mixture supported growth equal to that of casein hydrolysate and reportedly better than Evans peptone. However, a different synthetic mixture of amino acids based on analyses of Bacto peptone, which does not support growth, also failed to support growth.

2. Carbohydrate source

Sucrose (3%) was used initially (Williams *et al.*, 1966, 1967) but glucose (2–4%) has been used in most subsequent work as it appears to give more consistent results (Bushnell, 1968; Kuhl *et al.*, 1971). A limited survey of alternative carbohydrates showed that D-fructose, D-mannose or D-mannitol but not D-galactose could also support growth initiation from uredospores (Kuhl *et al.*, 1971). The following common, non-defined mycological media, which are rich in carbohydrates or oganic nitrogen, have not supported growth from uredospores: potato dextrose agar, potato carrot agar, oatmeal agar, beef extract agar and malt agar (Kuhl *et al.*, 1971).

3. Mineral nutrition and pH

Most workers starting from Williams *et al.* (1966) have used Czapek's minerals or simple modifications, such as the inclusion of 2–6 g/l of $Ca(NO_3)_2 \cdot 4H_2O$ (Bose and Shaw, 1974a, 1974b) which was thought to stimulate growth on chemically defined media. Trace minerals were thought to be necessary on defined media containing calcium (Bose and Shaw, 1974a, 1974b) but not necessary on defined media lacking calcium (Foudin and Wynn, 1972). In most experiments the pH has been adjusted to 6·0–6·4, although Bose and Shaw (1974a, 1974b) adjusted the pH to 5·0 to avoid the precipitation of calcium supplements.

4. Supplements to the basic medium

Although a number of supplements have been claimed to improve both the reproducibility and amount of growth from uredospores (see also page 47) some reports are in conflict presumably owing to the variable growth which seems intrinsic to uredospores as inoculum. Thus, different laboratories disagree on the efficacy of suspending uredospores in gelatin (Coffey *et al.*, 1969; Bushnell, 1976), or of adding pectin (Kuhl *et al.*, 1971; Wong and Willetts, 1970; Coffey *et al.*, 1969) or citrate (Scott and Maclean, 1969; Kuhl *et al.*, 1971; Coffey *et al.*, 1969; Bose and Shaw, 1974b). Nonetheless, the addition of 100 mM citrate to 10× concentrated stock solutions of minerals is useful because it prevents the precipitation of phosphates during storage. Added vitamins and cofactors do not appear to stimulate growth on defined media (Bose and Shaw, 1974a, 1974b) although sufficient vitamins may be present as "carryover" nutrients in dense uredospore inocula. However, except for the report of Bushnell (1978), there appears to be general consensus that 1% bovine serum albumin stimulates the growth of *P. graminis,* particularly at low inoculum densities of uredospores (Scott, 1968; Kuhl *et al.*, 1971; Coffey *et al.*, 1969; Wong and Willetts, 1970; Bose and Shaw, 1974a).

Recent work by Grambow and co-workers to be discussed in relation to sporeling development (see page 48), indicates that an aqueous extract prepared by boiling wheat leaves, and 3,3′-bis-indolylmethane (BIM)—a degradation product of indole-3-acetic acid (IAA), both stimulate saprotrophic growth of *P. graminis tritici* from uredospores (Grambow and Tücks, 1979; Grambow and Müller, 1978; Grambow *et al.*, 1977).

5. Preparation and application of uredospore inoculum

The prolonged period of incubation required for saprotrophic mycelium to arise from uredospores seeded onto nutrient media requires strict attention to aseptic techniques. Contaminant-free uredospores have been obtained either by setting up aseptic leaf cultures (Williams *et al.*, 1966; Bushnell, 1968, 1976; Katsuya *et al.*, 1978; Foudin and Wynn, 1972; Kuhl *et al.*, 1971), or by collection from freshly-opened pustules on intact, rust-infected plants (Kuhl *et al.*, 1971; Williams, 1971; Green, 1976; Grambow and Riedel, 1977). Whereas the latter method is more convenient and less time-consuming, the former method is better suited to collecting large quantities of uredospores with very low levels of contamination.

When the main objective is to obtain saprotrophic growth rather than observe the development of individual sporelings, a dense spot inoculum applied to nutrient agar with a convenient instrument (e.g. spoon, spatula, camel's hair brush, cotton swabs have been used), and streaked on one side for a rough evaluation of density effects and sporeling development, is sufficient. An elegant method for producing a very uniform uredospore density over a large surface area, is to spray on a suspension of uredospores in volatile fluorocarbons (Kuhl *et al.*, 1971; Williams, 1971; Grambow and Reidel, 1977).

6. Growth conditions

There is no unequivocal evidence that any particular temperature within the limits 13–23°C is optimal for the most consistent initiation of saprotrophic growth from uredospores, and different laboratories have incubated cultures at high humidity in the dark at the following routine temperatures: 13–15°C (Bushnell and Stewart, 1971; Bushnell, 1976; Foudin and Wynn, 1972), 17°C (Williams *et al.*, 1966, 1967; Maclean and Scott, 1970; Bose and Shaw, 1971), 20°C (Green, 1976), and 23°C (Williams, 1971; Hartley and Williams, 1971a). Lower temperatures may be advantageous for large-scale experiments to screen many isolates of *P. graminis*, because of slower growth, whereas higher temperatures (20–23°C) are recommended for studies of sporeling development. Discrepencies between different groups of workers illustrate again the unpredictability and inconsistency of growth initiation from uredospores, and cannot be explained by varietal differences because most of the work cited above was carried out using an Australian isolate, race 126-ANZ-6, 7 of *P. graminis tritici*. The failure of cultures to grow in some experiments has been explained by exposure to light during observation (Coffey *et al.*, 1969), but in our experience there are no consistent differences in growth between undisturbed and intermittently-exposed cultures (Kuhl, Maclean, Scott and Williams, unpublished).

7. Growth assessment

The most critical problem in assessing growth during the initiation of a saprotrophic mycelium, is to determine at which stage of sporeling development the fungus crosses the threshhold from a dependence on stored nutrients, to growth maintained wholly by exogenous nutrients taken up from the medium. Many experiments have been assessed by constructing a growth scale based on visual observations, such as the 0–3 scale used by Kuhl *et al.* (1971), a modification of which is described below.

Growth stage 0. Germ tube or infection hypha unbranched.

Growth stage I. Limited primary and secondary brancing of germ tube or infection hypha.

Growth stage II. Extensive secondary and some tertiary branching.

Growth stage III. Extensive tertiary branching to give a tight mat of mycelium; individual colonies just visible to naked eye (mycelial mat 0·2–0.4 mm thick in densely inoculated zones).

Growth stage IV. Formation of a dense, leathery stroma; individual colonies > 1 mm diameter (mycelium > 1 mm thick in densely inoculated zones).

Growth stage I can be attained on water agar, and accordingly represents an extension of the process of spore germination. Growth stage II requires exogenous nutrients, and encompasses the transition from endogenous to exogenous nutrients, but special experiments would have to be devised to determine whether or not this growth is totally independent of nutrients carried over from the spore. Thus, growth which terminates at stage II cannot unequivocally be regarded as saprotrophic.

For thinly seeded cultures, attainment of growth stages III or IV by colonies arising from a single spore or small group of spores clearly indicates the formation of a saprotrophic mycelium. However, for mass inocula it is almost impossible to trace the development of individual colonies from stage II onwards, and mycelium assessed at early stage III may represent somewhat equivocal saprotrophic growth. Only stage IV appears to be capable of supporting sporulation. Spores are borne on a thick, leathery mass of hyphae termed a stroma (cf. Williams *et al.*, 1967; Maclean and Scott, 1970). Stage III–IV growth is referred to as a "primary mycelium", and usually terminates growth by entering a terminal reproductive phase, or by becoming necrotic ("staling"). In general, it is not possible to maintain a primary mycelium of *P. graminis* by serial subculture (Maclean and Scott, 1970) (but see pages 49, 52 and 58).

8. Growth in relation to race and inoculum density

It is now firmly established that races of *P. graminis* f.sp. *tritici* differ in the ease with which they can be cultured (Kuhl *et al.*, 1971; Bushnell and Stewart, 1971; Bushnell, 1968, 1976; Bose and Shaw, 1974a; Green, 1976; Hartley and Williams, 1971a, 1971b). This variability is also shown by different form species of *P. graminis:* f.sp. *avenae* (oat stem rust) and f.sp. *secalis* (rye stem rust) (Kuhl *et al.*, 1971; Green, 1976). It was observed even in the early culture experiments that growth was best and most consistent in regions of high inoculum density (Williams *et al.*, 1966; Bushnell, 1968). Indeed, it is perhaps fortuitous that the first isolate cultured from uredospores was race 126-ANZ-6, 7, which we now know to have a high potential for initiating growth at relatively low inoculum densities compared to other races of *P. graminis* (Hartley and Williams, 1971b).

Early experiments with dilute peptone-yeast extract media used inoculum densities of race 126-ANZ-6, 7 within the range 100–2000 uredospores mm^{-2} (Williams *et al.*, 1966, 1967; Bushnell, 1968; Maclean and Scott, 1970; Bose and Shaw, 1971), although Coffey *et al.* (1969) obtained good growth at 20–40 uredospores mm^{-2} when inoculated as gelatin suspensions. Using media enriched in peptone, Hartley and Williams (1971b) have grown this race at spray-inoculated densities of 20 uredospores mm^{-2}, and isolated a strain of this race (culture No. 70165) which grew at densities as low as 2 uredospores mm^{-2}. Hartley and Williams (1971b) compared 11 Australian races for the minimum uredospore density required for growth, and noted that two races, 126-ANZ-1, 6, 7 and 126-ANZ-1, 4, 6, 7 did not grow even from a large, mass inoculum, whereas Green (1976) using essentially the same techniques failed to culture only one of 29 Canadian races tested with a mass inoculum, although a number of attempts were necessary for some races. In similar experiments, Bushnell (1976) found that one American race in particular, an isolate of race 17 grew very inconsistently, but that increasing the inoculum to 4–10 mg of uredospores ($\sim$2–4 $\times$ 10^6 uredo-spores) and decreasing the volume of enriched agar medium from 15 ml to 3 ml greatly improved both the reproducibility and amount of saprotrophic growth; however, some replicates still failed to grow. Supplements to the medium and procedures cited by other workers as being possibly beneficial and growth stimulatory, did not improve the growth of this isolate.

The effect of inoculum density on the frequency of uredospores of race 126-ANZ-6, 7 which initiate saprotrophic growth was examined critically by Kuhl *et al.* (1971), who used a spray-inoculation technique to obtain very uniform inoculum densities. These workers observed much variability, which appeared to result from a cessation of growth which overtook members of a population in increasing numbers, beginning at early stages of sporeling development 2–3 days after inoculation. Although it was clear from a number of experiments which included the use of "conditioned" media that sporelings mutally stimulated one another's growth, a maximum of only 4–6% of the initial population of sporelings could be induced to reach growth stage III (macroscopically-visible colonies).

Further information on mutual stimulation of sporeling development was obtained by Hartley and Williams (1971a, 1971b) who found that morphological diversity in a series of spray-inoculated cultures, ostensibly race 126-ANZ-6, 7, was caused by a contaminated inoculum consisting of a mixture of race 21-ANZ-1, 2, 3, 7 and two strains of race 126-ANZ-6, 7 (designated cultures 334 and 70165; culture 334 of 126-ANZ-6, 7 repre-sented the parental culture of this rust used originally by Williams *et al.*, 1966, 1967). Co-culture experiments set up by deliberately mixing the inoculum showed that saprotrophic colonies of race 21-ANZ-1, 2, 3, 7, which itself formed small stage III–IV colonies, stimulated sporelings of race 126-ANZ-6, 7 (culture 70165) to a higher frequency of stage III–IV colony formation than when grown alone. Furthermore, colonies of this strain of race 126-ANZ-6, 7 were larger when co-cultured with 21-ANZ-1, 2, 3, 7.

Further co-culture experiments showed that the two races of *P. graminis tritici*, 126-ANZ-1, 6, 7 and 126-ANZ-1, 4, 6, 7, which could not be induced to grow saprotrophically by themselves even from a mass-inoculum, were eventually observed to form a few large stage III–IV colonies when co-cultured with the "stimulator" strain, race 21-ANZ-1, 2, 3, 7 (which as noted above, characteristically forms small stage III–IV colonies). However, the genetic identity of the different size classes of colonies was not checked by reinfection experiments as in the other experiments cited above, and the evidence that the large colonies represented races 126-ANZ-1, 6, 7 and 126-ANZ-1, 4, 6, 7 was circumstantial (Hartley and Williams, 1971a).

In one series of experiments by Hartley and Williams (1971a), results were obtained consistent with the production of toxic substances by races 126-

ANZ-1, 6, 7 and 126-ANZ-1, 4, 6, 7, both of which as noted above are difficult to culture alone. When either of these races constituted 50% of a mixed inoculum with the "stimulator" race 21-ANZ-1, 2, 3, 7, both races in the mixture died before producing saprotrophic colonies, and the proportion of the stimulator race had to be raised to 85–90% of the inoculum to overcome the inhibitory effect of the other races.

C. Transition of P. Graminis from Sporeling to Saprotroph

1. General comments

Because saprotrophic growth is variable from an inoculum of uredospores, it was suggested by Scott and Maclean (1969) that a reorientation of metabolism occurs at some stage during sporeling development, and that this transition is necessary for the fungus to use exogeneous nutrients for sustained growth. Thus, a lag phase of 3–10 days from inoculation is necessary before stage I of growth is reached (cf. references cited on page 42). Kuhl et al. (1971) further suggested that a "random event" determines whether or not such a metabolic transition takes place; such a random event might for example be the presence or absence of particles of a stimulator or inhibitor of host origin, or differences in the microclimate of individual sporelings during incubation on nutrient medium.

It is useful to compare the transition from sporeling to saprotroph, with the transition from sporeling to biotroph in which infection structures are formed during penetration of host plants. Infection structures can also be formed in vitro by appropriate chemical or physical treatment of sporelings. Williams and co-workers and Grambow and co-workers have studied the morphology and cytology of germinating uredospores on nutrient media in the presence and absence of treatments and substances which induce the formation of infection structures or which stimulate the initiation of saprotrophic growth. Germination and maximum germ tube elongation (stage 0 of growth) is usually complete within 24 hours after inoculation (Scott and Maclean, 1969); if infection structures are formed, they do so during this 24 hour period.

Opinions differ as to whether or not a saprotrophic mycelium can develop only from infection structures, as is the case with biotrophic mycelium in a host plant (cf. Dickinson, 1949; Chakravarti, 1966), or whether a saprotrophic mycelium can only develop from undifferentiated germ tubes. Despite an earlier view that a dikaryotic, saprotrophic mycelium could develop from either germ tubes or infection structures (Williams et al., 1966), Williams (1971) later concluded that infection structures were necessary to organize the conjugate nuclear division essential to maintain a dikaryotic hypha, and that only dikaryotic hyphae were capable of sustained saprotrophic growth. On the contrary, Grambow and Müller (1978) concluded that dikaryotic, saprotrophic hyphae could develop either from infection structures or from undifferentiated germ tubes. The observations of each of these groups of workers require further critical consideration.

2. Cytology of sporeling development on nutrient agar

Williams (1971) determined the positions of nuclei and septa in mixed populations of differentiated and undifferentiated sporelings on nutrient medium. Septa were formed irregularly in undifferentiated germ tubes, resulting in cells containing 0, 1, 2, or 3 nuclei with uninucleate cells being most common. Monokaryotic hyphae which grew from some of these cells were narrow and highly branched, and ceased growth at an early stage III. In contrast, branched hyphae growing from infection structures were always dikaryotic and were thicker and less highly branched than the monokaryotic hyphae. Hartley and Williams (1971a, 1971b) described the subsequent development of dikaryotic hyphae to form substantial stage IV colonies, and observed that some races of *P. graminis* differed in development patterns.

Williams (1971) suggested that infection structures represent the location for changes in gene expression necessary to regulate the metabolism of vegetative cells. However, it was not demonstrated that incubation conditions which induce high frequencies of sporelings to form infection structures, overcame inconsistent growth. In this regard, the use of heat shock or cuticle wax to induce the formation of infection structures, did not overcome inconsistent growth between replicate inocula of an American race 17 (Bushnell, 1976) or an English race 21 (D. J. Maclean, unpublished) of *P. graminis tritici.*

Grambow and co-workers investigated the time-course of development of sporelings in the presence and absence of treatments which induce the formation of infection structures, and additives which stimulate the initiation of saprotrophic growth. An aqueous extract prepared by boiling wheat leaves, and 3, 3'-bis-indolylmethane (BIM), each stimulated saprotrophic growth from undifferentiated sporelings of race 32 of *P. graminis tritici* (Grambow and Müller, 1978; Grambow *et al.*, 1977). The presence of these stimulants in nutrient medium increased the fraction of germ tubes which after a lag phase of three days branched to form hyphae, and also increased the number of branches per sporeling (Grambow and Müller, 1978). The hyphae which grew from these undifferentiated germ tubes were dikaryotic, in contrast to the monokaryotic hyphae observed by Williams (1971).

Mild heat shock in the presence or absence of BIM or leaf extract induced the formation of infection structures within 24 hours of inoculation (Grambow and Müller, 1978). Nuclear division and branching of the infection hyphae began soon after differentiation, resulting in the growth of a stage I dikaryotic mycelium within three days of inoculation. Hyphae originating either from infection structures or directly from germ tubes were very similar in morphology, and four to six days after inoculation 93–97% of hyphal cells were dikaryotic. Monokaryotic hyphae appeared to be absent, although at one day after inoculation unbranched germ tubes sometimes produced abnormal cells containing 0, 1, 3 or higher numbers of nuclei.

The differences in nuclear condition of undifferentiated sporelings observed by Williams (1971) and Grambow and Müller (1978) are not yet resolved, and may well be related to differences in the nutrient medium (Williams did not add BIM or leaf extract to his medium), the higher inoculum densities used by Grambow and Müller, or by the different races of rust used in each laboratory, resulting in a higher incidence of abnormal sporeling development in Williams' experiments. Neither of these groups of workers observed anastomoses between germ tubes, a mechanism proposed by Bose and Shaw (1971) to be a necessary prerequisite for the formation of vegetative colonies.

3. Compounds which affect sporeling morphogenesis

The discovery by Grambow et al. (1977) of the effect of BIM and leaf extracts on the morphogenesis of undifferentiated germ tubes, proceeded from a general search by Grambow and Reisener (1976) for active compounds of host origin which affect sporeling development. BIM is a degradation product of indole-3-acetic acid (IAA) metabolism in wheat leaves (Grambow and Tücks, 1979), and was identified in autoclaved preparations of gramine, an indole derivative (Grambow et al., 1977). At concentrations frequently used for plant growth regulators ($1–6 \times 10^{-2}$M), BIM; (i) shortened the onset of branching of germ tubes, (ii) increased the number of hyphal branches per germ tube, (iii) increased the propotion of sporelings in a population which branched, and (iv) increased the yield of saprotrophic mycelium obtained three weeks after inoculation (Grambow and Tücks, 1979; Grambow and Müller, 1978; Grambow et al., 1977). On differentiated germ tubes, BIM stimulated the rate of growth of hyphae originating from infection structures (Grambow and Müller, 1978).

Further information on factors which affect the formation of infection structures has been obtained by testing various volatile fractions, solvent and aqueous extracts, and hydrolysis products obtained from leaf material

(Grambow and Riedel, 1977; Grambow and Grambow, 1978; Grambow, 1977), and by testing spores obtained from pustules of different ages (Williams, 1971). Two chemical stimulants of host origin: a volatile component plus a non-volatile compound (possibly phenolic) need to be used simultaneously to induce the formation of infection structures in the absence of heat shock (Grambow and Riedel, 1977; Grambow and Grambow, 1978; Grambow, 1978). Further work is necessary to establish whether or not any of the above compounds or treatments which affect sporeling development, can also increase the final yield of saprotrophic mycelium, or can eliminate the inconsistency in growth observed between replicate experiments.

D. Origin of Variant Strains of P. Graminis

Variant strains of *P. graminis tritici* distinguishable from the wild-type by their nuclear content, growth rate, gross morphology and pathogenicity, have been isolated on random occasions. On each occasion the experiments were set up to obtain a primary mycelium, which as noted previously (page 44) is difficult to maintain by serial subculture. The variant strains were generally looser in texture and less pigmented than primary mycelium, and because they could be maintained readily by serial transfer were termed "persistently vegetative colonies" (Maclean and Scott, 1970), "vegetative cultures" (Green, 1976), or "continuously subculturable, uninucleate lines" (Williams and Hartley, 1971).

Maclean and Scott (1970) first reported the isolation of variant strains in experiments where a mass inoculum of $\sim 10^6$ uredospores of race 126-ANZ-6, 7 germinated but failed to form a primary mycelium on liquid medium. These "failed" cultures were put aside in a cupboard and forgotten until about two months later, when they were found to contain 2 to 30 small white mycelial tufts per culture flask, floating on the liquid surface amidst the necrotic remains of sporelings (Fig. 1). Maclean (1971) also observed variants to arise after extended incubation on agar-solidified media, again after sporelings had failed to form a primary mycelium. In contrast, Williams and Hartley (1971) isolated variants as sectors from primary mycelium and subcultures of race 126-ANZ-6, 7. North American races of *P. graminis* have also produced variants as sectors from subcultures of primary mycelium (Green, 1976; Bushnell, 1976; Bushnell and Stewart, 1971) although Bushnell (1976) thought his subculturable lines had "adapted" to the medium rather than being genetically variant. Most of the variants isolated by Maclean and Scott (1970) and Williams and Hartley (1971) were monokaryotic when first examined cytologically, although the variant of Green

Fig. 1. Colonies of variant strains of *P. graminis* floating on the surface of liquid medium amidst masses of necrotic sporelings, 60 days after inoculation with uredospores. (From Maclean and Scott, 1970.) Magnification, about actual size.

(1976) and most of the cultures of Bushnell (personal communication) were dikaryotic.

No reports are available on observations of a variant arising from a single sporeling or single cell in a primary culture, probably because variants are formed at very low frequencies as suggested by the liquid culture experiments of Maclean and Scott (1970).

E. Other Rust Species Cultured from Uredospores

As well as *P. graminis*, the following other rust fungi have also been cultured from uredospores (a substantial mycelium was obtained by those authors indicated by an asterisk):

Melampsora lini (flax rust): Turel (*1969a, *1969b, *1971, *1972, *1973); Coffey *et al.* (*1970); Coffey and Shaw (*1972); Coffey and Allen (*1973); Bose and Shaw (*1974a, *1974b); Quick and Cross (*1971).

Uromyces dianthi (carnation rust): Jones (*1972a, *1972b, *1973a, *1973b).

Puccinia coronata (crown rust of oats): Kuhl *et al.* (1971); Siebs (1971) Ingram and Tommerup (1973); Jones (*1974); Green (1976); Katsuya *et al.* (*1978).

Puccinia helianthi (sunflower rust): Coffey and Allen (*1973).

Puccinia recondita (brown leaf rust of wheat): Kuhl *et al.* (1971); Ingram and Tommerup (1973); Green (1976); Raymundo and Young (1974; Katsuya *et al.* (*1978).

Phragmidium mucronatum (rose rust): Rahbar Bhatti and Shattock (*1980a).

Phragmidium violaceum (blackberry rust) and also *P. tuberculatum* and *P. rubi-idaei:* Rahbar Bhatti and Shattock (*1980b).

Puccinia striiformis (yellow stripe rust of wheat): Williams (1976); Ingram and Tommerup (1973).

Melampsora populnea (pine stem rust fungus): Vasil'ev and Saplina (1971).

Melampsora allii-populina: Riou *et al.* (1975).

Attempts to culture the maize rust *Puccinia sorghi* (Kuhl *et al.*, 1971) and the coffee rust *Hemileia vasatrix* (D. J. Maclean, unpublished) by modifications of the methods used for *P. graminis*, have failed. Presumably, attempts have been made to culture other rust species, but failures have not been published.

The major features associated with the culture of the above species are variability in the amount and type of growth obtained between experiments, and the need for an adequately balanced organic nitrogen source. Chemically defined media with organic constituents of a carbohydrate, a sulphur amino acid, and at least one other amino acid, have supported the initiation of growth of *M. lini* (Turel, 1973; Coffey and Shaw, 1972; Coffey and Allen, 1973; Bose and Shaw, 1974a, 1974b), *P. coronata* (Seibs, 1971) and *P. helianthi* (Coffey and Allen, 1973); otherwise, various peptones or yeast extract have served as nitrogen and sulphur source. Recent experiments on *Phragmidium* species (Rahbar Bhatti and Shattock, 1980a, 1980b) suggest that saprotrophic growth of this genus may be initiated more consistently than with most other rust genera. Nonetheless, all species grew better at higher inoculum densities, although problems with self-inhibition of germination by uredospores of *M. lini* (Turel, 1969; Coffey *et al.*, 1970) had first to be overcome using liquid media (Bose and Shaw, 1974a, 1974b). Extremely long lag phases (up to 13 weeks) were observed in some experiments before macroscopically-visible growth was obtained of the following species: *M. lini* (e.g. Coffey *et al.*, 1970), *U. dianthi* (Jones, 1972), *P. coronata* (Jones, 1974) and *P. helianthi* (Coffey and Allen, 1973).

Melampsora lini was particularly temperature-sensitive (Turel, 1969b; Coffey *et al.*, 1970) and was not grown at temperatures above 17°C, and was subcultured using chilled instruments. Care had to be taken not to mutilate mycelium to be used as inoculum for subculture, and even so inoculum pieces larger than 3–10 mm diameter were necessary to ensure successful subculture. Czapek's mineral solution as used for *P. graminis* was unsuitable for *M. lini*, which grew on modified Knop's minerals (Coffey and Shaw, 1972). Vigorous and consistent growth with prolific sporulation was reported on a chemically defined liquid medium supplemented with additional calcium (Bose and Shaw, 1974a, 1974b).

Continued growth of subcultures has been obtained with *M. lini* (Turel, 1969a, 1969b; Coffey and Shaw, 1972), *U. dianthi* (Jones, 1973a), *P. coronata* (Jones, 1974); *P. helianthi* (Coffey and Allen, 1973) and *Phragmidium* spp. (Rahbar Bhatti and Shattock, 1980a, 1980b). The formation of a stroma and consequent sporulation has been observed in aged cultures of *M. lini* (Coffey *et al.*, 1970; Turel, 1969), *P. coronata* (Katsuya *et al.*, 1978; Jones, 1974), *P. recondita* (Katsuya *et al.*, 1978) and *Phragmidium* spp. (Rahbar Bhatti and Shattock, 1980a, 1980b), whilst spore-like bodies have been observed in cultures of *U. dianthi* (Jones, 1972a, 1972b).

III. Axenic Culture from Mycelium in Infected Plants

A. From Dual Tissue Cultures

1. Development of the technique

The concept of using dual tissue cultures as an approach to isolating axenic cultures of otherwise intransigent biotrophic fungi, was first suggested by Morel (1944, 1948; cf. Brian, 1967 for a short review), and successfully applied to rust fungi by Cutter and co-workers. Cutter's approach was to set up pieces of rust-infected plants on tissue culture media under conditions which would encourage balanced growth of the fungus on host callus. Because some plant-rust combinations might allow the callus to outgrow the fungus (e.g. Turel and Ledingham, 1957), or the fungus to overgrow the host and thus destroy its source of nutrients (cf. Ingram and Tommerup, 1973),

Species and citations	Media and growth conditions	Observations and comments
(a) From dual tissue cultures		
1. *Gymnosporangium juniperi-virginianae* Hotson and Cutter (1951) Cutter (1959)	Modified Gautheret (1942) medium; includes 3% glucose or sucrose, 15% coconut milk, 1% yeast extract (?), pH ~5·3, 20°C/constant illumination at 350 foot candles	7 axenic isolates from 2617 primary or secondary cultures
2. *Uromyces ari-triphylli* Cutter (1952) Cutter (1960a)	Medium similar to above, but yeast extract omitted. Conditions as in 1 above	5 axenic isolates from 1603 primary or secondary cultures
3. *Puccinia malvacearum* Cutter (1960b)	Media and conditions as in 2 above	1 axenic isolates from 571 primary or secondary cultures
4. *Cronartium ribicola* Harvey and Grasham (1974)	Modified Czapek or tissue culture minerals, plus: 1% glucose, 0·1% yeast extract, 0·1% peptone (Evans or Bacto), 1·0% bovine serum albumin, 16 h light (21°C)/8 h dark (5°C), pH 5·5	15 axenic isolates from 15 dual cultures; 29 axenic isolates from 50 mycelial inocula grown in dual culture
5. *Cronartium fusiforme* Hollis *et al.* (1972)	Tissue culture minerals, plus: 3% glucose, 0·1% yeast extract, 0·1% peptone (Difco), pH 6·2–6·4	1 axenic isolate from about 140 cultures
(b) From senescing plant tissues		
6. *Puccinia coronata* 7. *P. striiformis* 8. *P. recondita* Ingram and Tommerup (1971)	Czapek's minerals, plus 0·3% sodium citrate, 4% glucose, 1% peptone (Evans), pH 6·0, 15°C, 16 h low light/8 h dark Inoculate with 1·5 cm segments of infected leaves	1 week: host cells yellow; 2 weeks: host cells dead; 3 weeks: aerial hyphae on leaf surface; 2 months: large stromata 5–10 cm diameter
9. *Melampsora occidentalis* Lane and Shaw (1974)	Minerals plus 5% sucrose, 0·6% aspartic acid, 0·056% cysteine, 17°C in the dark Inoculated with single pustules excised from poplar leaves	1 week: stromata develop in pustules; 2 weeks: mycelium emerging from pustules; 4 months: 30% of inocula produce saprotrophic mycelium
(c) From parasitic mycelium isolated from plants		
10. *Melampsora lini* Lane and Shaw (1972), and personal communication	Minerals plus 4% sucrose, 0·1% peptone (Evans), 0·1% yeast extract and 1% defatted bovine serum albumin; 17°C in the dark Inoculum: mycelium isolated from flax cotyledons by enzyme digestion	Slow phase of ~ one month, followed by saprotrophic growth; 100% of inocula grew

Cutter achieved most success with naturally-balanced systems as starting material, in which the host was already systemically infected by the fungus.

In an unpublished account of his work, Cutter* catalogued 23 151 attempts at obtaining dual cultures from a total of 41 rust species infecting 60 vascular plants. However, only seven infected callus systems were capable of propagation by subculture: *Gymnosporangium juniperi-virginianae* in *Juniperus* spp, *Uromyces ari-triphylli* in *Arisaema triphyllum*, *Gymnoconia peckiana* in *Rubus canadensis*, *Puccinia malvacearum* in *Althea roses*, *P. peridermiospora* in *Fraxinus americana*, *P. rubigo-vera* in *Clematis virginiana* and *P. bolleyana* in *Sambucus canadensis*. Only the first three of the above systems resulted in the isolation of axenic cultures of the fungal partner (Table I). Other workers have subsequently succeeded in isolating axenic cultures of the fungal partner from two other infected callus systems: *Cronartium ribicola* in *Pinus monticola* (Harvey and Grasham, 1974), and *C. fusiforme* in *Pinus elliottii* (Hollis *et al.*, 1972) (cf. Table I).

The problems posed by Cutter's approach are illustrated by the published failures of other workers to isolate saprotrophic mycelium of the following rust fungi from infected callus systems: *G. juniperi-virginianae* (Rossetti and Morel, 1958; Constabel, 1957; Turel and Ledingham, 1957), *Puccinia antirrhini* and *P. minutissima* (Rossetti and Morel, 1958), *P. helianthi* (Nozzolillo and Craigie, 1960), *P. tatarica* (Bauch and Simon, 1957; Witkowski and Grümmer, 1960) and *Melampsora lini* (Turel and Ledingham, 1957). The experiments on *M. lini* in flax tissue cultures are of particular interest because this fungus was eventually cultured from uredospores; media which supported growth from uredospores did not support growth of mycelium obtained from dual tissue cultures (Turel, 1969a). Details of experiments which led to the isolation of axenic cultures of rust fungi from dual tissue cultures as listed in Table I, are summarized below.

2. Gymnosporangium juniperi-virginianae

The dikaryophase of this fungus is present in telial galls ("cedar apples"), produced by systemic infections on *Juniperus* spp. On seven independent occasions four to nine months after callus initiation, mycelium from one of the primary cultures and six of the many secondary cultures grew out onto the agar medium (Table I), from whence it was maintained independently of the host callus by transfer and subsequent serial subculture (Hotson and Cutter, 1951; Cutter, 1959). Immediately prior to the emergence of saprotrophic mycelium from the dual culture, Cutter observed necrosis in surface tissues of the host, prompting the suggestion (Cutter, 1951) that the fungus

---

* Unpublished manuscript provided posthumously by Dr Lois Cutter.

changed from a biotroph to a necrotroph prior to becoming a saprotroph. Later however, Cutter (1959) speculated that a prior change in host callus could select for variant fungal nuclei with saprotrophic tendencies. At the time of their isolation, four strains were dikaryotic while the other three strains were monokaryotic.

All media contained 3% sucrose with 15% coconut milk in a modified Gautheret's agar (Gautheret, 1942; Cutter, 1959), and 1% yeast extract was added to media during isolation of the first axenic isolate (Hotson and Cutter, 1951) as it was thought to stimulate callus growth. It was not clearly stated if yeast extract was used during isolation of the other six strains (Cutter, 1959), but it was presumably omitted, as Cutter noted that only coconut milk of the many supplements tested (including yeast extract) was found after further experimentation to stimulate callus growth (Cutter, unpublished results).

3. Uromyces ari-triphylli

This fungus (previously named *U. caladii*) was isolated from infected callus obtained from systemically-infected, subterranean corms of *Arisaema triphyllum* (Cutter, 1952, 1960a) (Table I). On five independent occasions, mycelium grew out onto the agar medium under circumstances resembling the isolation of *G. juniperi-virginianae*, in that prior deterioration of host callus tissues was observed. The five axenic cultures were all dikaryotic, and grew readily after subculture without morphological or nuclear variation being evident.

4. Puccinia malvaearum

A single saprotrophic isolate of this fungus was obtained from explants of rust-infected *Althea rosea* on tissue culture media (Cutter, 1960b, and unpublished results; cf. also Scott and Maclean, 1969). The saprotrophic mycelium probably arose directly from a teliospore or basidiospore derived from the explant.

5. Cronartium ribicola

The haploid stage of *C. ribicola* was isolated in axenic culture by Harvey and Grasham (1974) after much perseverence with dual cultures. These workers initiated infected callus from systemically-infected cankers of *Pinus monticola*, western white pine (Harvey and Grasham, 1969), and conducted a series of experiments on the characteristics of the fungal partner in this two-membered system (e.g. Harvey and Grasham, 1970, and other papers cited by Harvey and Grasham, 1974). A host-free inoculum of the fungus was prepared by growing mycelium on a dialysis membrane overlaying host callus (Harvey and Grasham, 1970). When transferred to tissue culture media supplemented with an organic nitrogen source (Table I), 58% of these membrane-grown mycelial inocula continued to grow as saprotrophs (Harvey and Grasham, 1974). The high consistency of initiation of saprotrophic growth by this fungus was illustrated by other experiments: when infected callus was placed on the supplemented medium, saprotrophic mycelium grew out onto the medium in all (100%!) replicates. No morphological variants were observed during maintenance of axenic cultures by serial subculture.

6. Cronartium fusiforme

This fungus was isolated from dual cultures set up from segments of branch galls of *Pinus elliottii* (slash pine) systemically-infected with *C. fusiforme* (Hollis *et al.*, 1972). Of over 200 gall segments set up, about two-thirds survived contamination and formed rust-infected callus; five of these cultures produced mycelial outgrowths onto the medium after about six weeks incubation. Only one of these outgrowths was successfully subcultured when freed of host callus. This isolate grew from a necrotic host explant, in a manner reminiscent of the mycelium obtained from dual cultures by Cutter (cf. pages 54–55). Subcultures were maintained on a yeast extract-peptone medium (Table I).

B. From Senescing Plant Tissues

This approach was first reported by Ingram and Tommerup (1973), who placed segments of rust-infected cereal leaves on a nutrient medium supplemented with 1% Evans peptone, and incubated in low light at 15°C. These conditions kill the host "gently" by accelerating senescence. *Puccinia coronata* in first leaves of oat, and *P. striiformis* and *P. recondita* in first leaves of wheat, were tested in the above manner by using leaf segments harvested immediately prior to the opening of uredosori (Table I). Host tissues became yellow within one week, and appeared to be dead after two weeks. After three weeks incubation, aerial mycelium was visible on the surface of the dead host tissue, but did not spread onto the culture medium. However, the mycelial mass continued to expand, and after two months incubation had formed large, sporulating stromata, 5–10 mm in diameter. This slow but continuous fungal growth suggests that after the death of host cells, most of the mycelial increase was supported by exogenous nutrients originally present in the medium, rather than by unmodified or autolysed host materials. No attempts were made to subculture the stromata.

Lane and Shaw (1974) used methods similar to those of Ingram and Tommerup (1973), to culture the poplar rust *Melampsora occidentalis* (Table I). Single rust pustules plus a 1 mm surrounding ring of uninfected tissue, were excised from surface-sterilized leaves of black cottonwood, *Populus trichocarpa*. The leaf pieces were placed pustules up on a defined agar medium (Table I). Mycelium in the pustules continued growth, and four months after inoculation had grown out onto the medium from 30% of the pustules. This saprotrophic mycelium was maintained in axenic culture by serial transfer at 3–5 week intervals. Some three-month-old colonies formed stromata which bore uredospores.

C. From Parasitic Mycelium Isolated from Plants

Lane and Shaw (1972) cultured the flax rust *M. lini* by employing an enzyme digestion method to isolate mycelial colonies from infected flax cotyledons (Table I). The cotyledons were harvested when infection flecks were clearly visible, but before the sporogenous hyphae had formed a dense yellow stroma. After appropriate treatment, infected flax tissue was incubated in a mixture of hydrolytic enzymes which differentially digested plant but not fungal cell walls. The host mesophyll cells were liberated as protoplasts, leaving the fungal colonies intact but sometimes attached to pieces of host epidermis and vascular tissue. When transferred to nutrient agar, all (100%) of the isolated colonies resumed growth, and after a slow period of about one month, expanded into the large spreading colonies characteristic of *M. lini* isolated from uredospores (cf. page 52). This technique is not subject to the problems of inoculum density and inconsistent growth experienced with uredospores as inoculum, presumably, as suggested by Lane and Shaw (1972), because the transition stage between sporeling and vegetative growth has been by-passed.

IV. Characteristics of Established Cultures

A. Cultural Stability

1. General considerations

Sections II and III of this chapter have dealt with the problems of establishing rust fungi in axenic culture. However, the "primary cultures"

observed immediately after establishment represent an organism in transition: growth as a saprotroph presents a new set of selection pressures to
which these normally biotrophic fungi are not subjected under natural
conditions. Therefore, random genetic variants which grow more vigorously
than the parental genotype will be at a selective advantage during the
establishment and maintenance of axenic cultures. However, clearly-
identified genetic variants have only been observed to arise from primary or
established cultures of *P. graminis* and *G. juniperi-virginianae*, as demonstrated by changes in the number of nuclei per cell. Other species which have
possibly formed variants include *M. lini* (Coffey and Shaw, 1972; Turel,
1971), *P. malvacearum* (Cutter, unpublished results), *U. dianthi* (Jones,
1972a, 1972b, 1973a) and *P. coronata* (Jones, 1974), as suggested by
extremely long lag periods before growth was initiated from uredospores, by
progressive improvements in the viability and growth rate of subcultures
during maintenance of established cultures, or by changes in (or loss of)
pathogenicity after extended periods of growth in axenic culture.

2. *Puccinia graminis*

Variants of this species have been readily recognized, possibly because
primary cultures are more difficult to maintain by serial subculture than
other rust species so far grown in axenic culture. Most variants were monokaryotic when first examined cytologically (Maclean and Scott, 1970; Williams and Hartley, 1971) although one variant was dikaryotic (Green, 1976).
Other monokaryotic variants produced dikaryotic strains during maintenance by serial subculture (Maclean, 1974; Green *et al.*, 1978). Variant
strains were themselves unstable, and during maintenance by serial subculture produced morphologically and culturally distinct strains [Maclean and
Scott, 1970; Maclean, 1974; Rawlinson and Maclean. 1973 (cf. Fig. 2)]. When
propagated saprotrophically by serial subculture, each strain tended to
retain its distinctive characteristics, often for one to two years or more,
before morphological heterogeniety again became evident. Some strains
actually decreased in growth rate (Maclean, 1974). It is interesting that 13
distinct variants all isolated from race 126-ANZ-6, 7, each contained virus-
like particles (Rawlinson and Maclean, 1973). Very similar virus-like particles were also seen in uredospores of an English isolate of race 21 and in
wheat leaves infected with this race, but not in uninfected leaves or in
rust-free portions of infected leaves. It is possible that genetic changes in
viruses could explain some of the observed variability (Rawlinson and
Maclean, 1973).

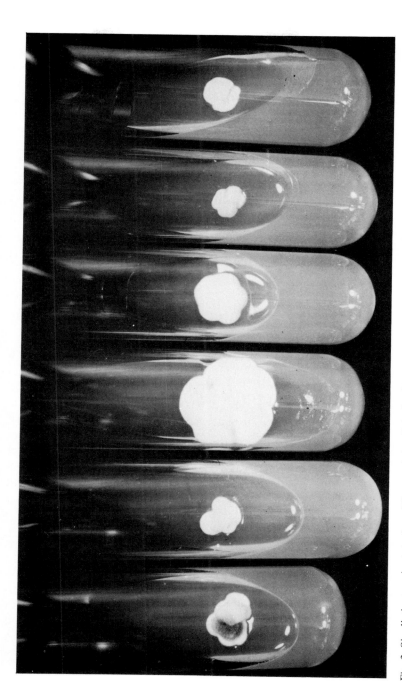

Fig. 2. Six distinct variant strains of *P. graminis*, all derived from strain V1 of Maclean and Scott (1970). Subcultures were grown at 20°C for 48 days. (Unpublished results of D. J. Maclean.) Magnification, about actual size.

Table II. Results of inoculation of saprotrophically-grown mycelium of *Gymnosporangium juniperi-virginianae* on host tissues.[a]

| | | No. nuclei per cell | | [f]Interaction with alternate hosts | | | | [f]Interaction with primary host, Juniperus |
| | | within 30 days of isolation | later | Pyrus | | Crategus | | |
Strain	Date isolated			[g]Infected Aecia	Spermogonia	[g]Infected Aecia	Spermogonia	[g]Infected Telia
1. C-31-49	Feb. 1950	2	1, 2[c] (Oct. 1952)	+	−/+	+	−/+	+
2. C-94-50	Apr. 1951	2	1 (>Oct. 1952)	+	+[d]	−	−	+/−
3. NY-150-50[b]	Mar. 1951	1	1 (>Oct. 1952)	+	−	−	−	−
4. C-121-51	Apr. 1951	1	1 (>Oct. 1952)	−	+	−	−	−
5. C-286-51	Sept. 1952	1	1 (>Oct. 1952)	+/−	−	−/+	−	−
6. NC-43-52	Apr. 1953	2	1 (>Oct. 1952)	+	+	−	−	+
7. NC-219-54	May 1955	2	[e]n.d.	+/−	−	−	−	+/−

[a]Data taken from Cutter (1959, and unpublished); [b] Culture difficult to maintain, lost 18 months after isolation; [c]Mostly dikaryotic with occasional monokaryotic sectors; [d]Tested after mycelium had become monokaryotic; [e]n.d. = not determined; [f]Symbols, + = successful, − = unsuccessful, +/− = success varied on different occasions, the most frequent being indicated first; [g]For brevity, no distinction is made between inoculations on tissue cultures, and greenhouse or field-inoculated plants.

3. Gymnosporangium juniperi-virginianae

Each of the seven strains of G. *juniperi-virginianae* isolated by Cutter (1959) was derived from a different telial gall, obtained from *Juniperus* plants growing in different localities in the eastern U.S.A. Because there is no repeating (uredial) stage of this rust fungus the strains were presumably of genetically diverse origin, each ultimately derived from a different aeciospore. Although mycelium of G. *juniperi-virginianae* is dikaryotic in telial galls, four of the strains were predominantly dikaryotic within 30 days of isolation whereas the other three strains were completely monokaryotic (cf. Table II and page 55). After serial subculture, two of the initially dikaryotic strains became monokaryotic, whereas a third strain (C-31-49) although mostly dikaryotic in young colonies, produced sectors of predominantly monokaryotic mycelium in older colonies. When grown on a range of mycological media, this strain formed pale pink to orange colonies during the first 25 days of growth, after which red, purple, brown, grey and pure white sectors frequently appeared. One of the initially monokaryotic strains, C-121-51, grew much more slowly than the others and eventually died out 18 months after its isolation during which time it had been serially transferred six times.

Fruiting bodies or structures which resembled them were produced by axenic cultures of some strains. Teliospores were produced only by strains C-31-49 and NC-219-54 but very irregularly. Strain C-31-49 sometimes formed structures resembling aecial primordia, and strain C-121-51 occasionally formed mycelial masses resembling spermogonial initials, but none of these structures ever matured.

B. Nutritional Requirements

1. Experimental systems

Experiments using uredospores as inoculum have shown that minimal nutrients required to initiate saprotrophic growth include a carbohydrate as bulk carbon source, a sulphur-containing amino acid as sulphur source, and another amino acid as bulk nitrogen source (cf. Section II, this chapter). However, because growth from uredospores is variable, further experiments with established axenic cultures are necessary to determine the full range of alternative carbon, sulphur and nitrogen sources which can support growth of rust fungi. Subcultures of M. *lini*, U. *dianthi* and P. *graminis* have been tested on a wide range of potential nutrients in agar-solidified media, whilst G. *juniperi-virginianae* has been tested in liquid cultures (Tables III and IV). There have been no reports of a vitamin requirement by any rust, and the minerals used for growth from uredospores are adequate for continued growth of established cultures.

2. Carbon source

Most rust species were first cultured saprotrophically on media containing a complex carbon source consisting of sucrose or glucose, together with natural products containing organic nitrogen, such as peptone (a peptic digest of animal muscle), yeast extract, acid hydrolysate of casein, or coconut milk. Variant strains of *P. graminis* did not grow on peptone in the absence of added carbohydrate (Maclean, 1974), and the extent of such growth reported for *U. dianthi* (Jones, 1973a) was virtually negligible and could have been due to carry-over carbohydrates present in the inoculum. Although not stated specifically, *M. lini* was clearly unable to grow in the absence of a carbohydrate (e.g. Turel, 1973; Bose and Shaw, 1974a; Coffey and Shaw, 1972; Coffey and Allen, 1973). Unfortunately, Cutter did not report any attempts to grow *G. juniperi-virginianae* on media lacking a carbohydrate.

3. Alternative carbohydrates

Although it is apparent that rust fungi can use a wide range of alternative carbohydrates as carbon source (Table III), some carbohydrates, e.g. ribitol, ribose, starch, which supported prolific growth of some species appeared to be totally unable to support growth of other species or variant strains. However, failure to use a particular carbohydrate may merely represent failure to adapt on a particular occasion, as evidenced by the erratic growth of variants of *P. graminis* on sugar alcohols, especially ribitol (and also D-ribose) on which a particularly long lag phase was observed (Maclean, 1974). A dikaryotic variant of *P. graminis* grew particularly poorly on sugar alcohols (Maclean, 1974). Table III is not exhaustive. Thus, Cutter observed growth of *G. juniperi-virginianae* on the following carbohydrates which either were not tested or did not support growth of *P. graminis*, *U. dianthi* or *M. lini:* D-xylose, L-sorbose, L-rhamnose, inulin, melibiose, melezitose, dextrin, salicin. The following sugars also failed to support growth when tested with *P. graminis, U. dianthi* or *M. lini:* xylitol, erythritol, *myo*-inositol, sedoheptulose, galactitol, L-arabitol, D-lyxose, D-galacturonic acid, polygalacturonic acid.

Experiments designed to devise optimal media showed that rust fungi can

Table III. Carbohydrates which support saprotrophic growth[a] of established cultures of rust fungi.

Carbohydrate	SPECIES				
	1. *Puccinia*[b] *graminis* (variants)	2. *Uromyces*[c] *dianthi*	3. *Melampsora*[d,e] *lini*	4. *Puccinia*[f] *helianthi*	5. *Gymnosporangium*[d] *juniperi-virginianae*
Concentration (% w/v)	1%	3%	4%	4%	3%
D-Glucose	++	++	(++)	(++)	(++)
D-Fructose	++	++	(++)	(++)	(++)
D-Mannose	++	++	(++)	(++)	(++)
Sucrose	++	++	(++)	(++)	(++)
Raffinose	++		(++)	(+)	(+)
Trehalose	+ to ++*		(++)	(0)	
D-Mannitol	+ to ++*	++	(++)	(++)	(++)
D-Glucitol	+ to ++*		(+)	(0)	(++)
D-Arabitol	0 to ++*		(++)		
Ribitol	0 to ++*†		(++)	(0)	
D-Ribose	0 to ++*†		(0)		
Maltose		++	(+)	(0)	(++)
Amylose, starch or glycogen	++	0	(0)		(+)
Glycerol			(++)		(+)
L-arabinose	0		(0)		(++)
D-arabinose	0		(0)		(+)?
D-Galactose	0	0	(0)		(++)
Lactose	0	0		(0)	(++)

[a]Assessment of growth:

++ = large increase in colony mass
+ = small increase in colony mass
0 = no increase in colony mass
(++) = good growth (visual estimate)
(+) = relatively poor growth (visual estimate)
(0) = very slight or no growth (visual estimate)
* = growth very variable between replicate inocula
† = extended lag phase (compared to sucrose or glucose) for new growth
to appear after inoculation.

[b]Maclean (1974): test of 10 variant strains; [c]Jones (1973a); [d]Coffey and Shaw (1972); [e]Coffey and Allen (1973); [f]Coffey and Allen (1973): uredospore inoculum; [g]Cutter, V. M. Jr. (unpublished results, in manuscript referred to in Section III A:2).

tolerate high sugar concentrations. Although the optimal concentration of D-glucose was about 4% w/v (i.e. 0·22 M) for *P. graminis*, vigorous growth was still obtained at 12% w/v (0·67 M), the highest concentration tested

(Maclean, 1974). *U. dianthi* grew sub-optimally on 10% w/v sucrose (0·3 M), and could still grow on 20% w/v sucrose (0·6 M) (Jones, 1973a).

Maclean (1974) made the potentially useful observation that variant strains of *P. graminis* differed in the amount of amylase they excreted after growth on amylose (soluble starch) medium. An inverse correlation was found between amylase excretion and pathogenicity: strains which excreted most amylase were least pathogenic. Growth on amylose might be useful for screening variants for pathogenic strains. It is possible that a contributing cause to loss of pathogenicity is a loss of control over the amount of hydrolytic enzymes excreted.

4. *Alternative sulphur sources*

Of the six species of rust fungi which have been investigated for their sulphur nutrition in defined media, only *G. juniperi-virginianae* was capable of using inorganic sulphate (2 mM) as a sole source of sulphur (Cutter, unpublished results). The other species: *P. graminis tritici* (Scott and Maclean, 1969; Kuhl *et al.*, 1971; Howes and Scott, 1972; Bose and Shaw, 1974a), *M. lini* (Coffey and Shaw, 1972; Coffey and Allen, 1973; Turel, 1973; Bose and Shaw, 1974a), *P. helianthi* (Coffey and Allen, 1973), *P. coronata* (Siebs, 1971) and *C. ribicola* (Harvey and Grasham, 1974), all appear to require a reduced, organic source of sulphur as the amino acids cysteine, cystine, homocysteine or methionine, or the tripeptide glutathione.

Alternative sulphur sources have been investigated most extensively for *P. graminis tritici*. Scott and Maclean (1969) first reported that this fungus required an organic source of sulphur, based on experiments (cf. Kuhl *et al.*, 1971) which showed that uredospores initiated growth on defined media containing 1·65 mM cysteine, cystine or glutathione (but not methionine), in the presence of 22·5 mM aspartic acid as bulk nitrogen source. In further work using subcultures of a variant strain of *P. graminis,* Howes and Scott (1972) and Howes (1972) showed that methionine was an effective sulphur source at lower concentrations, and that the optimal and toxic concentrations of methionine varied greatly with the supplementary bulk nitrogen source (cf. Table IV). Methionine is toxic at particularly low concentrations when aspartic acid is present, thus explaining why Kuhl *et al.* (1971) obtained no growth from uredospores with methionine. The variant strain could also grow on media containing DL-homocysteine (1 mM) but not any of the following alternative sulphur compounds: thiolacetic acid, cysteic acid, thiosulphate or glutathione (Howes and Scott, 1972; Howes, 1972); inorganic sulphide was taken up and metabolized but did not support nett growth (Howes and Scott, 1973). The inability to grow on glutathione was unex-

Table IV. Optimal and toxic concentrations of sulphur amino acids for growth[f] of strain V1 of *Puccinia graminis* f.sp. *tritici* in the presence of various bulk nitrogen sources.

Bulk nitrogen source		mM Methionine		mM Cysteine[d]
		Optimal[b]	toxic[c]	
L-Asparate	30 mM[b]	0·08(++)	>0·32(−)	3[b](++), >6[c](−)
Ammonium	32 mM[b]	0·32(++)	>0·7 (−)	4 (++)
L-Glutamine	6 mM[b]	0·2 (++++)	>1·0 (−)	3[b](++++)
L-Asparagine	8 mM[b]			3 (+++)
L-Gutamate	8 mM[b]			3 (+++)
	32 mM[d]	0·15(++)	>0·6 (−)	
L-Alanine	4 mM[b]			3 (+++)
	32 mM[d]	>1·0 (+++)	n.d.[e]	
L-Serine	4 mM[b]			3 (++)
Glycine	4 mM[b]			3 (++)

[a]Data summarized from Howes and Scott (1972) and Howes (1972); [b]Approximate optimal concentration (a range was tested); [c]Minimum toxic concentration (a range was tested); [d]Only the stated concentration was tested (unless indicated otherwise); [e]n.d. = not determined; [f]Growth assessment: increase in protein (μg per colony) after 3–4 weeks growth:

(−) = no increase
(+) = 0–100 μg increase
(++) = 100–300 μg increase
(+++) = 300–500 μg increase
(++++) = >500 μg increase.

pected in view of the ability of uredospores to grow on this compound, although it must be noted that uredospores formed only small colonies when using glutathione as sole source of sulphur (Bose and Shaw, 1974a). Howes and Scott (1973) noted that the requirement of rust fungi for a reduced, organic source of sulphur is also shared by some water molds (Saprolegniales and Blastocladiales—cf. Cantino, 1950, 1955).

5. Alternative nitrogen sources

Although most fungi can use inorganic nitrate or ammonium salts as a sole nitrogen source, *G. juniperi-virginianae* and *P. graminis* appear to be the only rusts of those cultured axenically that have been carefully tested on inorganic nitrogen sources. *G. juniperi-virginianae* had the simplest requirements, and could use either nitrate or ammonium as alternative nitrogen sources, as well as a wide range of amino acids and other organic nitrogen sources (V. M. Cutter, unpublished results cited by Scott and Maclean, 1969). Provided that a sulphur amino acid (e.g. 4 mM cysteine or 0.3 mM methionine) was present (Howes and Scott, 1972), *P. graminis* could use ammonium ion (8–80 mM) but not nitrate or nitrite as bulk source of nitrogen.

Organic sources of nitrogen have been investigated more extensively than inorganic sources, although the requirement of most rust species for a sulphur amino acid has complicated matters. Thus, neither cysteine nor methionine has supported growth of established cultures of either *P. graminis* (Howes and Scott, 1972) or *M. lini* (Coffey and Allen, 1973) when offered as a sole source of reduced nitrogen. However, *C. ribicola* grew on a medium containing either 8 mM cystine or 6·5 mM glutathione as sole organic nitrogen source (Harvey and Grasham, 1974), thus suggesting that this fungus can use sulphur amino acids to provide both nitrogen and sulphur.

Howes and Scott (1972) and Howes (1972) compared various amino acids for their relative ability to serve as bulk nitrogen source for the growth of *P. graminis* on defined media supplemented with a sulphur amino acid (Table IV). Best growth was obtained on 6 mM glutamine in the presence of either 3 mM cysteine or 0·2 mM methionine. The relative concentrations of an amino acid and the sulphur supplement were important; for example, growth on 32 mM alanine was best with 1 mM methionine, but this methionine concentration was toxic in combination with 32 mM aspartate. On the basis of these experiments, Howes and Scott (1973) used the following minimal defined medium for growing *P. graminis* in studies of sulphur and nitrogen metabolism: 4% glucose, 10 mM trisodium citrate, Czapek's minerals, 6 mM glutamine and either 0·1 mM methionine, or 3 mM cysteine, pH 6·0.

Coffey and Allen (1973), Coffey and Shaw (1972) and Turel (1973) grew *M. lini* on a number of amino acids supplemented with cysteine or methionine. The ability to use some amino acids depended on whether the inoculum had been pre-grown on a peptone-based medium, or on defined combinations of amino acids. Growth failed on some amino acids in the presence of methionine but not cysteine, possibly because the high concentration of the methionine supplement (0·7–0·8 mM) caused toxicity similar to that observed with *P. graminis* (Table IV). The apparent requirement for alanine in some experiments could have been due to the ability of this amino acid to overcome methionine toxicity.

C. Pathogenicity and Virulence

1. General considerations and experimental systems

It is important to determine whether or not axenic cultures, particularly after extended periods of saprotrophic growth, retain the ability to infect their host plants. The term "pathogenicity" is used in the general sense of ability to parasitise at least some cultivars (e.g. "universal suscepts") of the host, whereas "virulence" denotes an ability of the fungus to overcome specific genes for resistance in particular host cultivars (cf. Watson, 1970).

A number of techniques have been employed to infect host tissues with mycelium or spores produced in axenic culture. Infection of whole plants or excised portions of them is often best accomplished by injuring surface tissues to allow mycelium access to cells within the interior of the host. For example, wheat leaves have been infected after applying saprotrophically-grown mycelium to mesophyll exposed by prior removal of portion of the abaxial epidermis (Williams *et al.*, 1967; Bushnell, 1968; Bose and Shaw, 1971, 1974a; Maclean *et al.*, 1971; Maclean and Scott, 1974; Green *et al.*, 1978; Green, 1976; Hartley and Williams, 1971a, 1971b), or after injection of mycelial suspensions into developing leaf rolls (Mussell and Staples, 1974). Hartley and Williams (1971a, 1971b) use the term "implant" to indicate that mycelium was placed in direct contact with exposed mesophyll. Similarly to *P. graminis*, mycelium of *M. lini* was able to infect flax via injured tissue, in this case the cut ends of cotyledons in tissue culture (Turel, 1971, 1972). However, uredospores of *M. lini* produced in axenic culture were able to infect intact plants of flax directly (Bose and Shaw, 1974a), whereas similarly produced uredospores of *P. graminis* were only infective if placed on exposed mesophyll of wheat (Bose and Shaw, 1971, 1974a; Williams *et al.*, 1967).

In contrast to *P. graminis* and *M. lini,* saprotrophically-grown mycelium of *G. juniperi-virginianae* was able to infect intact, uninjured plants of both the primary and alternate hosts. A paste of a mycelial homogenate in carbowax was applied to leaf axils; a mycelial homogenate was also used to infect host callus (Cutter, 1959). Similar techniques were used to infect both plants and callus of *Arisema triphyllum* with saprotrophically-grown mycelium of *U. ari-triphylli* (Cutter, 1960a), and callus of *Althea rosea* with saprotrophically-grown *P. malvacearum* (Cutter, 1960b).

Axenic cultures of *Cronartium fusiforme* growing on a yeast extract-peptone medium produced aeciospores which established apparently normal infections on intact plants of the primary host, *Quercus nigra* (Hollis *et al.*, 1972). Similarly, homogenates of spore-producing axenic cultures of *C. ribicola* established infections on leaves of the primary host *Ribes nigrum* (Harvey and Grasham, 1974). Mycelium of *C. ribicola* was able to infect callus cultures of the alternate host *Pinus monticola* directly, but seedlings of this host could only be infected via wounds made in the stem (Harvey and Grasham, 1974).

Because re-infection experiments using axenic cultures of *P. graminis* and *G. juniperi-virginianae* have provided most information on genetic variability during maintenance as saprotrophs, further details of experiments with these two species will be discussed below.

2. *Implant experiments with* Puccinia graminis

After inoculation of exposed mesophyll of wheat leaves, successful infection was indicated by the fungus growing through the leaf and rupturing the epidermis opposite the implant, 10 to 12 days after inoculation with primary cultures (Williams *et al.*, 1967; Bushnell, 1968; Bose and Shaw, 1971), and 8 to 26 days after inoculation with variants (Maclean and Scott, 1974). However, primary cultures differed from variants both in the ease of infection and in fungal morphology at the infection site.

(a) Primary cultures. Wheat leaves successfully infected by implants of mycelium from primary cultures induced the fungus to produce large, friable masses of dry uredospores, which can be compared with the sparse sporulation which occurred in ageing axenic cultures of primary mycelium. In contrast to the uredospores produced *in vitro*, the implant-induced uredospores were directly pathogenic (Hartley and Williams, 1971a, 1971b; Bose and Shaw, 1971). Infection experiments with these uredospores on differential cultivars of wheat have established that primary cultures retain the virulence characteristics of the parental isolate of rust from which the primary cultures were initiated (Hartley and Williams, 1971a, 1971b; Green, 1976).

(b) Variant strains. Variant strains derived from race 126-ANZ-6, 7 differed widely between themselves both in pathogenicity (Maclean *et al.*, 1971; Maclean and Scott, 1974; Williams and Hartley, 1971; Green, 1976; Green *et al.*, 1978; cf. also page 64) and in the mode of fungal development in infected leaves (Maclean and Scott, 1974). These differences were seen most clearly in the experiments of Maclean and Scott (1974), who tested the pathogenicity of one dikaryotic and nine monokaryotic variants on a "universally susceptible" cultivar of wheat. Four of the variants including the dikaryon were highly pathogenic (55–72% success rate from implants), three variants were weakly pathogenic (2–5% success rate) and the other two variants showed no detectible pathogenicity. The weakly pathogenic strains took 18–26 days after implantation to rupture the host epidermis, and then produced only sterile hyphae rather than spores. However, the highly pathogenic strains took only 8–16 days to rupture the epidermis, and formed sori containing mixtures of uredospores, teliospores and sterile hyphae, the relative proportions of which differed between strains. A dikaryotic variant was isolated from a Canadian strain of race 11 (designated "race 11 yellow" because it produced yellow rather than the normal rust red uredospores)

(Green, 1976). This variant was much less pathogenic than strain V1C, and in a series of 778 implants made during January to May, 1974, only four leaves (0·5%) became infected.

3. Isolation of strain V1C of P. graminis

This monokaryotic variant has made a rather unique contribution to our knowledge of somatic variation in *P. graminis,* and originated from strain V1, which was one of 13 distinct variants isolated from uredospores in May, 1967 (Maclean and Scott, 1970). As summarized in Fig. 3, V1 was pathogenic when implanted, but the uredospores so produced failed to infect intact plants until an experiment in March, 1969, which resulted in the formation of a single pustule (Maclean *et al.*, 1971). The progeny of this pustule were designated strain V1C.

The first uredial culture of V1C was lost after four transfers on wheat, owing to the low infectivity of uredospores (Maclean, 1971). Nonetheless, sufficient uredospores were produced to re-establish axenic cultures (Maclean *et al.*, 1971), although replicate inocula on agar media gave the same inconsistent growth patterns in initiating saprotrophic growth as did uredospores of the dikaryotic parent (Maclean, 1971). However, primary mycelium of V1C retained the typical characteristics of variants rather than forming the dense stromata typical of primary cultures of the parent. Subsequent failures and successes in obtaining uredial cultures from saprotrophic mycelium of V1C are summarized in Fig. 3.

4. Progency of uredial cultures of variants of P. graminis

Only two variant strains, V1C and the Canadian variant of race 11 yellow, have been successfully propagated as uredial cultures by uredospore inoculation on wheat (or barley). The return from the selection pressures of nutrient medium to those of the host–parasite interaction resulted in further variation being observed, as outlined below.

(a) Morphological differences with the parent. In general, pustules of the parental race 126-ANZ-6, 7 developed more rapidly, were more elongated, and sporulated more profusely than did pustules of V1C, individual colonies of which formed circular pustules which spread more or less slowly through the leaf (cf. Fig. 4; Maclean and Scott, 1974; Green *et al.*, 1978). Some monokaryons which grew at the normal rate were isolated from V1C (Green *et al.*, 1978). Whereas the parent rarely formed teliospores on seedling

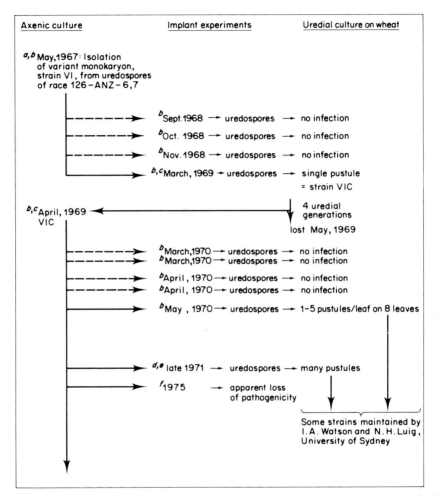

Fig. 3. Isolation and maintenance of axenic and uredial cultures of strain V1C of *Puccinia graminis* f.sp. *tritici.* [a]Maclean and Scott (1970); [b]Maclean (1971); [c]Maclean *et al.* (1971); [d]Series of unpublished experiments by D. J. Maclean, G. J. Green, I. A. Watson, N. H. Luig and P. G. Williams; [e]Green *et al.* (1978); [f]D. J. Maclean, unpublished experiments.

leaves, pustules of V1C often did so two to four weeks after inoculation (Maclean *et al.*, 1971), although the tendency to produce teliospores diminished the longer V1C was maintained by uredial transfer (Maclean, 1971).

(b) Infectivity. Uredospores of V1C and the Canadian variant were

Fig. 4. Abaxial surfaces of wheat leaves 14 days after inoculation with uredospores of *P. graminis.* A = strain V1C (diploid monokaryon); B = race 126-ANZ-6, 7 (parental dikaryon). (From Maclean and Scott, 1974.) Magnification, about × 6.

initially much less infective than uredospores of their parents (Maclean, 1971; Green, 1976; Green *et al.*, 1978). During uredial propagation, uredospores derived from different pustules varied widely in infectivity, and eventually isolates were selected which retained the slowly-spreading pustule morphology and the monokaryotic condition, but which produced uredospores as infective as the dikaryotic parent (Green *et al.*, 1978; unpublished results of I. A. Watson cited by Williams, 1975b).

(c) Chlorosis and necrosis. On some occasions, uredospore infections produced chlorotic and necrotic reactions on host cultivars fully susceptible to the parental dikaryon. These low infection types were observed with both V1C (Green *et al.*, 1978; I. A. Watson and N. H. Luig, personal communication) and the Canadian variant dikaryon (Green, 1976).

(d) Uredospore colour. The colour of uredospores varied in the progeny of uredial cultures of both V1C and the Canadian variant. Thus, V1C produced colonies which varied from Sudan Brown (the colour of the parent) to lighter shades of brown, orange and yellow. The Canadian variant, derived from a yellow parent, produced colonies which were normal rust red and various shades of orange, but none with the original yellow (Green, 1976). Uredospore colour is frequently used as a marker in genetical studies on rusts.

(e) Dikaryotization. Dikaryotic colonies were isolated from amongst the progeny of uredial transfers of the originally monokaryotic V1C (Green *et al.*, 1978).

(f) Virulence, and segregation of progeny of V1C into two groups. Green *et al.* (1978) selected 20 representative uredial cultures derived from V1C, which differed from one another by the parameters described above. Two groups of cultures were recognized based on their nuclear condition and reactions on differential varieties of wheat. One group of 13 cultures which varied only slightly in virulence from the parental dikaryon were all dikaryotic, and also resembled the parent in spore colour, rate of growth, infectivity and ease of maintaining uredial cultures. The remaining group of seven cultures were all monokaryotic, and differed strikingly between one another and the parent in spore colour, virulence, growth rate, infectivity and ease of maintaining uredial cultures. However, all these monokaryons, even those with growth rates approaching that of the parent, formed characteristic poorly-sporulating, spreading pustules which readily distinguished them from the parent.

(g) Virulence of progeny of Canadian variant. The four cultures derived from the variant differed between themselves and from the parent in virulence, and all had lost the parental yellow colour [cf. (e) above] (Green, 1976). In contrast, uredospores derived from implants of primary mycelium

of the parent, produced infections that were characteristic of the parent both in yellow uredospore colour and in virulence.

5. Gymnosporangium juniperi-virginianae

Cutter (1959) tested the pathogenicity of all seven of his strains of *G. juniperi-virginianae* on both callus and intact plants of the primary host *Juniperus virginiana* (red cedar) and the expected alternate host *Pyrus* spp. (native and cultivated apple), together with *Crategus* spp. (hawthorn), the alternate host of closely related species of *Gymnosporangium*. Successful infection was judged by the formation of characteristic fruiting bodies of the fungus or, in the case of callus, the formation of fungal haustoria (fruiting bodies did not always form on infected callus). The results of Cutter's inoculations are summarized in Table II.

Cutter (1959) found that the ability of any strain to infect particular host tissues and the type of fruiting bodies formed, was related to the nuclear condition of the mycelium. Under natural conditions the monokaryotic haplophase exists only in the alternate host and produces spermogonia, whereas the dikaryophase is generated in the alternate host producing aecia, and is continued in the primary host where it forms telia. Thus, strains 3, 4, and 5 (Table II, cf. page 60) which were completely monokaryotic within 30 days of isolation, were unable to infect *Juniperus* but could infect *Pyrus* or *Crategus*. The other strains, which were dikaryotic or of mixed nuclear condition, were able to infect both *Juniperus* and one of the alternate hosts. Although the data on fruiting body formation were incomplete, those bodies which did form were also consistent with the nuclear condition of the fungal strain (Table II). However, no cross-spermatization was carried out to verify the putative haploid status of those monokaryotic strains which formed spermogonia.

Strain C-31-49 was consistently able to infect *Crategus,* a finding which created taxonomic difficulties because *G. juniperi-virginianae* is usually restricted to *Pyrus* spp. as its alternate host (Cutter, 1959). The aecial morphology on some species of *Crategus* resembled the hawthorn rust *G. globosum,* and on other species of *Crategus* resembled the quince rust *G. claripes,* but on *Pyrus* spp. was characteristic of *G. juniperi-virginianae.* All three *Gymnosporangium* species can grow on *Juniperus* as primary host. It was not known whether C-31-49 originated from a *Gymnosporangium* isolate of wider host range than previously assumed, or whether host specificity was broadened by axenic culture. In either case, these observations suggest that details of aecial morphology are determined by or are at least influenced by the host.

D. *Genetic Status of Variants of* Puccinia Graminis

1. *Ploidy of nuclei*

Before mechanisms for the origin and further variation of variant strains can be postulated, it is important to compare the ploidy levels of variant and parental nuclei. The number of chromosomes in fungal nuclei has traditionally been determined by counting the number of chromatinic particles in fixed and stained cells, or by determining the number of linkage groups by genetic analysis. Both of these methods, as will become clear below, are of limited application to the variants of *P. graminis*. Therefore, other methods have also been used to approach the problem, and it is now almost certain that the variant monokaryons are somatic diploids. The various lines of evidence are listed below.

(a) *Chromosome counts in fixed and stained cells.* There is disagreement about the number of staining particles which represent distinct chromatinic bodies in nuclei of the monokaryotic variants. McGinnis (1953) established a haploid number of $n = 6$ in nuclei of germinating basidiospores of a wild isolate of *P. graminis*. In figures interpreted as late prophase of mitosis, Williams and Hartley (1971) saw a mean number of 12 particles in nuclei of monokaryotic variants, and six particles in the haploid nuclei of the dikaryotic parent. In contrast, Maclean *et al.* (1971) saw approximately six particles in nuclei of both their monokaryotic variants and the dikaryotic parent, a finding which they considered supported haploidy. In a later paper to resolve the difference with Williams and Hartley (1971), Maclean *et al.* (1974) confirmed their counts of $n = 6$ but presented other evidence consistent with the variants being diploid (see below). It was concluded that mitotic chromosome counts could not distinguish between haploid and diploid nuclei in somatic cells of *P. graminis*. This conclusion is consistent with cytological studies on the fungi *Aspergillus nidulans* (Robinow and Caten, 1969) and *Ustilago violacea* (Day and Jones, 1972), both of which showed no apparent difference in the number of stained chromatinic particles in haploid and diploid nuclei of strains characterized independently by genetic analysis.

It appears that homologous chromosomes are closely paired during mitosis in diploid fungal nuclei. Williams (1975b) suggested that his fixation, staining and mounting techniques enabled the homologous chromosomes to be seen. However, Maclean *et al.* (1974) compared the different techniques, and suggested that Williams and Hartley (1971) obtained counts of $n = 12$ by viewing stained nuclei at resting to early prophase, and that in drying their preparations prior to mounting, the chromatin became constricted into discrete bodies which gave an impression of individual chromosomes at late prophase of mitosis.

(b) Size of spores and spore nuclei. The volume of cytoplasm per genome is generally constant in spores of haploid and diploid strains of fungi (Clutterbuck, 1969; Ishitani *et al.*, 1956; Tinline and MacNeill (1969), and the size of spore nuclei is also directly related to the size of the genome (Esser and Kuenen, 1967; Tinline and MacNeill, 1969). Observations and measurements on the size of uredospores and teliospores, the number of nuclei per spore cell, and the size of nuclei in uredospores and teliospores, were consistent with nuclei in monokaryotic variants of *P. graminis* being diploid (Maclean, 1971; Maclean *et al.*, 1974; Williams, 1975b), and nuclei in dikaryotic variants being haploid (Maclean *et al.*, 1974). The observations and measurements on teliospores (Maclean *et al.*, 1974) are of particular significance because these structures are the site of karyogamy (nuclear fusion) in wild isolates of *P. graminis*, thus providing a ready means of directly comparing authentic haploid and diploid nuclei. Nuclei in binucleate teliospore cells of the parental dikaryon and a variant dikaryon were small (~3·5 μm diameter) compared to the large fusion nucleus which they formed (~5 μm diameter) (cf. Fig. 5). However, all nuclei in teliospore cells of variant monokaryons were large (~5 μm diameter), even though up to 30% of these cells were binucleate presumably owing to faulty septation during their formation.

(c) DNA in nuclei of sporelings. Williams and Mendgen (1975) compared the DNA content of nuclei in sporelings of a variant monokaryon and the parental dikaryon. The fluorescence from Feulgen-stained nuclei of the monokaryon was twice that from individual nuclei of the dikaryon, an observation which clearly supports diploidy of the variant. It was presumed that the nuclei in germ tubes were at the G1 stage of the cell cycle, so that the DNA values represented 1C and 2C levels in the nuclei of the dikaryon and monokaryon, respectively.

2. Mechanisms of somatic variation

The variability observed in axenic cultures of *P. graminis* must be considered in relation to somatic variation observed in natural populations of this fungus, and to mechanisms of somatic variation demonstrated in other fungi.

(a) The parasexual cycle in natural populations? A parasexual cycle has

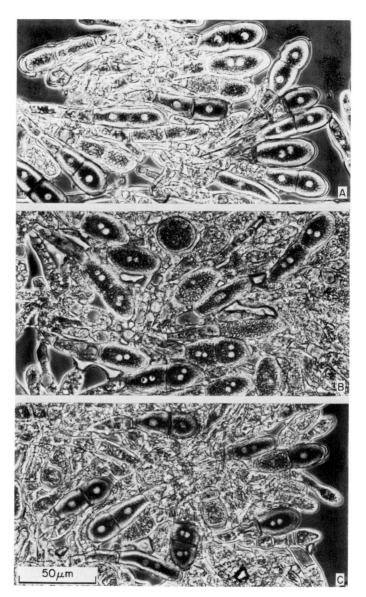

Fig. 5. Teliospores of variant strains of *P. graminis* mounted in water and visualized by phase contrast microscopy. A = large nuclei in teliospore cells of strain V1B (diploid monokaryon); B = small nuclei of strain V3B (dikaryon) prior to karyogamy; C = large fusion nuclei in some cells of strain V3B after karyogamy. (After Maclean *et al.* 1974.) Magnification shown by bar (50 μm).

been demonstrated in other fungi to provide a means of genetic recombination via vegetative diploid nuclei (cf. Roper, 1966). Such a cycle was postulated to operate in rust fungi to explain the origin of hybrids obtained by co-culturing mixtures of races of rusts on host plants. The new races could not have arisen by simple nuclear exchange between the co-cultured dikaryons (e.g. Watson and Luig, 1958; Bridgmon, 1959; Ellingboe, 1961; Sharma and Prasada, 1969; Bartos et al., 1969). The demonstration that somatic diploids can exist in axenic culture provides a potential "missing link" to explain the origin of new races in natural populations. In search of such a link, Williams (1978) presented evidence that only about 0·1% of a natural population of uredospores of race 126-ANZ-6, 7 of P. graminis contained putatively diploid nuclei, but did not find a fully diploid colony, thus confirming the results of earlier searches by Little and Manners (1969a, 1969b) and Bartos et al. (1969). Presumably, diploid colonies occur very rarely in nature.

(b) The parasexual cycle in diploids isolated from axenic cultures? It now remains to interpret observations on the origin of variants and their further variation in terms of parasexual phenomena. Presumably, variant diploids arose from the fusion of the two haploid nuclei in a dikaryotic cell of the parental strain. The other possibility, that diploids arose by chromosome doubling in a single haploid nucleus of the dikaryon, was rejected by Green et al. (1978) because such a diploid would be totally homozygous, and could not account for the observed recovery from diploids of dikaryons with the phenotype of the parent.

Because the parental phenotype was recovered when the uredial cultures of the diploid strain V1C spontaneously produced dikaryons, Green et al. (1978) suggested that these dikaryons arose by somatic meiosis (cf. Ellingboe, 1961, 1964). Such a mechanism would conserve the parental genome, but the distribution of genes between the two haploid nuclei of the reconstituted dikaryon would differ from the parent; this different gene distribution might be expected to cause minor changes in phenotype (cf. Hartley and Williams, 1971c). An alternative mechanism of reconstituting a dikaryon from a diploid requires nondisjunction at mitosis (Käfer, 1961; Roper, 1966). Such a mechanism would presumably require at least two separate nondisjunction events to generate the (+) and (−) haploid (or hyperhaploid) mycelia necessary to anastomose and regenerate a dikaryon. Such reconstitued dikaryons would be expected to demonstrate greater genetical diversity than was observed by Green et al. (1978), because whole chromosomes are lost after nondisjunction with the progeny tending towards homozygosity. However, Green et al. (1978) suggested that because of the apparent genetic diversity of pathogenic diploids recovered from strain V1C on wheat, the diploids probably arose from hyperdiploids formed during nondisjunction. Recombinant diploids could also arise from crossing-over during mitosis of diploid nuclei.

A problem is posed by Green's (1976) finding that a dikaryotic variant of a yellow isolate of a Canadian race 11 when propagated uredially on wheat, continued to throw off further variants (the nuclear condition of uredial cultures was not reported). Theoretically, a dikaryon should be more stable than a diploid. The simplest explanation is that the original saprotrophic culture of the variant was genetically heterogeneous, and perhaps even contained diploid cells (Green assessed nuclear condition from thin sections in the electron microscope). This explanation is consistent with the loss of pathogenicity of the saprotrophic culture soon after starting infection trials. The four uredial cultures which varied in virulence were each derived from a separate implant of saprotrophic mycelium.

(c) Potential uses for somatic diploids. Diploid strains of rust fungi isolated from axenic cultures have a number of potential uses for genetical work, but some technical problems must first be solved. The most obvious use is in the "selfing" of dikaryotic races, particularly those races which do not readily form teliospores or whose teliospores are difficult to germinate. Such selfing experiments can reveal heterozygous loci (Green *et al.*, 1978) and hence demonstrate if a natural population of a race is prone to lose a dominant avirulent gene, thus creating a new race with a wider host range.

The most pressing technical problem is to develop efficient and reproducible methods of synthesizing somatic diploids from uredial cultures. It is not yet known why diploids arise at relatively high frequencies in some experiments but not in others. Green (1976) can testify to this problem, as he obtained only one variant (a dikaryon?) despite deliberate and repeated attempts to re-create the conditions of Scott and Maclean (1970) and Williams and Hartley (1971) which resulted in the isolation of diploids. Reproducible techniques of synthesizing diploids are also necessary so that implants can be made as soon as possible after they arise, and before pathogenicity is lost. In this regard, it is perhaps fortuitous that the diploid strain V1C retained its pathogenicity for such an extended period of time in axenic culture compared to other variants, thus making possible the experiments of Green *et al.* (1978).

The scope of diploids for extensive parasexual analysis might depend on whether or not haploid mycelia can be isolated and maintained saprotrophically. The only report of haploid mycelium of *P. graminis* (Williams, 1971) indicates that growth ceases at early stages of saprotrophic development. Other species of rust fungi might provide better experimental systems than *P. graminis.*

V. Concluding Remarks and Summary

In retrospect, we now know that the major problem associated with axenic cultures of rust fungi is their initial isolation. Once established, most (but not all) of those rusts which have been cultured can form a quite substantial mass of mycelium, and can be maintained as saprotrophs by serial subculture on relatively simple media. The variability and inconsistency encountered in establishing axenic cultures requires further consideration, as this restricts their usefulness as experimental tools.

A. Why is Growth Initiation Variable?

To answer this question, we must recognize that rust fungi have developed a remarkable degree of adaptation to a biotrophic mode of existence. First, rusts must undergo a co-ordinated sequence of morphogenetic changes to establish infection (spore germination, formation of infection structures as an aid to penetration, formation of infection hyphae, formation of haustorial mother cells and finally haustoria). Secondly, the physical and chemical microenvironment of a living host tissue is very much more heterogeneous and complex than the surface or substratum of an artificial agar or liquid medium. When growing biotrophically, the intercellular hyphae and the haustoria of rusts are presumably each bathed in a host fluid of somewhat different chemical composition. Compounds of host origin present on the leaf surface and in these bathing fluids must therefore play two major roles (a) to induce or assist in the control of fungal morphogenesis, and (b) to provide all the nutrients necessary for growth.

(a) The role of the host in controlling fungal morphogenesis has been discussed in Section IIC of this chapter. The direction that Grambow and co-workers' research seems to be leading us is that the transition from sporeling to vegetative hypha is under the control of host metabolites (cf. Brian, 1967). Fungal growth regulators of host origin such as BIM and stimulators of infection structure formation (cf. pages 47–49) may have to be identified and added if vegetative growth is to be initiated consistently on nutrient media.

(b) Host fluids provide a balance of nutrients to which rust fungi have adapted over a long evolutionary time span. The relative concentrations of essential organic nutrients in artificial media must represent an approximation to these host fluids. Methionine toxicity when in imbalance with other amino acids (cf. pages 64–66) is one consequence of a bad approximation.

Another consequence of a bad approximation is related to the selective "leakiness" of amino acids from rust hyphae (cf. Part B of this Chapter), which can potentially result in metabolic disorganization if essential leaked metabolities are not replaced from the medium. Presumably, the growth stimulatory effects of high inoculum density, and of "conditioned" media (cf. pages 44–46) result from countering this leakage (Scott and Maclean, 1969).

The low frequency of isolation of saprotrophic mycelium from dual tissue cultures might well be related to the nutrient balance in the medium. In most of these experiments, more thought was given to providing optimal nutrients for the host callus than for the fungal partner. For example, the coconut milk added by Cutter to simulate callus growth provided the major source of mixed organic metabolities available for independent fungal growth. It is possible that Cutter selected saprotrophic strains of *G. juniperi-viriginianae* and *U. ari-triphylli* that were better able than the normal population to cope with the perhaps uncongenial nutrient balance of coconut milk. Harvey and Grasham (1974) were unable to isolate *C. ribicola* until they added yeast extract and peptone to their tissue culture medium. However, Turel (1969) was not able to isolate *M. lini* by adding dilute (0·1%) yeast extract to media supporting dual tissue cultures. Perhaps an enriched or better balanced source of amino acids would have proved more successful, e.g. by countering any leakage of nutrients from hyphae growing out onto the medium.

B. What Should be Done with Axenic Cultures?

The two avenues of greatest potential use appear to be the study of rust metabolism and genetics. Part B of this Chapter discusses the advances that have been made to our knowledge of the vegetative metabolism of rust fungi. Liquid suspension cultures of a variant of *Puccinia graminis* have proved particularly useful in these investigations. The demonstration of somatic diploidy in rust fungi has come only from axenic culture experiments on *P. graminis*. Although this finding is useful in itself in understanding how natural populations of rust fungi can produce new races with altered virulence patterns, it would be more useful if somatic diploids could be used for parasexual analysis. However, as discussed on page 78, many difficult technical problems may have to be solved before such parasexual analysis can complement or compete with standard techniques of sexual analysis (cf. McIntosh and Watson, Chapter 3, this book).

References

Allen, P. J. (1954). *Ann. Rev. Plant Physiol.* **5**, 225–248.
Allen, P. J. (1959). *In* "Plant Pathology, Problems and Progress 1908–1958" (C. S. Holton *et al.*, eds), pp. 119–129. University of Wisconsin Press, Wisconsin.
Allen, P. J. (1965). *Ann. Rev. Phytopathol.* **3**, 313–342.
Arthur, J. C. (1929). "The Plant Rusts (Uredinales)." John Wiley, New York.
Bartos, P., Fleischmann, G., Samborski, D. J. and Shipton, W. A. (1969). *Can. J. Bot.* **47**, 1383–1387.
Bauch, R. L. and Simon, V. (1957). *Ber. Dtsch. Bot. Ges.* **70**, 145–156.
Bose, A. and Shaw, M. (1971). *Can. J. Bot.* **49**, 1961–1964.
Bose, A. and Shaw, M. (1974a). *Can. J. Bot.* **52**, 1183–1195.
Bose, A. and Shaw, M. (1974b). *Nature* **251**, 646–648.
Brian, P. W. (1967). *Proc. R. Soc. London Ser. B.* **168**, 101–118.
Bridgmon, G. H. (1959). *Phytopathology* **49**, 386–388.
Bushnell, W. R. (1968). *Phytopathology* **58**, 526–527.
Bushnell, W. R. (1976). *Can. J. Bot.* **54**, 1490–1498.
Bushnell, W. R. and Rajendren, R. B. (1970). *Phytopathology* **60**, 1287.
Bushnell, W. R. and Stewart, D. M. (1971). *Phytopathology* **61**, 376–379.
Cantino, E. C. (1950). *Q. Rev. Biol.* **25**, 269–277.
Cantino, E. C. (1955). *Q. Rev. Biol.* **30**, 138–149.
Chakravarti, B. P. (1966). *Phytopathology* **56**, 223–229.
Clutterbuck, A. J. (1969). *J. Gen. Microbiol.* **55**, 291–299.
Cochrane, V. W. (1960). *In* "Plant Pathology, An Advanced Treatise" (J. G. Horsfall and A. E. Dimond, eds), Vol. II, pp. 169–202. Academic Press, New York.
Coffey, M. D. (1975). *Symp. Soc. Exp. Biol.* **29**, 297–323.
Coffey, M. D. and Allen, P. J. (1973). *Trans. Br. Mycol. Soc.* **60**, 245–260.
Coffey, M. D. and Shaw, M. (1972). *Physiol. Plant Pathol.* **2**, 37–46.
Coffey, M. D., Bose, A. and Shaw, M. (1969). *Can. J. Bot.* **47**, 1291–1293.
Coffey, M. D., Bose, A. and Shaw, M. (1970). *Can. J. Bot.* **48**, 773–776.
Constabel, F. (1957). *Biol. Zentralbl.* **76**, 385–413.
Cutter, V. M., Jr. (1951). *Trans. N.Y. Acad. Sci.* **14**, 103–108.
Cutter, V. M., Jr. (1952). *Phytopathology* **42**, 479.
Cutter, V. M., Jr. (1959). *Mycologia* **51**, 248–295.
Cutter, V. M., Jr. (1960a). *Mycologia* **52**, 726–742.
Cutter, V. M., Jr. (1960b). *ASB (Assoc. Southeast Biol.). Bull.* **7**, 26.
Day, A. W. and Jones, J. K. (1972). *Can. J. Microbiol.* **18**, 663–670.
Dickinson, S. (1949). *Ann. Bot. (Lond.) NS* **13**, 219–236.
Dougherty, E. C. (1953). *Parasitology* **42**, 259–261.
Ellingboe, A. H. (1961). *Phytopathology* **51**, 13–15.
Ellingboe, A. H. (1964). *Genetics* **49**, 247–251.
Esser, K. and Kuenen, R. (1967). "Genetics of Fungi". Springer-Verlag, New York.
Foudin, A. S. and Wynn, W. K. (1972). *Phytopathology* **62**, 1032–1040.
Gautheret, R. J. (1942). "Manual Technique de La Culture des Tissus Végétaux". Masson and Cie, Paris.
Grambow, H. J. (1977). *Z. Pflanzenphysiol.* **85**, 361–372.
Grambow, H. J. (1978). *Z. Pflanzenphysiol.* **88**, 369–372.
Grambow, H. J. and Grambow, G. E. (1978). *Z. Pflanzenphysiol.* **90**, 1–9.
Grambow, H. J. and Müller, D. (1978). *Can. J. Bot.* **56**, 736–741.

Grambow, H. J. and Reisener, H. J. (1976). *Ber. Dtsch. Bot. Ges.* **89**, 555–561.
Grambow, H. J. and Riedel, S. (1977). *Physiol. Plant Pathol.* **11**, 213–224.
Grambow, H. J. and Tücks, M. T. (1979). *Can. J. Bot.* **57**, 1765–1768.
Grambow, H. J., Garden, G., Dallacker, F. and Lehmann, A. (1977). *Z. Pflanzenphysiol.* **82**, 62–67.
Green, G. J. (1976). *Can. J. Bot.* **54**, 1198–1205.
Green, G. J., Williams, P. G. and Maclean, D. J. (1978). *Can. J. Bot.* **56**, 855–861.
Hartley, M. J. and Williams, P. G. (1971a). *Trans. Br. Mycol. Soc.* **57**, 129–136.
Hartley, M. J. and Williams, P. G. (1971b). *Trans. Br. Mycol. Soc.* **57**, 137–144.
Hartley, M. J. and Williams, P. G. (1971c). *Can. J. Bot.* **49**, 1085–1087.
Harvey, A. E. and Grasham, J. L. (1969). *Can. J. Bot.* **47**, 663–666.
Harvey, A. E. and Grasham, J. L. (1970). *Can. J. Bot.* **48**, 71–73.
Harvey, A. E. and Grasham, J. L. (1974). *Phytopathology* **64**, 1028–1035.
Hollis, C. A., Schmidt, R. A. and Kimbrough, J. W. (1972). *Phytopathology* **62**, 1417–1419.
Hotson, H. H. and Cutter, V. M. Jr. (1951). *Proc. Nat. Acad. Sci., U.S.A.* **37**, 400–403.
Howes, N. K. (1972). Ph.D. Thesis, University of Queensland, Australia.
Howes, N. K. and Scott, K. J. (1972). *Can. J. Bot.* **50**, 1165–1170.
Howes, N. K. and Scott, K. J. (1973). *J. Gen. Microbiol.* **76**, 345–354.
Ingram, D. S. and Tommerup, I. C. (1973). *In* "Fungal Pathogenicity and the Plant's Response" (R. J. W. Byrde and C. V. Cutting, eds), pp. 121–140. Academic Press, London and New York.
Ishitani, C., Uchida, K. and Ikeda, Y. (1956). *Exp. Cell Res.* **10**, 737–740.
Jones, D. R. (1972a). *Trans. Br. Mycol. Soc.* **58**, 29–36.
Jones, D. R. (1972b). Ph.D. Thesis, University of Keele, U.K.
Jones, D. R. (1973a). *Physiol. Plant Pathol.* **3**, 379–386.
Jones, D. R. (1973b). *Trans. Br. Mycol. Soc.* **61**, 227–235.
Jones, D. R. (1974). *Trans. Br. Mycol. Soc.* **63**, 593–594.
Käfer, E. (1961). *Genetics* **46**, 1581–1609.
Katsuya, K., Kakishima, M. and Sato, S. (1978). *Ann. Phytopathol. Soc. Jpn.* **44**, 606–611.
Kuhl, J. L., Maclean, D. J., Scott, K. J. and Williams, P. G. (1971). *Can. J. Bot.* **49**, 201–209.
Lane, W. D. and Shaw, M. (1972). *Can. J. Bot.* **50**, 2601–2603.
Lane, W. D. and Shaw, M. (1974). *Can. J. Bot.* **52**, 2228–2229.
Lewis, D. H. (1973). *Biol. Rev. Cambridge Philos. Soc.* **48**, 261–278.
Link, G. K. K. (1933). *Phytopathology* **23**, 843–862.
Little, R. and Manners, J. G. (1969a). *Trans. Br. Mycol. Soc.* **53**, 251–258.
Little, R. and Manners, J. G. (1969b). *Trans. Br. Mycol. Soc.* **53**, 259–267.
McGinnis, R. C. (1953). *Can. J. Bot.* **31**, 522–526.
Maclean, D. J. (1971). Ph.D. Thesis, University of Queensland, Australia.
Maclean, D. J. (1974). *Trans. Br. Mycol. Soc.* **62**, 333–349.
Maclean, D. J. and Scott, K. J. (1970). *J. Gen. Microbiol.* **64**, 19–27.
Maclean, D. J. and Scott, K. J. (1974). *Can. J. Bot.* **52**, 201–207.
Maclean, D. J., Scott, K. J. and Tommerup, I. C. (1971). *J. Gen. Microbiol.* **65**, 339–342.
Maclean, D. J., Tommerup, I. C. and Scott, K. J. (1974). *J. Gen. Microbiol.* **84**, 364–378.
Mains, E. B. (1917). *Amer. J. Bot.* **4**, 179–220.

Morel, G. (1944). *C. R. Acad. Sci. Paris,* **218**, 50.
Morel, G. (1948). *Ann. Épiphyt. (II)* **14**, 1–112.
Mussell, H. W. and Staples, R. C. (1973). *Phytopathology* **63**, 653–654.
Nozzolillo, C. and Craigie, J. H. (1960). *Can. J. Bot.* **38**, 227–233.
Quick, W. A. and Cross, S. L. C. (1971). *Can. J. Bot.* **49**, 187–188.
Rahbar Bhatti, M. H. and Shattock, R. C. (1980a). *Trans. Br. Mycol. Soc.* **74**, 595–600.
Rahbar Bhatti, M. H. and Shattock, R. C. (1980b). *Trans. Br. Mycol. Soc.* **75**, 327–331.
Rawlinson, C. J. and Maclean, D. J. (1973). *Trans. Br. Mycol. Soc.* **61**, 590–593.
Raymundo, S. A. and Young, H. C. Jr. (1974). *Phytopathology* **64**, 262–263.
Riou, A., Harada, H. and Taris, B. (1975). *C. R. Acad. Sci. Paris Ser. D.* **280**, 2765–2767.
Robinow, C. F. and Caten, C. E. (1969). *J. Cell Sci.* **5**, 403–431.
Roper, J. A. (1966). *In* "The Fungi, an Advanced Treatise" (G. C. Ainsworth and A. S. Sussman, eds), Vol. 2, pp. 589–617. Academic Press, New York and London.
Rossetti, V. and Morel, G. (1958). *C. R. Acad. Sci. Paris* **247**, 1893–95.
Scott, K. J. (1968). *In* "Abstract of Papers. First International Congress of Plant Pathology", London.
Scott, K. J. (1972). *Biol. Rev. Cambridge Philos. Soc.* **47**, 537–572.
Scott, K. J. (1976). Encyclopedia of plant physiology new series, Volume 4. *In* "Physiological Plant Pathology" (R. Heitefuss and P. H. Williams, eds), pp. 719–742. Springer-Verlag, Berlin, Heidelberg, New York.
Scott, K. J. and Maclean, D. J. (1969). *Ann. Rev. Phytopathol.* **7**, 123–146.
Sharma, S. K. and Prasada, R. (1969). *Aust. J. Agric. Res.* **20**, 981–985.
Shaw, M. (1963). *Ann. Rev. Phytopathol.* **1**, 259–294.
Shaw, M. (1964). *Phytopathol. Z.* **50**, 159–180.
Shaw, M. (1967). *Can. J. Bot.* **45**, 1205–1220.
Siebs, E. (1971). *Phytopathol. Z.* **72**, 97–114.
Staples, R. C. and Wynn, W. K. (1965). *Bot. Rev.* **31**, 537–564.
Thrower, L. B. (1965). *Phytopathol. Z.* **52**, 319–334.
Tinline, R. D. and MacNeill, B. H. (1969). *Ann. Rev. Phytopathol.* **7**, 147–170.
Turel, F. L. M. (1969a). *Can. J. Bot.* **47**, 821–823.
Turel, F. L. M. (1969b). *Can. J. Bot.* **47**, 1637–1638.
Turel, F. L. M. (1971). *Can. J. Bot.* **49**, 1993–1997.
Turel, F. L. M. (1972). *Can. J. Bot.* **50**, 227–230.
Turel, F. L. M. (1973). *Can. J. Bot.* **51**, 131–134.
Turel, F. L. M. and Ledingham. G. A. (1957). *Can. J. Microbiol.* **3**, 813–819.
Vasill'ev, O. A. and Saplina, V. I. (1971). *Mikol. Fitopatol.* **5**(2), 182–183 (cited in *Rev. Plant Pathol.* (1972) **51**, 125).
Watson, I. A. (1970). *Ann. Rev. Phytopathol.* **47**, 510–511.
Watson, I. A. and Luig, N. H. (1958). *Proc. Linn. Soc. N.S.W.* **83**, 190–195.
Williams, P. G. (1971). *Phytopathology* **61**, 994–1002.
Williams, P. G. (1975a). *In* "Advances in Mycology and Plant Pathology" (S. P. Raychaudhuri, A. Varma, K. S. Bhargava and B. S. Mehrotra, eds) pp. 67–82. Professor R. N. Tandon's Birthday Celebration Committee, New Delhi.
Williams, P. G. (1975b). *Trans. Br. Mycol. Soc.* **64**, 15–22.
Williams, P. G. (1976). *Arch. Microbiol.* **110**, 173–175.
Williams, P. G. (1978). *Trans. Br. Mycol. Soc.* **70**, 293–294.

Williams, P. G. and Hartley, M. J. (1971). *Nature (Lond.) New Biol.* **229**, 181–182.
Williams, P. G. and Mendgen, K. W. (1975). *Trans. Br. Mycol. Soc.* **64**, 23–28.
Williams, P. G., Scott, K. J. and Kuhl, J. L. (1966). *Phytopathology* **56**, 1418–1419.
Williams, P. G., Scott, K. J., Kuhl, J. L. and Maclean, D. J. (1967). *Phytopathology* **57**, 326–327.
Witkowski, R. and Grümmer, G. (1960). *Z. Allg. Mikrobiol.* **1**, 79–82.
Wolf, F. T. (1974). *Can. J. Bot.* **52**, 767–772.
Wong, A. L. and Willetts, H. J. (1970). *Trans. Br. Mycol. Soc.* **55**, 231–238.
Yarwood, C. E. (1956). *Ann. Rev. Plant Physiol.* **7**, 115–142.

Note added in proof

Recent work of Boasson and Shaw (1979, 1981), indicates that carbon dioxide is necessary for growth to initiate from sporelings of *M. lini*. This finding is of much potential value if it can be shown that carbon dioxide levels above ambient can improve the consistency of growth from uredospores of other rust species *in vitro*.

Boasson, R. and Shaw M. (1979). *Can. J. Bot.* **57**, 2657–2662.
Boasson, R. and Shaw M. (1981). *Can. J. Bot.* **59**, 1621–1622.

Part B: Metabolism of Axenic Cultures*

CONTENTS OF PART B

---

* A list of abbreviations can be found at the end of the chapter.

I. Introduction

The metabolism of rust fungi can be considered from a number of viewpoints, including (i) vegetative growth (e.g. the uptake of exogenous nutrients and their conversion into new cellular materials), (ii) morphology and morphogenesis (e.g. translocation between different types of cell; ontogeny of fruiting bodies; spore germination) and (iii) host-parasite interactions (e.g. metabolite exchange between host cells and parasitic hyphae; control of compatibility between fungal and host tissues). Once established, axenic cultures of rust fungi consist mostly of vegetative hyphae rather than sporogenous cells, except in some aged cultures (cf. Part A of this chapter). Therefore, the biochemistry of axenic cultures is essentially the biochemistry of vegetative cells, and is relatively uncomplicated by considerations of morphogenesis or host compatibility.

This review will be mostly restricted to intermediary metabolism, i.e. those metabolic pathways necessary to direct the flow of carbon, nitrogen, sulphur etc. from exogenous substrates or storage pools to metabolities required by cells, and to organize energy production. Much current research is also directed to understanding gene expression via DNA, RNA and protein metabolism in rust-infected plant tissue, using saprotrophically-grown mycelium as a control; this work will not be reviewed here (but see Chakravorty, Chapter 5, this book).

Before discussing experiments carried out on axenic cultures, other materials which have been exploited (or which can potentially be exploited) for metabolic studies will be considered and compared to axenic cultures.

II. Systems Available for Metabolic Studies

A. Uredospores and Sporelings

The ready availability of uredospores has resulted in many investigations being carried out on the metabolism of germinating uredospores (cf. Staples and Wynn, 1965; Wolf, Chapter 4 of this book). Because washed uredospores germinate readily in distilled water (Yarwood, 1956; Scott and Maclean, 1969), and do not require an exogenous nutrient source for sporeling development as do many other fungi (Allen, 1965; Burnett, 1976), rust fungi exist as self-contained metabolic units mobilizing stored nutrients during germ tube elongation. However, uredospores rapidly lose up to 50% of their soluble metabolites by leakage when hydrated in free water (Staples and Wynn, 1965; Daly et al., 1967), and take up and reutilize some of these metabolites (especially trehalose and acyclic polyols) during subsequent sporeling development (Daly et al., 1967).

Although the intermediary metabolism of germinating uredospores is primarily concerned with the mobilization of stored metabolites, many radioisotope experiments have demonstrated that sporelings retain a capacity to assimilate exogenous nutrients (cf. Staples and Wynn, 1965; Pfeiffer et al., 1969; Burger et al., 1972). Thus, [^{14}C]-labelled short-chain fatty acids are readily taken up by sporelings, and Ziegler and Reisener (1975) used exogenous [^{14}C]-valerate as a "probe" to demonstrate that citric acid was compartmentalized in two separate metabolic pools. In other experiments, Burger et al. (1972) demonstrated that exogenous [^{14}C]-D-glucose after uptake was selectively diverted to the synthesis of cell wall carbohydrates in the elongating germ tube, and only a small proportion of the assimilated label was converted into soluble metabolites such as polyols and amino acids. However, it has not yet been demonstrated that exogenous nutrients leached either from the spores or from host surfaces, assist sporelings on the leaf prior to penetration.

B. Mycelium in Infected Host Tissues

Dynamic aspects of the movement of host metabolites into parasitic mycelium and the subsequent assimilation into fungal metabolites can only be determined indirectly, and the following experimental approaches have been adopted:

(a) [^{14}C]-carbon dioxide offered to rust-infected plant tissues has been converted to host photosynthate, and the movement of label into characteristic fungal metabolites such as trehalose and acyclic polyols has been followed (reviewed by Smith et al., 1969; Scott, 1972).

(b) Specifically-labelled amino acids and sugars have been taken up by the vascular system of sporulating, rust-infected wheat leaves. Uredospores subsequently collected have been analysed to determine the distribution of label between fungal metabolities (e.g. Jäger and Reisener, 1969; Pfeiffer et al., 1969; Reisener et al., 1970). Such experiments, for example, have demonstrated that ^{14}C from exogenously-supplied glucose, glutamate, alanine or glycine was widely distributed amongst all fungal sugars and amino acids, whereas the carbon skeletons of exogenous lysine and arginine showed little randomization of ^{14}C when incorporated into uredospores.

(c) The movement of metabolites from host to parasite has been investigated by the "inhibition technique" originally devised by Drew and Smith (1967) for lichens, and applied to leaf discs of Tussilago farfara infected with

D

the rust *Puccinia poarum*. Results were consistent with sucrose being released by the host, and inverted to glucose and fructose by enzymes present in spaces between host and fungal cells, prior to uptake by the fungus (experiments of P. M. Holligan and C. Yuen, cited by Lewis, 1976; cf. also Long *et al.*, 1975). A similar pattern of carbohydrate movement was found from *Vigna sesquipedalis* to *Uromyces appendiculatus* (So and Thrower, 1976). Modification of the technique to follow the movement of amino acids from *T. farfara* to *P. poarum*, yielded results consistent with serine and alanine being absorbed more readily from the host than aspartate, glutamate or glutamine (Burrell and Lewis, 1977).

The above approaches are subject to a number of severe reservations. Substrates fed to infected tissues may be extensively altered by host tissues before reaching the fungus (cf. Burrell and Lewis, 1977). The number of different metabolites moving simultaneously between host and parasite may be many, and their relative fluxes may be difficult to determine. Exogenous application of metabolites, especially if presented in high concentrations, may perturb the "normal" metabolic processes which occur in the host-parasite complex.

C. Mycelium Isolated from Infected Host Tissues

Dekhuijzen *et al.* (1967, 1968) isolated hyphae of *Uromyces phaseoli* from infected bean leaves by grinding, sieving and settling in sucrose gradients. This technique is of limited value, because it is difficult to determine the extent to which fungal metabolism will be perturbed by the trauma resulting from mechanical damage and fractionation in sucrose solutions. Enzymic digestion to separate colonies of *Melampsora lini* from flax cotyledons has been achieved by Lane and Shaw (1972). This gentle isolation would appear to be superior to the mechanical methods of Dekhuijzen *et al.*, and may have merit for future experiments, provided that the fungus can recover from any disruption to steady-state metabolism reasonably soon after isolation.

D. Mycelium Grown on Host Tissue Cultures

Host-free vegetative mycelium for investigating metabolism was first obtained by excising tufts of aerial hyphae from infected flax cotyledons in tissue cultures (Turel and Ledingham, 1957). These hyphae took up ^{14}C-labelled D-glucose, pyruvate and D-mannitol (Williams and Shaw, 1968), and assimilated the label into amino acids, acyclic polyols and trehalose

(Mitchell and Shaw, 1968). The limitations of aerial hyphae for metabolic studies must be recognized, however. Nutrients are initially gained from host cells by the intercellular mycelium within the interior of the leaf culture, and metabolites of this mycelium are translocated to the aerial hyphae. Thus, the metabolism of aerial hyphae is adapted to the utilization of translocated fungal metabolites (e.g. trehalose, polyols), rather than being adapted to assimilating exogenous nutrients of host origin (e.g. glucose, fructose, sucrose).

E. Axenic Cultures

Axenic cultures present a ready source of free-living vegetative mycelium, with the advantage that the mycelium can be prepared under uniform, steady-state conditions, especially if it can be grown in liquid suspension culture. Liquid suspension cultures have been described for *Gymnosporangium juniperi-virginianae* (Cutter, unpublished results; Wolf, 1956), *Uromyces ari-triphylli* (Cutter, 1960, and unpublished results) and variants of *Puccinia graminis tritici* (Maclean 1971; Howes and Scott, 1972, 1973). Surface-grown cultures of rust fungi form tight mycelial masses and are frequently morphologically heterogeneous, and are therefore less suitable than liquid suspension cultures for time-course radioisotope experiments which require rapid changes of medium or rapid washing of mycelium.

F. Gymnosporangium juniperi-virginianae

Wolf (1956) grew liquid cultures of *G. juniperi-virginianae* on a rotary shaker, resulting in the production of small, brownish pellets, quite hard in texture. Analysis of culture filtrates using paper chromatography and chemical methods of detection, demonstrated the synthesis of indole acetic acid (IAA) from tryptophane. Tryptamine and indole acetaldehyde appeared to be metabolic intermediates of IAA synthesis.

G. Variants of Puccinia graminis tritici

It was the fortuitous isolation of diploid and dikaryotic variants which grow more evenly and reproducibly than primary cultures of *P. graminis tritici* (cf. Part A of this chapter), which has made it possible to carry out effectively the experiments on sulphur, nitrogen and carbohydrate metabolism which are discussed in the remainder of this chapter. The use of variant strains to study metabolism should not be a cause for concern however. Variants are extremely unlikely to gain new metabolic pathways not found in parental wild-types, although they might differ in some aspects of regulation of metabolism. Metabolic pathways eluicidated with relative ease in variants, can form a basis for designing experiments to elucidate metabolic regulation in wild-type strains, both growing axenically (where this is possible) and parasitically.

Experiments to investigate the nutrition and metabolism of variant strains of *P. graminis,* need to be designed with the following background information in mind. First, liquid suspension cultures seem to require enriched nutrients compared to agar-solidified media for significant growth to occur, possibly because leakage of fungal metabolities (cf. Sections III and IV below) is greater from the diffuse hyphae in suspension cultures. For this reason, the nutritional experiments of Howes (1972) and Howes and Scott (1972) to define minimum sulphur and nitrogen requirements, and the experiments of Maclean (1974) on carbohydrate nutrition, used agar-solidified media. Secondly, submerged non-shaken suspension cultures were used for all radioisotope feeding experiments, even though growth is much less profuse compared to shaken cultures or surface-grown cultures. As well as allowing a better exchange of nutrients with the medium (see Section II E above), liquid suspension cultures can be readily handled by pipetting, and are more uniform morphologically than either shaken liquid cultures (which form pellets) or surface cultures (which consist of aerial hyphae dependent on metabolites translocated from nutrient-absorbing hyphae). A morphologically (and metabolically) diverse mycelium can

Table I. Comparative growth[a] of strain VI of *Puccinia graminis* in different liquid media (compiled from data of Howes, 1972).

Media	μg mycelial protein/ml culture	
Preincubation → growth	At time of transfer	5 days after transfer
[b]Growth medium → growth medium	60	110
Growth medium → [c]basal medium	70	48
Basal medium → basal medium + 3 mM cysteine	57	82

[a]Growth at 23°C in submerged, non-shaken culture (cf. Howes and Scott, 1973). Data are selected to indicate relative growth rates from similar inocula.

[b]Growth medium: 1% peptone, 10 mM citrate, 2·1 mM histidine HCl, 4% glucose, in Czapek's minerals, pH 6·0.

[c]Basal medium: 6 mM glutamine, 10 mM citrate, 4% glucose, in Czapek's minerals, pH 6·0.

give mixed kinetics with experiments designed to follow radioactivity from exogenous substrates through successive pools of metabolites.

The liquid suspension cultures used for the experiments described in Sections III, IV and V were grown on an enriched peptone-based medium (growth medium, cf. Table I) to which histidine had been added as it enhanced growth rates (N. K. Howes, unpublished observations). A defined medium was necessary for experiments on nitrogen and sulphur metabolism; the basal medium used for these experiments did not support nett protein synthesis in the absence of a suitable concentration of a sulphur amino acid such as 3 mM cysteine (Table I).

III. Sulphur Metabolism in *Puccinia Graminis*

A. Uptake and Assimilation of Inorganic Sulphur

In Part A of this chapter, it was noted that a number of rust fungi including *P. graminis* appear to have an essential requirement for a reduced, organic source of sulphur. Howes and Scott (1972) showed that mycelium of *P. graminis* was unable to grow on media containing inorganic sulphate, sulphite, thiosulphate, or sulphide as a sole source of sulphur. Most fungi can use inorganic sulphate as a sole source of sulphur by the pathway of reductive assimilation summarized in Fig. 1. The first step in this pathway is transport into the cell, followed by activation of the sulphate first to APS and then to PAPS (steps (1), (2) and (3) of Fig. 1). The activated sulphate (PAPS in Fig. 1) is then reduced to a disulphide which can donate sulphide into intermediary metabolism. Note that APS is sulphur donor for step (4) in green plants, and future research might show this to be so for some fungi.

Howes and Scott (1973) investigated the uptake and assimilation of both [^{35}S]-labelled inorganic sulphate and hydrogen sulphide (H_2S) to determine the location of any metabolic block in *P. graminis*. Inorganic sulphate was transported into cells of *P. graminis* at 38% of the rate of transport into *Neurospora crassa*, a much faster-growing fungus which can assimilate and reduce sulphate. However, none of the sulphate taken up by *P. graminis* was subsequently reduced and incorporated into cysteine (either free or in mycelial protein). In contrast, H_2S was readily incorporated into cysteine and other organic compounds by *P. graminis* (e.g. Table II). It can therefore be surmised that a metabolic block must occur somewhere between sulphate transport (step (1) of Fig. 1) and sulphide assimilation (step (6) of Fig. 1).

Because PAPS is required by living organisms as a sulphating agent for synthesizing sulphate esters, Howes and Scott (1973) reasoned that *P. graminis* must carry out steps (2) and (3) of Fig. 1. The location of the

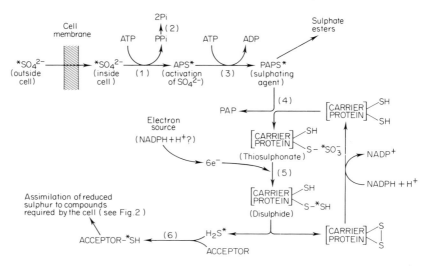

Fig. 1. Scheme for the reductive assimilation of inorganic sulphate by microorganisms (based on Schiff and Hodson (1973) and Siegel (1975). Enzymes: (1) = ATP-sulphurylase; (2) = pyrophosphatase; (3) = APS-kinase; (4) = PAPS: thiol sulphotransferase; (5) = sulphite reductase; (6) = O-acetylserine sulphydrylase.

metabolic block in sulphate assimilation would therefore lie at step (4) or (5) of Fig. 1, i.e. the fungus is unable to reduce PAPS to thiosulphonate, or thiosulphonate to sulphide, or both. The exact nature of this metabolic block is yet to be determined.

B. *Leakage of Assimilated Products of H_2S*

The most interesting feature of the assimilation of [^{35}S]-H_2S by *P. graminis*, was that more assimilated label leaked out into the medium as small molecular weight organo-sulphur compounds, than was incorporated into mycelial protein (Table II). This leakage was demonstrated on two independent occasions with different strains of the fungus (Table II), and occurred both on basal medium which does not support nett growth (increase in mycelial protein) and on cysteine-supplemented medium which does support growth. A further, highly significant finding was that only a small fraction of the leaked label was present in methionine (0·5–2% of total assimilated ^{35}S), the remainder of the leaked label being present in cysteine or cysteine derivatives (74–89% of total assimilated ^{35}S). These sulphide assimilation experiments thus pose a number of questions: (1) what are the pathways of

Table II. Assimilation of $[^{35}S]$-H_2S into organic compounds by *Puccinia graminis*.

| | % distribution of assimilated label [d] (nmoles of ^{35}S/mg mycelial protein) | | | |
| | [a,c]Experiment 1 | | [b,c]Experiment 2 | |
Compound	A. Basal medium	B. Basal medium +3mM cysteine	C. Basal medium	D. Basal medium +3mM cysteine
Mycelial protein				
Cysteine	18·4(0·34)	7·6(0·06)	16 (2·0)	8·6(0·43)
Methionine	6·5(0·12)	1·6(0·01)	7·5(0·91)	1·7(0·09)
Total	24·9(0·46)	9·2(0·07)	23·5(2·91)	10·3(0·52)
Culture filtrate				
Cysteine	14 (0·26)	36 (0·29)	14 (1·67)	34 (1·73)
Methionine	0·5(0·01)	1·1(0·01)	0·8(0·10)	2·1(0·10)
Glutathione	12 (0·23)	13 (0·10)	13 (1·60)	15 (0·77)
Cysteinylglycine	44 (0·80)	38 (0·30)	44 (5·40)	36 (1·8)
S-methylcysteine	4 (0·08)	2 (0·02)	5 (0·56)	2·4(0·12)
Total	74·5(1·38)	90·1(0·72)	76·8(9·33)	89·5(4·52)
. · . Total $[^{35}S]$-H_2S assimilated	100 (1·84)	100 (0·79)	100 (12·2)	100 (5·0)

[a]Calculated from data of Howes and Scott (1973): strain V1 incubated for five days.
[b]Calculated from unpublished data of Bullock and Scott: strain V1C incubated for nine days.
[c]Liquid suspension cultures containing ~0·5 mM $[^{35}S]$-H_2S (115 mCi mmole) in basal medium (cf. Table I).
[d]Values in parentheses represent absolute amount (nmole) of assimilated $[^{35}S]$.

assimilation and interconversion of reduced sulphur compounds, (2) why is much of the cysteine leaked out as glycine-containing peptides and (3) what is the biological significance of the leakage of sulphur-containing amino acids and peptides?

C. Pathways of Assimilation and Interconversion of Sulphur Compounds

1. General considerations

A general scheme for the assimilation and interconversion of reduced sulphur compounds in fungi is summarized in Fig. 2. Note that both cysteine

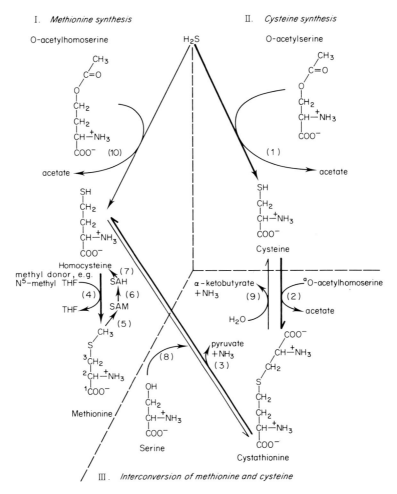

Fig. 2. General scheme for the assimilation of inorganic sulphide (cf. Siegel (1975), Schiff and Hodson (1973), Giovanelli *et al.* (1974), Datko *et al.* (1974). Bold arrows indicate apparent pathway of assimilation of sulphide by *Puccinia graminis.* [a]Note that some organisms appear to use O-phosphohomoserine or O-succinylhomoserine as homoserine donor. Enzymes: (1) = O-acetylserine sulphydrylase; (2) = Cystathionine-γ-synthase; (3) = Cystathionine-β-lyase (or β-cystathionase); (4) = Methyltransferase; (5) = Methionine adenosyltranferase; (6) = Methyltransferase; (7) = Adenosylhomocysteinase; (8) = Cystathionine β-synthase; (9) = Cystathionine-γ-lyase (or γ-cystathionase); (10) = O-acetylhomoserine sulphydrylase.

and methionine can be synthesized independently from inorganic sulphide (H_2S or the carrier disulphide in Fig. 1), and that these two amino acids can be interconverted via cystathionine. Although Howes and Scott (1973) did not suggest which pathways actually operate in *P. graminis*, some clues can be gleaned from their data. For example, the two-way movement of sulphur between cysteine and methionine is demonstrated by the ability of the fungus to use either amino acid as a sole source of sulphur (Howes and Scott, 1972). Other experiments are discussed below.

2. Inhibition of $[^{35}S]$-H_2S assimilation by cysteine

It can be seen from Table II that less ^{35}S was assimilated from $[^{35}S]$-H_2S into amino acids (mycelial protein + culture filtrate) in the presence of 3 mM unlabelled cysteine than in the absence of cysteine, despite the growth promoting effect of exogenous cysteine (Howes and Scott, 1972). This inhibition was selective, with assimilation of ^{35}S into methionine being proportionately less in the presence than in the absence of cysteine. These results are best explained if all or most of the ^{35}S from H_2S was assimilated via cysteine (pathway II of Fig. 2), with the exogenous "cold" cysteine supplying precursor molecules for methionine synthesis, and causing partial inhibition of cysteine synthesis. The alternative route of assimilation of H_2S via homocysteine (pathway I of Fig. 2) would lead to selective enrichment of ^{35}S in methionine compared to cysteine, which is the reverse of what was observed.

The proposal that *P. graminis* assimilates sulphide via cysteine is consistent with sulphide assimilation in other organisms such as *Neurospora crassa* and *Salmonella typhimurium*, in which the direct pathway of methionine synthesis from H_2S via homocysteine does not appear to be of physiological significance (Siegel, 1975; Kerr and Flavin, 1970). In contrast, the direct synthesis of methionine from H_2S appears to be important in yeast (Siegel, 1975; Cherest *et al.*, 1969).

3. Conversion of methionine to cysteine

Puccinia graminis is able to grow on media containing low concentrations of methionine as sole source of sulphur (Howes and Scott, 1972), showing that a pathway exists for the conversion of methionine to cysteine. The fate of the carbon skeleton of methionine was determined after growth on media containing 0·1 mM $[1-^{14}C]$-L-methionine in the absence of both cysteine and inorganic sulphide (Howes and Scott, 1973). After five days incubation,

41% of the radioactivity remaining in the culture filtrate was present in methionine, 2% in cystathionine, and 57% in an unidentified compound thought to be α-ketobutyrate derived from the spontaneous γ-elimination of cystathionine. No label was detected in homocysteine, although this compound is presumed to be an obligatory intermediate in the conversion of methionine to cysteine via cystathionine (Fig. 2); intracellular metabolites were not analysed.

The excretion of cystathionine (and a possible derivative of it) only after growth on methionine, but not apparently after growth on H_2S or cysteine as sulphur source, might reflect differences in pool sizes of intermediates depending on the direction of sulphur flow between methionine and cysteine. In no experiment was labelled homocysteine excreted (Howes and Scott, 1973).

IV. Nitrogen and Amino Acid Metabolism in *Puccinia Graminis*

A. *Sources of Nitrogen*

Nutritional experiments described in Part A of this chapter showed that a reduced source of nitrogen such as ammonium ion or an amino acid, was necessary for a net increase in mycelial protein. No growth was observed on nitrate or nitrite as sole nitrogen source (Howes and Scott, 1972). Nitrate is assimilated in three steps by organisms able to use it (Hewitt, 1975; Miflin and Lea, 1976): (1) reduction of nitrate to nitrite, (2) reduction of nitrite to ammonium and (3) reductive assimilation of the ammonium into either glutamate or glutamine. The first step is generally considered to be rate limiting.

Howes (1972) investigated nitrate reductase activity in mycelium of *P. graminis,* and compared it with the activity in mycelium of a cysteine-requiring mutant of *Neurospora crassa* able to use nitrate as a sole source of nitrogen. Nitrate reductase is subject to induction and repression in many organisms. Therefore, each fungus was pre-grown on agar-solidified medium containing 0·2% peptone (reduced nitrogen) as sole nitrogen source, then transferred to a liquid "induction medium" containing no reduced sulphur source such as cysteine or methionine (i.e. nutritionally inadequate for each species, but necessary so as not to introduce reduced nitrogen). The induction medium contained 24 mM nitrate as sole nitrogen source, together with 1 mg/l of MoO_3 (nitrate reductase is a molybdenum-containing enzyme).

No constitutive nitrate reductase activity was detected in *P. graminis* growing on peptone, but activity was induced by nitrate, reaching a maximum of 1·5 units of activity after two days incubation in the induction medium (1 unit of activity = 1 μmole nitrite produced/mg protein/hour). In contrast, *N. crassa* had a constitutive activity of 80 units, which rose to a maximum of 150 units of activity after five hours incubation in the induction medium. Thus, nitrate reductase activity was induced 100 fold higher in *N. crassa* than in *P. graminis*.

Howes (1972) calculated that to provide adequate reduced nitrogen for protein synthesis, *P. graminis* would require activities of nitrate reductase in excess of 30 units (as defined above), i.e. 20 times higher than that observed. This sluggish rate of nitrate reduction readily explains the need of *P. graminis* for reduced sources of nitrogen, but also demonstrates that rust fungi have not totally lost the ability to reduce nitrate. If species of rust fungi differed in the ease and extent to which nitrate reductase can be induced, this would explain the observed ability of *Gymnosporangium juniperivirginianae* to grow on nitrate as sole nitrogen source (see Part A of this chapter). Many fungi other than *P. graminis* cannot grow with nitrate as sole nitrogen source, particularly many of the higher basidiomycetes (Cochrane, 1958), and also the Blastocladiales and many Saprolegniales (Cantino, 1955). The latter Phycomycetes also share with *P. graminis* an ability to reduce sulphate (Section III, this chapter).

B. Metabolism of Glycine

The metabolism of glycine and excretion of glycine-containing peptides is of interest, because incubation of *P. graminis* on media containing $[^{35}S]$-H_2S resulted in about half of the assimilated radioactivity being excreted as glycine-containing peptides (36–44% as cysteinylglycine and 12–15% as glutathione—cf. Howes and Scott, 1973; Table II, Section III B). The direct role of glycine in metabolism was investigated by incubating the fungus with $[^{14}C]$-labelled glycine in media containing 6 mM glutamine as bulk nitrogen source, and either cysteine or methionine as sulphur source (cf. Table III). Serine and cysteine were the only protein amino acids synthesized from glycine; both of these amino acids were also excreted free into the medium. However, much of the excreted radioactivity was in the form of glutathione and cysteinylglycine; this is especially evident in experiment 2 of Table III. Hydrolysis of glutathione, cysteinylglycine, and two unidentified peptides from experiment 1 of Table III (Howes and Scott, 1973) showed that the radioactivity was mostly located in the glycine residues. These experiments confirmed the excretion of cysteine- and glycine-containing peptides seen after incubation with $[^{35}S]$-H_2S, and demonstrated that this excretion or leakage occurred with either cysteine or methionine as sole sulphur source in

Table III. Metabolism of exogenous [^{14}C]-glycine by *Puccinia graminis*.

Compound	% distribution of ^{14}C into compounds other than glycine			
	[a]Experiment 1		[b]Experiment 2	
	[c]0·1 mM methionine	[c]3 mM cysteine	[c]0·1 mM methionine	[c]3 mM cysteine
Mycelial protein				
Cysteine	0·2(0·03)[d]	0·7(0·04)[d]	0·06(0·06)[d]	0·16(0·06)[d]
Serine	1·0(0·14)	3·7(0·22)	0·30(0·27)	1·0 (0·40)
Culture filtrate				
Cysteine	5·5(0·77)	3·5(0·20)	1·9(1·9)	1·1 (0·44)
Serine	29 (4·1)	7·7(0·46)	8 (7·9)	2·6 (1·0)
Glutathione	6 (1·0)	5 (0·29)	19 (19)	16 (6·4)
Cysteinylglycine	18 (2·5)	22 (1·3)	61 (60)	64 (26)
S-methylcysteine	18 (2·6)	24 (1·4)	4 (4·0)	5 (2·1)
Other peptides	21 (3·0)	33 (2·0)	5 (5·0)	9 (3·7)
. · . Total	100 (14)	100 (5·9)	100 (98)	100 (40)

[a]Calculated from data of Howes and Scott (1973): strain V1 incubated for five days in 0·15 mM [2-^{14}C]-glycine (7·5 μCi/5 ml) in basal medium (Table I).

[b]Calculated from unpublished data of Bullock and Scott: strain V1C incubated for nine days in 0·1 mM [U-^{14}C]-glycine (4·5 μCi/5 ml) in basal medium (Table I).

[c]Supplement to the medium.

[d]Values in parentheses represent nmoles of [^{14}C]-glycine incorporated into each compound per mg mycelial protein.

the medium. Indeed, 2–8 times as much label was excreted in these peptides after growth on 0·1 mM methionine compared to 3 mM cysteine (cf. Table III).

C. *Full Extent of Amino Acid Excretion*

The full extent of metabolite excretion was not determined after incubating *P. graminis* on media containing [^{35}S]-H$_2$S, [^{14}C]-glycine or [^{14}C]-methionine, because these experiments were designed to follow the movement of radioactivity into specific classes of compounds. Metabolites derived from unlabelled substrates would not have been detected. To assess the full extent of amino nitrogen loss into the medium, Howes (1972) and Bullock and Scott (unpublished) incubated *P. graminis* in media containing [^{14}C]-labelled glucose, and isolated all amino acids from the culture filtrates

Table IV. Synthesis and excretion of amino acids (cationic substances) from [^{14}C]-D-glucose by *Puccinia graminis*.

| | [a]Results expressed as the ratio of: | |
| | mg glucose assimilated into amino acids $\diagup$ mg mycelial protein | |
	Basal medium	Basal medium +3 mM cysteine
Mycelial protein	0·163[b](17%)	0·578 (38%)
Culture filtrate	0·801 (83%)	0·942 (62%)

[a]Previously unpublished data of Bullock and Scott: growth of strain V1C for nine days on Basal medium (Table I) containing 4% [U-^{14}C]-D-glucose of specific activity 0·056 mCi/mmole. Very similar results were obtained with strain V4 of *P. graminis*.

[b]Values in parentheses represent the per cent of the total ^{14}C assimilated into cationic substances.

as a cation fraction. Individual amino acids in the cation fraction were separated by thin layer chromatography, and some were quantified by the ninhydrin reaction. Concurrent radiochemical analysis made it possible to assess whether or not the excreted amino acids had been synthesized from the glucose present in the medium.

More of the glucose assimilated into amino acids was excreted into the medium rather than being incorporated into mycelial protein (Table IV). Thus, after incubation for nine days on basal medium (which does not support nett protein synthesis), 83% of the carbon in amino acids synthesized from glucose was released into the culture filtrate, compared to 62% released after incubation in cysteine-supplemented medium (which supports growth).

The individual amino acids identified in culture filtrates are listed in Table V; assessment of radioactivity in these amino acids was only carried out after incubation in basal medium. The following salient features can be noted:

(a) A wide range of amino acids was excreted (Table V). Two other compounds were also present: one was γ-aminobutyric acid (thought to be a degradation product of glutamic acid, cf. Howes, 1972), while the other was thought to be γ-glutamyl-glutamate because it yielded variable mixtures of glutamic acid and γ-aminobutyric acid after acid hydrolysis (Howes, 1972).

(b) Despite the lack of a reduced sulphur source to promote net protein synthesis, the relative amounts of amino acids released after incubation in basal medium were very similar to the amounts released during growth in cysteine-supplemented medium (Table V). Thus, the relative amounts of

Table V. [a]Amino acids excreted into culture filtrates by *Puccinia graminis,* after incubation for nine days with [^{14}C]-D-glucose.

Amino acid	Basal medium		Basal medium + 3 mM cysteine
	[b]nmoles excreted	[c][^{14}C]	[b]nmoles excreted
Glutamate	1470	+ +	579
Glycine	647	+ +	372
Alanine	538	+ +	452
Arginine	107	trace	153
Lysine	231	trace	260
Glutamine[d]	Det.[d]	+ +	Det.[d]
γ-Glutamylglutamate(?)	Det.	+	Det.
Serine	Det.	+ +	Det.
Leucine + isoleucine	Det.	trace	Det.
Phenylalanine	Det.	trace	Det.
Valine	Det.	N. Det.	Det.
Threonine	Det.	N. Det.	Det.
Asparagine	Det.	N. Det.	Det.
Proline	trace	N. Det.	trace
Cysteine	trace	N. Det.	Det.[d]
Methionine	trace	N. Det.	trace

[a]Further analysis of culture filtrates described in Table IV.
[b]Total nmole excreted (ninhydrin assays) per mg mycelial protein.
[c]Radioactivity detected on thin-layer chromatograms.
[d]Supplied in the medium.
Det. = detected by ninhydrin reaction, but not quantified.
N. Det. = not detected.
trace, +, + + = relative amounts.

alanine, arginine and lysine excreted into each medium were comparable, although somewhat more glycine and glutamate were excreted in the absence of cysteine.

(c) The cysteine- and glycine-containing peptides cysteinyl-glycine and glutathione were not detected in significant quantities in comparison to the free amino acids listed in Table V (they should not, of course, have been present after incubation in basal medium, but might have been expected to be present in cysteine-supplemented medium). The reason for the above becomes clear when Table V is compared with Tables II and III which were compiled from experiments in which these peptides were detected by incorporated radioactivity, after feeding [^{35}S]-H$_2$S and [^{14}C]-glycine respectively. Fewer nmoles of these peptides were excreted per mg mycelial protein (Tables II and III) than the free amino acids (Table V).

(d) Radioactivity from the [^{14}C]-glucose present in the medium was assymetrically distributed between the amino acids, with most being present in glutamate, glycine, alanine, glutamine, serine and glutamylglutamate. Only traces were present in arginine, lysine, leucine/isoleucine and phenylalanine, although these compounds were present in large amounts. The unlabelled amino acids which were excreted were presumably synthesized from the glutamine or citrate supplied in the medium.

D. Nutritional Significance of Metabolite Leakage

1. Loss of amino acids

A large part of the synthetic capacity of *P. graminis* appears to be wasted by excreting amino acids as end products of metabolism (cf. Sections III B, III C, IV B, IV C). The extent to which these excreted amino acids are continuously being taken up and reutilized is not known, although turnover has been demonstrated with glutamine; culture filtrates accumulated labelled glutamine after growth on [^{14}C]-glucose, even though unlabelled glutamine was initially added to the medium as bulk nitrogen source (Table V).

A useful way of viewing the leakage is to consider the culture medium as an external "reservoir" pool of metabolities, with a two-way exchange occurring between internal and external pools. Provided that the external pool contains a well-balanced mixture of amino acids, any component lost by leakage can immediately be replaced by uptake, and indeed, in the presence of an enriched medium there may well be a nett uptake of all or most amino acids. When the rust fungus is growing parasitically, the plant fluids which bathe the haustoria and intercellular hyphae presumably represent such a well-balanced, enriched medium, to which the fungus has adapted during co-evolution with its host. Minimal or unbalanced media in axenic culture would therefore result in a nett loss of amino acids, thus imposing an excessive drain on metabolism and depleting internal metabolite pools, which in turn would result in lesser growth rates.

It can be concluded that leakage of amino acids is presumably the cause of the requirement of *P. graminis* for enriched media for optimal growth (cf. Section II G), and provides evidence for the hypothesis that uredospores of rust fungi grow better on nutrient media in regions of high inoculum density, owing to leakage and reutilization of metabolites (cf. Part A of this chapter).

2. Nett sulphur loss

The leakage of sulphur-containing amino acids and peptides from *P. graminis,* although not quantitatively as great as the leakage of other free amino acids, nonetheless appears to represent a considerable drain on the sulphur resources of the organism. The inability of hydrogen sulphide to support nett protein synthesis when offered as sole sulphur source is probably caused by this leakage, because 75% of the assimilated sulphur was released into the culture filtrate rather than being incorporated into mycelial protein (Table II). Because most of this sulphur was lost as cysteine residues, Howes and Scott(1973) suggested that the rate of cysteine synthesis from sulphide was insufficient to maintain the internal pool of cysteine at adequate levels. A depleted cysteine pool would thus be inadequate either as a precursor pool for methionine biosynthesis, as a source of cysteine for protein synthesis, or as an intermediate required for the operation of the γ-glutamyl cycle (see Section E below).

The 30-fold lower exogenous concentration of methionine (0·1 mM) necessary to support nett protein synthesis compared to cysteine (3 mM), appears to be related to the relative amounts of each of these amino acids excreted by the fungus. After growth with H_2S as sole sulphur source, the amount of methionine (synthesized via cysteine) excreted was 1% or less of the amount of cysteine and cysteine derivatives excreted (Table II). However, when methionine served as sole sulphur source, extensive leakage was observed of cysteine precursors (cystathionine and a possible derivative of cystathionine, detected after growth on [^{14}C]-methionine—cf. Section III C), and cysteine residues (detected after growth with [^{14}C]-glycine—cf. Section IV B). Apparently, the rate of conversion of methionine to cysteine via cystathionine is sufficiently efficient from small exogenous methionine concentrations, to maintain an adequate pool of cysteine for nett protein synthesis, with excess cysteine leaking from the mycelium.

E. The γ-Glutamyl Cycle

After growth on either [^{35}S]-H_2S or [^{14}C]-glycine, considerable amounts of radioactivity were excreted as cysteinylglycine and glutathione. As noted by Howes and Scott (1973), these compounds are intermediates of the "γ-glutamyl cycle", a cyclic set of reactions first proposed by Orlowski and Meister (1970) as a mechanism for the active transport of amino acids into mammalian cells (Meister, 1973, 1974, 1978; Meister and Tate, 1976) and into yeast (Mooz and Wigglesworth, 1976; Mooz, 1979).

During operation of the γ-glutamyl cycle as presently proposed for amino

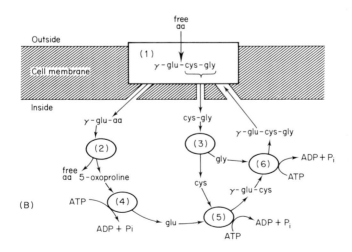

Fig. 3. Reactions of the γ-glutamyl cycle based on Meister (1973) and Meister and Tate (1976). A = enzymes (1) and (2). B = full cycle. Enzymes: (1) = γ-glutamyltranspeptidase; (2) = γ-glutamylcyclotransferase; (3) = peptidase; (4) = 5-oxoprolinase; (5) = γ-glutamylcysteine synthetase; (6) = glutathione synthetase.

acid transport, a free amino acid outside the cell reacts with glutathione (γ-glutamylcysteinyglycine, Fig. 3A) in a reaction catalysed by γ-glutamyl transpeptidase, an enzyme situated on or accessible to the outside of the cell membrane (Fig. 3B). The products formed by the reaction (a γ-glutamyl amino acid and cysteinylglycine) are translocated into the cell. Further reactions, shown to occur in the cystosol in Fig. 3B, involve liberation of the transported amino acid by a γ-glutamyl cyclotransferase, cleavage of cysteinylglycine by a peptidase and three ATP-requiring reactions to regenerate glutathione.

As far as I am aware, no reports are available on the excretion of glutathione and cysteinylglycine during malfunction of the γ-glutamyl cycle, except for the report by Howes and Scott (1973) on *Puccinia graminis*. Howes and Scott (1973) suggested that *P. graminis* possesses a defective γ-glutamyl cycle which lacks a peptidase capable of cleaving cysteinylglycine, resulting in the accumulation and excretion of cysteinylglycine and other intermediates of the cycle. However, this suggestion suffers from a number of problems, e.g. a total blockage of cysteinylglycine cleavage would lead to this dipeptide being excreted in equimolar proportion to transported amino acids, which would impose a severe drain on the supply of cysteine and glycine from intermediary metabolism, and such a mechanism would offer no apparent advantage to the fungus.

While the present information is too fragmentary to offer a precise model to explain the results of Howes and Scott (1973), a number of suggestions can be made which might aid future research. Although the scheme presented in Fig 3B shows the peptidase (enzyme (3)) being present in the cytosol, Meister (1973) and Meister and Tate (1976) have considered the possibility that γ-glutamyl transpeptidase is located in a membrane complex with other enzymes (including peptidases, cf. Hughey *et al.*, 1978), that functions in the general transport or degradation of peptides. If such a membrane complex exists in *P. graminis* and is intrinsically leaky, some of the substrates and products of reactions occurring in the complex would be excreted. A further point to be born in mind is that γ-glutamyl transpeptidase is an enzyme which shows a broad substrate specificity (Meister, 1973; Meister and Tate, 1976), and can catalyse a number of exchange reactions involving the γ-glutamyl bond and resulting in the release of free amino acids, e.g.: γ-glu aa$_1$ + aa$_2 \rightarrow \gamma$-glu aa$_2$ + aa$_1$, and γ-glu aa$_1$ + γ-glu aa$_2 \rightarrow \gamma$-glu-γ-glu aa$_2$ + aa$_1$.

Furthermore, the γ-glutamyl transpeptidase appears to be capable of catalysing reactions between intracellular amino acids and a γ-glutamyl donor such as glutathione or γ-glutamyl cysteine. It is therefore evident that a leaky or malfunctioning γ-glutamyl cycle could account for the excretion of intracellular amino acids as well as some of the intermediates of the cycle.

V. Carbohydrate Metabolism

A. Analysis of Sugars in Rust Fungi and Host Plants

1. Compounds identified

Knowledge of the major soluble sugars present in rust fungi and their host plants is necessary first, to suggest which sugars and their analogues should be used as substrates for *in vitro* experiments on nutrition and metabolism, and secondly, to assist in designing methods for the analysis of metabolic products formed during radiotracer experiments. The early literature on this subject is extensive and has been reviewed and updated on a number of occasions (e.g. Lewis and Smith, 1967; Smith *et al.*, 1969; Scott, 1972; Maclean and Scott, 1976), so that a number of generalizations can now be made.

Uredospores or mycelium of rust fungi have been shown to contain the disaccharide trehalose (α-D-glucopyranosyl-$(1 \rightarrow 1)$-α-D-glucopyranoside) and the acyclic polyols D-mannitol, D-glucitol (*syn.* sorbitol), D-arabitol, ribitol, erythritol and sometimes glycerol as major soluble sugars, with lower amounts of glucose and fructose, and traces of inositol and xylitol (cf. Maclean and Scott, 1976; Daly *et al.*, 1967; Lewis and Smith, 1967). The storage polysaccharide glycogen has been reported in some rust fungi (Holligan *et al.*, 1974a, 1974b), although most workers have not attempted to identify glycogen when analysing rust constituents. Sugars which can serve as a sole carbohydrate source for the growth axenic cultures of rust fungi are listed in Part A of this chapter (page 63).

The major sugars which accumulate in non-infected tissues of those host plants which have been studied during rust infection, are sucrose, glucose and fructose (Smith *et al.*, 1969). However, Maclean and Scott (1976) detected traces of compounds with the chromatographic mobilities of ribitol, arabitol and mannitol on chromatograms very heavily loaded with extracts of healthy wheat leaves, although these compounds could not be detected at chromatogram loadings used for the analysis of polyols in rust-infected wheat. In general, it is assumed that many host plants do not accumulate significant quantities of polyols. Some plants belonging to the Rosaceae accumulate D-glucitol as a major translocatory sugar (Lewis and Smith, 1967; Bieleski and Redgwell, 1980) but such glucitol-containing hosts have not yet been investigated during plant-rust interactions.

2. Carbohydrate levels in rust-infected plants

Carbohydrate levels in rust-infected tissues are related to the stage of disease development and the intensity of infection (cf. Smith *et al.*, 1969; Long *et al.*, 1975; Mitchell *et al.*, 1978 for reviews of earlier work). At low infection intensities, which represent a better nutritional situation for the fungus due to less competition between colonies for host metabolites, soluble host sugars at infection sites (sucrose, glucose and fructose) tend to increase or remain comparable in concentration with concentrations in uninfected tissues, despite increases in invertase activity at the infection site (e.g. Mitchell *et al.*, 1978; Long *et al.*, 1975). At higher infection densities, sucrose concentrations tend to drop while glucose and fructose concentrations often increase. Storage polysaccharides of host origin such as starch, fructans and glucomannans accumulate in host tissues surrounding or within infection sites (Holligan *et al.*, 1973, 1974a, 1974b; Long *et al.*, 1975; Mitchell and Roberts, 1973).

Fungal sugars (trehalose and polyols) increase at infection sites in proportion to the spread of the parasitic mycelium after inoculation, until after sporulation has reached a peak and both host and parasite undergo senescence (e.g. Smith *et al.*, 1969; Mitchell *et al.*, 1978). The relative proportions of different rust sugars in mycelium and spores was investigated by Maclean and Scott (1976), who showed that when sporogenous mycelium was just beginning to form beneath the intact epidermis of wheat leaves, D-glucitol was the predominant hexitol (D-glucitol = 0·29 mg/g fresh weight of leaves; D-mannitol = 0·20 mg/g), and ribitol was the predominant pentitol (ribitol = 0·16 mg/g; D-arabitol = 0·09 mg/g). Because uredospores and sporelings contained no detectible glucitol and only traces of ribitol, yet contained large amounts of mannitol and arabitol, Maclean and Scott (1976) suggested that glucitol and ribitol may be intermediates (or by-products) of the uptake and metabolism of exogenous sugars such as glucose, whereas mannitol and arabitol represent reserve or storage compounds.

3. Identification of glucitol and ribitol as rust metabolites

The identification of D-glucitol and ribitol as major metabolites of *P. graminis* (Maclean and Scott, 1976) requires further comment, because the presence of these acyclic polyols has been reported on very few occasions. Maclean (1971) found large amounts of both D-glucitol and ribitol in glucose-grown mycelium of *P. graminis*, and these polyols were purified from this source and positively identified by their infared spectra and melting points (Maclean and Scott, 1976); traces of xylitol were also detected and tentatively identified. These polyols have been reported in rust fungi on only three other occasions: (1) Daly *et al.* (1967) noted that significant amounts of ribitol were present in some (but not all) preparations of bean and wheat rust uredospores; (2) Wicker *et al.* (1976) showed that glucitol and ribitol were the major sugars present in pycnial (i.e. spermogonial) fluid of *Cronartium ribicola* together with much smaller concentrations of fructose and mannitol, and (3) Lewis (1976) cited unpublished work of A. K. Fung, who detected the accumulation of small quantities of glucitol in uredial infections of *Puccinia poarum* on *Poa pratensis*, a grass.

It is highly likely that many other workers failed for technical reasons to detect glucitol and ribitol in extracts of rust-containing materials. The first analyses of rust sugars were carried out on resting and germinating uredo-spores of *P. graminis tritici*, resulting in the identification of D-mannitol, D-arabitol, erythritol and glycerol (Prentice *et al.*, 1959; Reisener *et al.*, 1962). As noted previously, uredospores and sporelings of this fungus contain very little if any glucitol and ribitol, and subsequent workers prob-ably assumed that the hexitols and pentitols which accumulated in rust-infected plant tissues were those that had been identified in uredospores. Accordingly, analytical techniques were frequently used which are intrinsi-cally unable to resolve the isomeric hexitols or the isomeric pentitols, e.g. the solvent systems for paper chromatography used by Mitchell and Roberts, 1973; Mitchell and Shaw, 1968; Holligan *et al.*, 1974a, or the use of trimethylsilyl derivatives for gas–liquid chromatography (e.g. Mitchell and Roberts, 1973; Mitchell *et al.*, 1978; Holligan *et al.*, 1973). Unfortunately, Holligan *et al.*, (1973) used gas chromatography of polyol acetates (which resolves isomeric hexitols) only for hydrolysates of cell walls and not for soluble sugars.

Another problem arises with the use of the ethyl methyl ketone/acetic acid/boric acid/water solvent for paper chromatography of Rees and Reynolds (1958), which should separate mannitol from glucitol on paper chromatograms, and which has been applied to the analysis of extracts of rust-infected plant materials by, for example, So and Thrower (1976) and probably Holligan *et al.* (1973). In our laboratory, we have frequently found that some components of crude extracts retard the chromatographic mobil-ity in this solvent system of hexitols and pentitols and prevent their clean separation, especially if the chromatography paper has not been pre-washed (cf. Maclean and Scott, 1976). Thus, the direct chromatography of 80% ethanol extracts of glucose-grown mycelium of *P. graminis* even after ion-exchange treatment gives a single hexitol spot in the mannitol region (glucitol has a greater mobility than mannitol in this solvent system); upon eluting and rechromatographing this spot a satisfactory separation of gluci-tol and mannitol is achieved, with mobilities corresponding to adjacent standards (Manners *et al.*, 1982).

Table VI. Relative proportions of acyclic polyols in vegetative mycelium of various fungi grown in liquid culture (mixtures of surface and mostly submerged hyphae) (Unpublished data of D. J. Maclean).

Species	Carbohydrate source in medium	Percent of total acyclic polyols					
		Glucitol	Mannitol	Arabitol	Ribitol	Xylitol	Erythritol
A. Heterobasidiomycetes							
Puccinia graminis	glucose	49	41	3·5	6	0·7	trace
Sporobolomyces roseus	glucose	5	72	15	0	5	2
Sporobolomyces roseus	fructose	0·2	97	2	0	0	0·6
B. Homobasidiomycetes							
Schizophyllum commune	glucose	2	86	9	0	3	1
Schizophyllum commune	fructose	13	60	27	0	0·6	0·4
Armillariella mellea	glucose	2	88	6	0	1	3
Poria vincta	glucose	trace	43	56	0	0·5	0·3
Rhizoctonia sp.	glucose	0	78	19	0	3	0

C. Ascomycetes and imperfect: the following fungi contained mostly D-mannitol and D-arabitol, and negligible amounts of D-glucitol or ribitol:

   Emericella nidulans (glucose or fructose),
   Fusarium oxysporum (glucose),
   Sclerotium rolfsii (glucose or fructose),
   Claviceps paspali (glucose or fructose)

4. Survey of other fungi for glucitol and ribitol

Because inadequate techniques as outlined above may have been responsible for the paucity of reports of D-glucitol (and possibly ribitol) in fungi (Lewis and Smith, 1967), a limited number of fungal species were assayed to determine the relative proportions of different polyols present in mycelium after growth on glucose or fructose as sole carbohydrate source. Mycelium was harvested, the aldoses and ketoses present in a deionized 80% ethanol extract of the mycelium were removed by exhaustive digestion with strong base, and the residual polyols were determined by gas–liquid chromatography of their acetylated derivatives (D. J. Maclean, unpublished methods). Although the major mycelial polyols were mannitol and arabitol, some basidiomycetes contained small but significant amounts of glucitol (Table VI). The highest proportion was found in one strain of *Schizophyllum commune*, which accumulated glucitol as 13% of its total polyols after growth on fructose, although this was still much less than the 50% found in *P. graminis* after growth on glucose. No ribitol was detected in any of the fungi surveyed (except *P. graminis*, which only accumulates low proportions of pentitols in liquid submerged culture compared to surface-grown cultures). Surveys of other fungi (Pfyffer and Rast, 1980a, 1980b) confirm that very few species accumulate significant amounts of ribitol and glucitol. Further surveys of other rust fungi are necessary to determine if *P. graminis* is unique in accumulating such a high proportion of glucitol.

B. Sugar Uptake by Puccinia graminis

The uptake of sugars into vegetative mycelium of *Puccinia graminis* has been investigated using liquid suspension cultures of strain VIB of the fungus, and using the host sugars sucrose, glucose, fructose and some of their analogues as substrates. When sucrose is presented as sole carbohydrate source to either sucrose- or glucose-adapted mycelium, rapid inversion of the sucrose occurs (D. J. Maclean, T. Hoppner and K. J. Scott, unpublished results), consistent with the sucrose being hydrolysed to its constituent monosaccharides either prior to or during uptake. Other experiments indicate that free glucose is taken up preferentially to fructose by either glucose- or fructose-adapted mycelium (Henry *et al.*, 1977).

The kinetic characteristics for the transport of glucose and glucose analogues into glucose-adapted cells of *P. graminis* have been examined (J. M. Manners, D. J. Maclean and K. J. Scott, unpublished results). Both a high and low affinity transport system were found for D-glucose (apparent K_m values of 3·5 mM and 85 mM respectively) and for 3-0-methyl-D-glucose

(7·6 mM and 650 mM). An even higher affinity was shown for 2-deoxy-D-glucose (apparent $K_m = 1·25$ mM), although this glucose analogue was rapidly converted to other compounds after uptake and was eventually toxic after 2–5 hours, especially at higher concentrations thus precluding investigation of low affinity systems. However, 3-0-methyl-D-glucose accumulated linearly for at least five hours, was not modified inside the cell, and did not appear to be toxic over 24 hour periods and eventually reached intracellular concentrations well in excess of the medium, indicating that transport was active.

C. Pathways of Assimilation of Glucose by Puccinia Graminis

1. Entry of glucose into intermediary metabolism

The possibility suggested by Maclean and Scott (1976) that D-glucitol is a key intermediate of glucose assimilation in *P. graminis*, prompted a series of kinetic experiments to determine the pathway(s) of carbon flow from [U-^{14}C]-labelled D-glucose through pools of metabolic intermediates to end-products (Henry *et al.*, 1977; Manners *et al.*, 1982). Selected data from Manners *et al.*, are summarized below, to illustrate the movement of carbon into various classes of compounds under steady-state conditions from 1 or 20 mM glucose (both concentrations gave similar results) at various time intervals after adding label:

| Metabolites | % of total assimilated label | | |
	2 min	10 min	180 min
anions	16–19	16–19	12–13
cations	16–19	29–34	31–33
neutral	62–63	41	37–38
insoluble in 80% ethanol	2–3	5–9	15–17
lipids	1<<	1<	~1

Further analysis of the anion, cation and neutral fractions revealed that the cellular pools of glucose and amino acids, and the phosphate esters of trehalose, glucose, fructose and gluconic acid, were most rapidly labelled. Glucose was also readily converted to free fructose, and more slowly to free trehalose, glucitol and mannitol.

In a cold chase experiment, the fungus was grown for four days in labelled, 200 mM D-glucose, then rapidly washed and resuspended in cold medium of the same composition: analysis of the carbohydrate pools showed that radioactivity was first chased out of the glucose pool, followed by the fructose pool, with label subsequently being chased out of the glucitol and mannitol pools.

2. Generalized scheme of carbohydrate metabolism

Data from the above and other experiments of Manners *et al.* (1982) have led to the scheme presented in Fig. 4 as a working hypothesis to explain the major pathways of carbon flow during growth of *P. graminis* on glucose as sole carbohydrate source. According to this scheme, glucose transported into the cell is assimilated by three competing pathways, each with fructose-6-phosphate as a common end product: (1) glucose → glucose-6-phosphate → fructose-6-phosphate (it is possible that at least some of the glucose-6-phosphate is formed by phosphorylation of glucose during transport—i.e. group translocation, Jennings, 1974), (2) glucose → fructose → mannitol → mannitol-1-phosphate → fructose-6-phosphate, (3) glucose (and fructose from (2)) → glucitol → glucitol-6-phosphate → fructose-6-phosphate. Identification of a phosphate ester of gluconate suggests that the pentose phosphate pathway may offer an additional source of fructose-6-phosphate, together with triose phosphates. Carbon from fructose-6-phosphate can then reach the citric acid cycle via glycolysis; conversely, fructose-6-phosphate can be resynthesized by gluconeogenesis from citric acid cycle

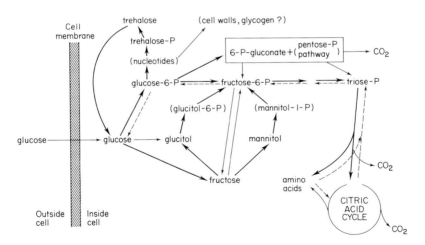

Fig. 4. Scheme for the assimilation and synthesis of glucose by *Puccinia graminis.* Parentheses indicate hypothetical intermediates; -P = phosphate ester,→ = postulated pathway(s) of glucose assimilation,--→ = postulated pathway of gluconeogenesis,→ = other possible or probable pathways.

intermediates, ultimately resulting in the formation of free glucose. Trehalose is synthesized from glucose-6-phosphate but breaks down to free glucose during turnover. Some of the novel features of this scheme are discussed below, in relation to supporting evidence.

The simultaneous operation of both gluconeogenesis and glycolysis was assessed by the following criteria: (i) the equilibrium specific activities of cellular pools of glucose, fructose, glucitol and probably mannitol, which were only 12% and 46% of the specific activity of exogenous $[^{14}C]$-glucose offered to cells at 1 mM and 20 mM respectively. This indicates that unlabelled carbon was entering these pools either from peptone or citrate provided in the medium, or by reutilization of storage reserves Manners *et al.* (1982) note that of lipid or glycogen. If the unlabelled carbon came from exogenous peptone or citrate, or from endogenous lipid, pathways involving gluconeogenesis would be necessary. (ii) A large and rapid counterflow of carbon, from glucose to amino acids derived from glycolytic or citric acid cycle intermediates, was seen two minutes after offering labelled glucose at 1 or 20 mM. (iii) Although little gluconeogenesis appeared to occur on 200 mM $[^{14}C]$-glucose, after four days growth 11% of accumulated radioactivity was present in amino acids, compared to 28% for sugars and polyols. A possible storage role for these amino acids was suggested by their very slow turnover compared to hexoses and polyols.

The interconversion of hexoses and hexitols by single step oxidation-reduction reactions is outlined in Fig. 4. Contrary to the possibility that fructose might be synthesized from glucose via glucitol as an intermediate, as is known to occur in some animal systems (Hers, 1960; Touster and Shaw, 1962), fructose was the first non-phosphorylated sugar or polyol to accept label from glucose, and also the first after glucose to lose label during a cold

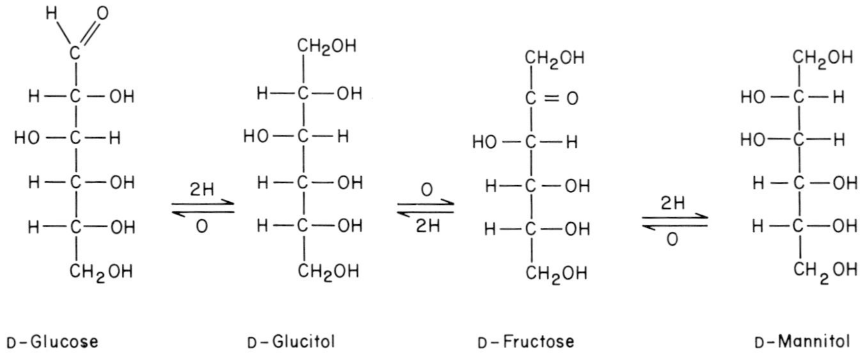

Fig. 5. Structural relationships between some hexoses and hexitols (2H = reduction step, O = oxidation step).

chase (Manners *et al.*, 1982). Kinetic and other data were consistent with the fructose being synthesized by the direct isomerization of glucose, a reaction known to be catalysed by some enzymes isolated from prokaryotes (Suekane *et al.*, 1978; Chen *et al.*, 1979), although an origin of fructose from small, rapidly turning-over pools of glucose and fructose phosphates was not ruled out. Kinetic data further suggested that flux rates through pools of metabolities were in the order glucose > glucitol > fructose > mannitol, and that the flux of carbon through fructose was surplus to that required for mannitol synthesis, but was insufficient to account for all the glucitol synthesis. On the basis of these and other considerations including the formation of 2-deoxyglucitol from 2-deoxyglucose (J. M. Manners, D. J. Maclean and K. J. Scott, unpublished data) it was concluded by Manners *et al.* (1982) that glucitol was synthesized from both glucose and fructose, but mannitol only from fructose. Because radioactivity in the glucitol and mannitol pools continued to decline after the glucose and fructose pools had become totally devoid of radioactivity during a cold chase experiment, Manners *et al.* (1982) suggested that these polyols were oxidized to sugar phosphates (probably fructose-6-phosphate) via very small pools of rapidly turning-over polyol phosphates.

Trehalose appeared to be synthesized in *P. graminis* by a pathway first elucidated in yeast (Cabib and Leloir, 1958), which involves the formation of glucose-6-phosphate and uridine nucleotides of glucose as intermediates, resulting in the synthesis of trehalose phosphate, which is then hydrolysed to free trehalose (Manners *et al.*, 1982). However, turnover of trehalose was faster at lower exogenous glucose concentrations (1–20 mM), suggesting wastage of energy by a futile cycle. Thus, two molecules of ATP are required for the synthesis of trehalose from glucose, but no ATP is recovered during its hydrolysis to free glucose by the enzyme trehalase (Elbein, 1974), and a saving of only one ATP would be effected if trehalose was broken down by trehalose phosphorylase (trehalose + P_i → glucose-1-phosphate + glucose; cf. Marechal and Belocopitow, 1972).

To verify schemes for metabolic pathways such as that presented in Fig. 4, it is necessary to demonstrate both (i) that the kinetics of carbon flow are consistent with the scheme, and (ii) that enzymes necessary to catalyse key reactions exist. The enzymes which require verification are those which result in the isomerisation of glucose to fructose, the reduction of glucose and fructose to glucitol and mannitol, the phosphorylation of these hexitols to hexitol phosphates, and the oxidation of hexitol phosphates to hexose phosphates which can then enter well-established metabolic pathways. Although a number of workers have detected activity in cell-free extracts of rust uredospores or mycelium, or rust-infected plant materials, which can couple the oxidation of hexitols or reduction of hexoses (or hexose phosphates) to $NAD(P)^+$ reduction or $NAD(P)H$ oxidation respectively, these experiments are incomplete because in only very few cases were products of the enzymatic reaction identified (Shu *et al.*, 1956; Hendrix *et al.*, 1964; Wynn, 1966; Maclean, 1971; Clancy and Coffey, 1980). Nonetheless, consistent with the scheme in Fig. 4 is the demonstration by Maclean (1971) of high activities of NADP-dependent glucitol oxidation and NAD-dependent mannitol oxidation in cell-free extracts of glucose-grown *P. graminis*, and the demonstration by Hendrix *et al.* (1964) of NADH-dependent reduction of fructose-6-phosphate in cell-free extracts of uredospores and *P. graminis*-infected wheat. Of course, the above activities represent enzymes assayed in the reverse direction to reactions postulated in Fig. 4.

3. Pentitol metabolism

The size of the pentitol pool (mostly ribitol and arabitol with a smaller proportion of xylitol—Maclean and Scott, 1976) fluctuated widely during growth on different carbon sources and under different conditions. Manners *et al.* (1982) found that in liquid suspension cultures, very little radioactivity entered the pentitol pools from exogenous [^{14}C]-glucose, suggesting a very minor role for pentitols in vegetative cells actively taking up glucose.

D. Regulation of Carbohydrate Metabolism in P. Graminis

Glucitol and trehalose, followed by mannitol, accumulate as the largest cellular pools of soluble carbohydrates during growth of *P. graminis* in liquid suspension culture using glucose as sole carbohydrate source (cf. Section V C). In addition, relatively large pools of free fructose and glucose were present. The role of these large pools of intracellular metabolites in regulating the flow of carbon through intermediary metabolism will now be considered, and a number of possible control mechanisms will be evaluated.

1. Role of glucitol and mannitol in storage and assimilation

Although the glucitol pool was much larger than the mannitol pool, the kinetics of movement of label from exogenous [^{14}C]-glucose into and out of these pools was very similar, suggesting that both of these hexitols play a similar role in vegetative metabolism (Manners *et al.*, 1982). Presumably, the hexitols are intermediates in alternative pathways for the movement of carbon into glycolysis and the HMP shunt. The quantitative importance of these pathways was shown during the cold chase in 200 mM glucose, at which concentration 68% of assimilated glucose was metabolized via hexitols (52% through glucitol with a turnover time of about 3½ hours, and 16% through mannitol with a turnover time of about 2 hours). However, these hexitol pools turned over very slowly compared to sugar (and putative polyol) phosphates. This relatively slow turnover of large pools suggests that the hexitols serve as a temporary storage of carbon during glucose assimilation, i.e. as a "buffer reserve" to hold assimilate pending its distribution to end products of metabolism. The hexitol metabolism of vegetative cells should be contrasted with that of uredospores, in which mannitol is the only hexitol present, and where it clearly serves as a long-term store of carbon and energy until germination is initiated (Maclean and Scott, 1976; Daly et al., 1967).

Manners et al. have suggested a mechanism to account for the observation that at higher exogenous glucose concentrations, a higher proportion of assimilated glucose entered a buffer pool of hexitols rather than entering metabolism via glucose-6-phosphate. The K_m values of fungal polyol dehydrogenases greatly exceed K_m values of fungal hexokinases. Therefore, at the higher influx rates forced by increasing the exogenous concentration of glucose, hexokinase would be saturated with substrate well before polyol dehydrogenases, thus directing a relatively greater flux of carbon into the hexitol pools. It should be noted that the movement of carbon from free glucose to fructose-6-phosphate (or to glucose-6-phosphate) via hexitols would involve no energy wastage, first because reducing power consumed during the synthesis of hexitols would be recouped during their oxidation (or oxidation of hexitol phosphates), and secondly, because the single phosphorylation step proposed is necessary for the entry of hexoses into further intermediary metabolism by any route.

2. Role of trehalose synthesis, glycolysis and gluconeogenesis in energy conservation

During growth on 1–20 mM glucose, the trehalose pool appeared to turn over in about three hours, compared to little or no turnover during growth on 200 mM glucose during which the trehalose simply accumulated within the mycelium (Manners et al., 1982). Thus, trehalose can be considered a long-term storage compound during growth on high glucose concentrations. Although a mechanism could not be offered for the regulation of trehalose turnover, Manners et al. (1982) noted that such turnover at low glucose concentrations would result in the operation of a futile cycle and thus waste ATP.

The simultaneous operation of glycolysis and gluconeogenesis which seem to function at low exogenous glucose concentrations (Manners *et al.*, 1982) would result in the operation of a number of other futile cycles (cf. Hue, 1981), e.g. (1) the interconversion of fructose-6-phosphate and fructose-1,6-diphosphate by phosphofructokinase and fructose-1,6-diphosphatase, (2) the interconversion of free glucose glucose-6-phosphate (glucose must be formed from glucose-6-phosphate and derived from gluconeogenesis, to account for the low specific activity of [^{14}C] observed in the mycelial pool of free glucose) and (3) the pyruvate/phosphoenolpyruvate cycle. Metabolic inefficiency resulting from a series of futile cycles could explain why *P. graminis* grows relatively poorly at low glucose concentrations, and requires relatively high glucose concentrations for optimal growth (cf. Maclean, 1974).

VI. Concluding Remarks and Summary

It is clear from the work summarized above, that our understanding of the metabolism of vegetative cells of rust fungi is still in its infancy. Nonetheless, a firm foundation has been laid for future work. The search for a "metabolic block" postulated to explain the apparent unculturability of rust fungi prior to 1966 (cf. Part A of this chapter), prompted many studies on the biochemistry of germinating uredospores, and succeeded in demonstrating that rusts possess the normal "housekeeping" pathways and enzymes of glycolysis, the pentose phosphate pathway, the citric acid cycle and lipid metabolism (cf. Staples and Wynn, 1965). Interest in the metabolism of rust fungi must now centre on discovering those details of the regulation of metabolism of these organisms which enable them to succeed as biotrophic parasites, and which distinguish them from other groups of fungi. Two areas of intermediary metabolism are of current interest in this regard.

In the area of amino acid metabolism, the major discoveries of interest are that at least some rust fungi are heterotrophic in their need for adequate supplies of reduced nitrogen and sulphur, and that massive leakage of amino acids can occur from vegetative mycelium (cf. Sections III and IV of this chapter). Inability to reduce sulphate or adequate nitrate should have little (if any) deleterious effect *per se* on the fungus when growing as a biotroph, because the host can reduce both of these anions and supply the needs of the fungus. What is not yet known however, is the extent to which parasitic mycelium leaks amino acids into its host—or whether such leakage is an "artifact" of axenic culture caused, for example, by the reversal of transport systems which normally take up host metabolites absent from (or in incorrect balance in) artificial media. It is possible that the membrane components (and associated intracellular enzymes?) which are responsible for the observed leakage carry out biochemical functions necessary for the viability of the organism.

In the area of carbohydrate metabolism, *Puccinia graminis* (and probably many if not all other rust species) appears to be rather unique amongst higher fungi in the large amount of glucitol and relatively small amount of mannitol which it synthesizes and accumulates during vegetative growth. Other fungi accumulate mannitol as their major hexitol, with only traces (if any) of glucitol (cf. Section V A). It is not yet clear that the ability to accumulate glucitol confers any particular advantages on rusts compared to other fungi, because both glucitol and mannitol appear to serve a similar role as metabolic "buffers" in an alternative pathway of carbon flow during sugar assimilation by *P. graminis*. Our knowledge of the pathway is still very incomplete however, and glucitol and mannitol may have different roles in coenzyme regulation for example, which are necessary for the viability of the organism. The operation of futile cycles at low exogenous glucose concentrations, resulting in loss of metabolic efficiency when it is least wanted, is another point of interest related to the viability of the organism, especially since growth does not occur on glucogenic amino acids in the absence of exogenous carbohydrates, and further investigation of the control of gluconeogenesis and trehalose turnover would appear to be warranted.

Although the points of metabolic interest cited above need to be investigated in rust fungi growing parasitically in host plants, such studies should be made easier by first defining biochemical pathways and mechanisms in the isolated, host-free fungus, as has been demonstrated by the work completed thus far. The phenomena discussed in this chapter are closely related to fungal nutrition, and are therefore of prime importance to a biotrophic parasite which has evolved specialized mechanisms of extracting adequate quantities of food from living cells of its host. If these mechanisms involve transport proteins and intracellular enzymes not present in the host, the potential exists to formulate specific inhibitors or antimetabolites which can be exploited for disease control.

Acknowledgements

I wish to thank Professor K. J. Scott and Mrs J. Owens (*née* Bullock) for permission to cite some of their previously unpublished work, and also Dr J. M. Manners for permission to refer to unpublished results.

The Wheat Industry Research Council, the Rural Credits Development Fund, and the Australian Research Grants Committee have provided partial financial assistance for much of the research reported herein.

References

Allen, P. J. (1965). *Ann. Rev. Phytopathol.* **3**, 313–342.
Bieleski, R. L. and Redgwell, R. J. (1980). *Aust. J. Plant Physiol.* **7**, 15–25.
Burger, A., Prinzing, A. and Reisener, H. J. (1972). *Arch. Mikrobiol.* **83**, 1–16.
Burnett, J. H. (1976). "Fundamentals of mycology", 2nd ed. Edward Arnold, London.
Burrell, M. M. and Lewis, D. H. (1977). *New Phytol.* **79**, 327–333.
Cabib, E. and Leloir, L. F. (1958). *J. Biol. Chem.* **231**, 259–275.
Cantino, E. C. (1955). *Q. Rev. Biol.* **30**, 138–149.
Chen, W. P., Anderson, A. W. and Han, Y. W. (1979). *Appl. Environ. Microbiol.* **37**, 785–787.
Cherest, H., Eichler, F. and De Robichon-Szulmajster, H. (1969). *J. Bact.* **97**, 328–336.
Clancy, F. G. and Coffey, M. D. (1980). *J. Gen. Microbiol.* **120**, 85–88.
Cochrane, V. M. (1958). "Physiology of Fungi". John Wiley, New York.
Cutter, V. M. Jr. (1960). *Mycologia* **52**, 726–742.
Daly, J. M., Knocke, H. W. and Wiese, M. V. (1967). *Plant Physiol.* **42**, 1633–1642.
Datko, A. H., Giovanelli, J. and Mudd, S. H. (1974). *J. Biol. Chem.* **249**, 1139–1155.
Dekhuijzen, H. M. and Staples, R. C. (1968). *Contr. Boyce Thompson Inst.* **24**, 39–51.
Dekhuijzen, H. M., Singh, H. and Staples, R. C. (1967). *Contr. Boyce Thompson Inst.* **23**, 367–372.
Drew, E. A. and Smith, D. C. (1967). *New Phytol.* **66**, 380–400.
Elbein, A. D. (1974). *Adv. Carbohydr. Chem. Biochem.* **30**, 227–256.
Giovanelli, J., Mudd, S. H. and Datko, A. H. (1974). *Plant Physiol.* **54**, 725–736.
Hendrix, J. W., Daly, J. M. and Livne, A. (1964). *Phytopathology* **54**, 895.
Henry, R. J., Maclean, D. J. and Scott, K. J. (1977). *Proc. Aust. Biochem. Soc.* **10**, 42.
Hers, H. G. (1960). *Biochim. Biophys. Acta.* **37**, 127–138.
Hewitt, E. J. (1975). *Ann. Rev. Plant Physiol.* **26**, 73–100.
Holligan, P. M., Chen, C. and Lewis, D. H. (1973). *New Phytol.* **72**, 947–955.
Holligan, P. M., Chen, C., McGee, E. M. M. and Lewis, D. H. (1974a). *New Phytol.* **73**, 881–888.
Holligan, P. M., McGee, E. M. M. and Lewis, D. H. (1974b). *New Phytol.* **73**, 873–879.
Howes, N. K. (1972). Ph.D. Thesis, University of Queensland, Australia.
Howes, N. K. and Scott, K. J. (1972). *Can. J. Bot.* **50**, 1165–1170.
Howes, N. K. and Scott, K. J. (1973). *J. Gen. Microbiol.* **76**, 345–354.
Hue, L. (1981). *Adv. Enzymol. Relat. Areas Mol. Biol.* **52**, 247–331.
Hughey, R. P., Rankin, B. A., Elce, J. S. and Curthoys, N. P. (1978). *Arch. Biochem. Biophys.* **186**, 211–217.
Jäger, K. and Reisener, H. J. (1969). *Planta* **85**, 57–72.
Jennings, D. H. (1974). *Trans. Br. Mycol. Soc.* **62**, 1–24.
Kerr, D. S. and Flavin, M. (1970). *J. Biol. Chem.* **245**, 1842–1855.
Lane, W. D. and Shaw, M. (1972). *Can. J. Bot.* **50**, 2601–2603.
Lewis, D. H. (1976). *In* "Perspectives in Experimental Biology" (N. Sunderland, ed.), Vol. 2 "Botany", pp. 207–219. Pergamon Press, Oxford and New York.
Lewis, D. H. and Smith, D. C. (1967). *New Phytol.* **66**, 143–184.

Long, D. E., Fung, A. K., McGee, E. M. M., Cooke, R. C. and Lewis, D. H. (1975). *New Phytol.* **74**, 173–182.
Maclean, D. J. (1971). Ph.D. Thesis, University of Queensland, Australia.
Maclean, D. J. (1974). *Trans. Br. Mycol. Soc.* **62**, 333–349.
Maclean, D. J. and Scott, K. J. (1976). *J. Gen. Microbiol.* **97**, 83–89.
Manners, J. M., Maclean, D. J. and Scott, K. J. (1982). (manuscript in preparation).
Marechal, L. R. and Belocopitow, E. (1972). *J. Biol. Chem.* **247**, 3223–3228.
Meister, A. (1973). *Science* **180**, 33–39.
Meister, A. (1974). *Life Sci.* **15**, 177–190.
Meister, A. (1978). *In* "Functions of Glutathione in Liver and Kidney" (H. Sies and A. Wendel, eds), pp. 43–59. Springer-Verlag, Berlin, Heidelberg and New York.
Meister, A. and Tate, S. S. (1976). *Ann. Rev. Biochem.* **45**, 559–604.
Miflin, B. J. and Lea, P. J. (1976). *Phytochemistry* **15**, 873–885.
Mitchell, D. T. and Roberts, S. M. (1973). *Physiol. Plant Pathol.* **3**, 481–488.
Mitchell, D. and Shaw, M. (1968). *Can. J. Bot.* **46**, 453–460.
Mitchell, D. T., Fung, A. K. and Lewis, D. H. (1978). *New Phytol.* **80**, 381–392.
Mooz, E. D. (1979). *Biochem. Biophys. Res. Commun.* **90**, 1221–1228.
Mooz, E. D. and Wigglesworth, L. (1976). *Biochem. Biophys. Res. Commun.* **68**, 1066–1072.
Orlowski, M. and Meister, A. (1970). *Proc. Nat. Acad. Sci., U.S.A.* **67**, 1248–1255.
Pfeiffer, E., Jäger, K. and Reisener, H. J. (1969). *Planta* **85**, 194–201.
Pfyffer, G. E. and Rast, D. M. (1980a). *New Phytol.* **85**, 163–168.
Pfyffer, G. E. and Rast, D. M. (1980b). *Exp. Mycol.* **4**, 160–170.
Prentice, N., Cuendet, L. S., Geddes, W. F. and Smith, F. (1959). *J. Am. Chem. Soc.* **81**, 684–688.
Rees, W. R. and Reynolds, T. (1958). *Nature* **181**, 767–768.
Reisener, H. J. Goldschmid, H. R., Ledingham, G. A. and Perlin, A. S. (1962). *Can. J. Biochem. Physiol.* **40**, 1248–1251.
Reisener, H. J., Ziegler, E. and Prinzing, A. (1970). *Planta* **92**, 355–357.
Schiff, J. A. and Hodson, R. C. (1973). *Ann. Rev. Plant Physiol.* **24**, 381–414.
Scott, K. J. (1972). *Biol. Rev. Cambridge Philos. Soc.* **47**, 537–572.
Scott, K. J. and Maclean, D. J. (1969). *Ann. Rev. Phytopathol.* **7**, 123–146.
Shu, P., Neish, A. C. and Ledingham, G. A. (1956). *Can. J. Microbiol.* **2**, 559–563.
Siegel, L. M. (1975). *In* "Metabolic Pathways", 3rd edn. (D. M. Greenberg, ed.), **7**, 217–286. Academic Press, New York, San Francisco and London.
Smith, D., Muscatine, L. and Lewis, D. (1969). *Biol. Rev. Cambridge Philos. Soc.* **44**, 17–90.
So, M. L. and Thrower, L. B. (1976). *Phytopathol. Z.* **86**, 302–309.
Staples, R. C. and Wynn, W. K. (1965). *Bot. Rev.* **31**, 537–564.
Suekane, M., Masaki, T. and Chikaka, T. (1978). *Agric. Biol. Chem.* **42**, 909–918.
Touster, O. and Shaw, D. R. D. (1962). *Physiol. Rev.* **42**, 181–225.
Turel, F. L. M. and Ledingham, G. A. (1957). *Can. J. Microbiol.* **3**, 813–819.
Wicker, E. F., Mosher, D. P. and Wells, J. M. (1976). *Phytopathol. Z.* **87**, 97–106.
Williams, P. G. and Shaw, M. (1968). *Can. J. Bot.* **46**, 435–440.
Wolf, F. T. (1956). *Phytopathol. Z.* **26**, 219–223.
Wynn, W. K. (1966). *Plant Physiol.* **41**, xxvi (Abstr.)
Yarwood, C. E. (1956). *Ann. Rev. Plant Physiol.* **7**, 115–142.
Ziegler, E. and Reisener, H. J. (1975). *Z. Pflanzenphysiol.* **75**, 307–321.

E

List of Abbreviations

ATP = adenosine 5'-triphosphate. ADP = adenosine 5'-diphosphate. P_i = inorganic phosphate. PPi = inorganic pyrophosphate. APS = adenosine 5'-phosphosulphate. PAPS = 3'-phosphoadenosine 5'-phosphosulphate. PAP = 3'-phosphoadenosine 5'-phosphate. NAD^+, NADH = oxidized and reduced forms respectively, of nicotinamide adenine dinucleotide. $NADP^+$, NADPH = oxidized and reduced forms respectively, of nicotinamide adenine dinucleotide phosphate. $NAD(P)^+$, NAD(P) H = oxidized and reduced forms respectively, of either NAD^+ or $NADP^+$. H_2S = hydrogen sulphide. SO_4^{2-} = inorganic sulphate ion. THF = tetrahydrofolate. SAM = S-adenosylmethionine. SAH = S-adenosylhomocysteine. gly = glycine. cys = L-cysteine. glu = L-glutamic acid. aa = unspecified L-amino acid. e^- = electron. HMP = hexose monophosphate.

3. Genetics of Host-Pathogen Interactions in Rusts

R. A. McINTOSH and I. A. WATSON

Plant Breeding Institute, University of Sydney, Australia

CONTENTS

I. Introduction

The rust fungi, the Uredinales, cover a wide range of obligate parasites which have parasitised some ferns and virtually all groups of higher plants. In considering the genetics of such fungi the first requirement is a knowledge of the life cycle and hence the role of sexual recombination and other recombinational processes as causes of the wide range of variability that may be encountered among them. Additionally, the dominant spore stage is important in determining the methods to be employed in studying a particular type and, if a pathogen of an agricultural or silvicultural crop plant, is important in determining the means of control.

Figure 1 presents a generalized life cycle for a macrocyclic rust pathogen such as *Puccinia graminis*. The first major characteristic of rust fungi is that some are *autoecious* producing all spore stages on a single host, whereas others are *heteroecious* and in order to complete the entire life cycle they must be capable of parasitising alternate hosts which are quite different species. Heteroecious species, therefore, must be equipped with the ability to infect the different hosts, often by different mechanisms (for example, stomatal entry of one host compared with direct penetration of the other), and presumably there are genetic systems controlling pathogenicity with respect to each host. Different species show variation in life cycle and in

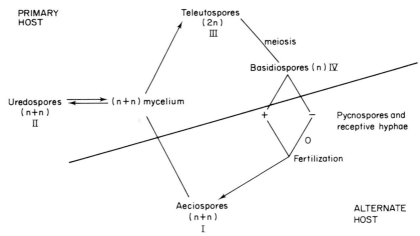

Fig. 1. Generalized life cycle of a typical macrocyclic heteroecious rust fungus. Roman numerals designate the spore types (McAlpine, 1906). Pycnia and aecia, O and I, occur on the alternate host species and uredospores and teleutospores occur on the primary host species. In autoecious rust fungi all spore stages occur on a single host.

some instances evolution is toward loss of certain spore forms. For example, the most extreme microcyclic forms appear to be those forming only teleutospores with or without spermagonia (Wilson and Henderson, 1966) and where some teleutospores (leptospores) will germinate without a resting stage. In some species only the uredial stage may be effective in the survival of the species. In *P. striiformis*, for example, no alternate host is known and the teleutospore stage—although vital for species identification—has no known role in survival. Where agricultural plants and heteroecious rust pathogens have been taken to areas where alternate hosts do not occur, or are resistant, survival is also dependent only upon a repeating uredial cycle. In such cases survival is enhanced by milder climates where the host, or closely related wild or weed species, can survive and maintain infections throughout the year.

A second characteristic is that the different species of rust fungi often show pathogenic variation that is usually not, or only poorly, detected by traditional methods of taxonomy. Populations within a species may be adapted to different related host species. For example, several *formae speciales* of *P. graminis* adapted to various graminacious hosts were described by Eriksson and Henning. Such forms are easily distinguished into groups with respect to various hosts but overlapping characteristics can be shown by host surveys and genetic manipulations of actual genotypes within host species, and by pathogenicity with respect to related common host species. In *P. graminis* where the various form species infect the common barberry, *Berberis vulgaris*, overlapping variation can be demonstrated by crossability studies and the genetic segregations produced as a result of sexual recombination on the alternate host. Similar variation with respect to graminacious hosts is apparent in *P. recondita*, but in this instance, a study of the genetic relationships between the forms is more difficult since different alternate hosts are involved. Wilson and Henderson (1966) list 13 form species adapted to nine grass genera and to at least ten alternate host genera. For two form species alternate hosts were unknown. Three non-crossable forms of *P. recondita* f. sp. *tritici* infect three entirely different alternate hosts (Anikster and Wahl, 1979).

Detailed genetic studies have been restricted to relatively few diseases of agricultural plants. In all instances more is known in regard to the genetics of reaction in the host than in regard to genetics of pathogenicity in the pathogen. Genetic analysis of a disease is conducted by genetic analysis of either the host or the pathogen while the genotype of the pathogen or host, respectively, is held constant. In order to study the genetics of the host, the host life cycle must be known. For ease of manipulation the host should be sexually reproducing, although clonal reproduction in addition can be useful under certain circumstances. Short generation times, ease in effecting crosses and large numbers of progenies following crossing, and/or self pollination are distinct advantages in genetic studies.

In rust pathogens variation may be generated by both sexual and asexual procedures. Rust fungi are often heterothallic and although mating type differences are usually assumed to be controlled by alleles at a single locus, recent work at the University of Adelaide with *Melampsora lini* indicates greater complexity, perhaps similar to the two-locus system in *Schizophyllum commune* (Lawrence, 1977). Determinations of mating types are made more difficult because the rusts are obligate parasites and axenic culture of some of them, despite early promise, has not contributed technically to our ability to conduct genetic studies. Where the uredospore stage is of primary interest large quantities of inoculum derived from single spore cultures can be readily obtained and stored in liquid nitrogen. This ensures the availability of identical inocula for successive experiments and reduces the risk of contamination. Moreover, it allows long-term storage of comprehensive culture collections in living historic type collections. One major problem in genetic studies of rust pathogens is germination of teleutospores under controlled conditions and, in the case of heteroecious pathogens, the availability of susceptible genotypes within the alternate host species. Tetrad analysis, the isolation and study of the individual ordered products of meiosis is not possible. Since each genetic recombinant is produced as an aecium and is usually maintained as a uredial culture, the generation and maintenance of large progeny populations without contamination is expensive, time consuming and tedious. However, with autoecious fungi there are possibilities for basidiospore selection of rare recombinant genotypes.

For the above reasons detailed genetic information is restricted mainly to the flax rust (*M. lini: L. usitatissimum*, autoecious) and various cereal rust (heteroecious) systems. In the remainder of this chapter most attention will be given to these.

II. Terminology

A consistent terminology for the description of characters in host:pathogen interactions that is acceptable to geneticists as well as plant pathologists and plant breeders has not been achieved. Most authors have therefore defined terms for a particular publication or have selected an earlier paper that best suited their purposes. Loegering (1966, 1972) has emphasized the need to consider variation at three levels *viz.*, host, interaction and pathogen, the expressions of which were described as *reaction, infection type* and *pathogenicity*, respectively. Each of these may show contrasting phenotypes *viz.* resistance or susceptibility, incompatibility or compatibility, and avirulence or virulence, respectively, depending upon actual genotypes of the organisms involved. The use of pathogenicity in the above sense has been widely accepted by geneticists interested in inheritance studies but precludes its use as defined by Watson (1970). In this chapter, pathogenicity will be used as above while *reproductive potential* will be used in the sense of Watson's 1970 definition. Factors which contribute to reproductive potential include infection efficiency, incubation period, lesion size, number of spores producted, etc. *Competitive ability* relates to the behaviour of pathogen genotypes in heterogeneous populations.

III. The Gene-for-Gene Concept

Current understanding of host:pathogen genetics, interorganismal genetics (Loegering, 1978), was made possible by Craigie (1927) who demonstrated the role of pycnia in sexual recombination and by studies of Flor (see 1956 review) who was first to study the inheritance of pathogenicity in the pathogen in conjunction with the inheritance of resistance in the host.

In studies of host:pathogen interactions there are several ways by which disease phenotypes can be measured (Ellingboe, 1978). With rusts the traditional methods have involved the scoring of growth (compatibility) versus no growth (incompatibility) of the pathogen as with studies of flax rust, or the scoring of restricted pathogen growth compared with conceptionalized or real compatible standards as with studies of the cereal rusts. For the latter, various infection type ratings have been published (Stakman *et al.*, 1962; Browder and Young, 1975). The choice of comparing inter-action phenotypes with compatible standards is justified by genetic analyses indicating that compatibility is a passive condition and that the genetically-determined deviations from compatibility are controlled by genes with positive function in both hosts and pathogens. Genetic investigations have not provided any information in relation to basic compatibility (Ellingboe, 1976) between host species and pathogen species but genetic analyses within host:pathogen systems have been built upon a situation of basic compatibility.

From about 1910, Stakman and co-workers at the University of Minnesota found that collections of *P. graminis tritici* often showed variation in pathogenicity on various wheat hosts. They chose 12 wheat genotypes on which many physiologic races could be distinguished (Stakman *et al.*, 1962). Likewise, with *M. lini* Flor eventually distinguished 239 physiologic races on as many as 27 host differentials each possessing a unique single gene for resistance (Flor, 1956). From his early studies it was apparent to Flor that genes controlling pathogenicity in *M. lini* must interact with those controlling reaction in the flax host. His inheritance studies in 1942 showed that in order for each unique resistance gene in the host to be overcome by the pathogen, the pathogen carried a matching virulence gene. Flor (1956) defined the gene-for-gene hypothesis . . . "for each gene conditioning rust reaction in the host there is a specific gene conditioning pathogenicity in the parasite". Subsequently the host genes and parasite genes became known as corresponding genes and the relationship strictly indicates that incompatibility occurs only when a host resistance gene and a corresponding pathogen avirulence gene participate in the same interaction. Table I shows the disease phenotypes arising from interactions of one and two corresponding gene pairs. These refer to host and pathogen phenotypes assuming homozygosity and/or full dominance of the corresponding genes. When two corresponding gene pairs are involved in the interaction, the phenotype is always as low as, or lower than, the more incompatible of the individual interactions and under some circumstances category IV interactions (Loegering, 1966) may be difficult to distinguish from a third corresponding gene pair.

Day (1974) includes several rusts among his list of disease systems for which there is evidence of gene-for-gene relationships. The greatest value of the relationship is its use in systems where it is impossible, or extremely difficult, to study the inheritance of host reaction and pathogen pathogenicity simultaneously. For example, the host may be a tree species requiring several years between generations, or may be an apomict or a clonally propagated species. Alternatively, the pathogen may have no known sexual stage despite demonstrable variation in pathogenicity. Secondly, the relationship serves as a model for the design of physiological and biochemical experiments where, due to variation in the participating organisms, both pathogen and host controls are essential for valid conclusions to be reached. This experimental requirement for host:pathogen studies has been called the quadratic check (Rowell et al., 1963).

In practice host variation is studied while pathogen genotypes are held constant as in surveys of host collections, or pathogen variation is studied while host genotypes are held constant as is common in race/strain or virulence surveys. The data matrices so generated are interpreted in relation to prior information. Identical arrays in the matrices indicate identical host or pathogen genotypes. When genetically known standards of one organism are included, unknowns of the same organism can be typed relative to those standards and, moreover, corresponding genotypes for the other organism can be inferred. Analysis of large masses of host pathogen survey data can be carried out by computer (Loegering et al., 1971; Dinoor and Peleg, 1972; Browder and Eversmeyer, 1976).

Pathogen genotypes are usually described in terms of races, strains or pathotypes of which there are several systems of nomenclature (see Day,

Table I. Interactions involving corresponding genes in hosts and pathogens.

(a) 1-gene model. R and r = alleles for resistance
and susceptibility.
A and a = corresponding avirulence and
virulence alleles. Phenotypes of *Rr* and *Aa* assumed
to be similar to *RR* and *AA*, respectively.
I = incompatibility, C = compatibility.

		Pathogen genotypes		
		AA		aa
Host genotypes	RR	$\dfrac{AA}{RR} = I$		$\dfrac{aa}{RR} = C$
	rr	$\dfrac{AA}{rr} = C$		$\dfrac{aa}{rr} = C$

(b) 2-gene model. R1 and R2 = resistance alleles at
different host loci.
Assumed to be homozygous.
A1 and A2 = corresponding alleles for
avirulence. Assumed to
be homozygous.
I1 and I2 = incompatibilities expressed.

		Pathogen genotypes			
		A1A2	A1a2	a1A2	a1a2
Host genotypes	R1R2	$\dfrac{A1A2}{R1R2} = I1 + I2^{a}$	$\dfrac{A1a2}{R1R2} = I1$	$\dfrac{a1A2}{R1R2} = I2$	$\dfrac{a1a2}{R1R2} = C$
	R1r2	$\dfrac{A1A2}{R1r2} = I1$	$\dfrac{A1a2}{R1r2} = I1$	$\dfrac{a1A2}{R1r2} = C$	$\dfrac{a1a2}{R1r2} = C$
	r1R2	$\dfrac{A1A2}{r1R2} = I2$	$\dfrac{A1a2}{r1R2} = C$	$\dfrac{a1A2}{r1R2} = I2$	$\dfrac{a1a2}{r1R2} = C$
	r1r2	$\dfrac{A1A2}{r1r2} = C$	$\dfrac{A1a2}{r1r2} = C$	$\dfrac{a1A2}{r1r2} = C$	$\dfrac{a1a2}{r1r2} = C$

[a]If I1 and I2 are phenotypically distinguishable, I1 + I2 is similar to the more incompatible interaction or may show increased incompatibility. Acknowledgement (McIntosh, 1978).

1974). Irrespective of the system of naming, the information encoded is an avirulence:virulence formula. This was recognized by Green (1965a) but due to the cumbersome means of conveying entire formulae, Green introduced a system of C (Canada) numbers immediately reintroducing the original problem that the designation is a file reference which must be decoded in terms of the known, and unknown, genes present in the host tester set.

Clearly, as Van der Plank (1975) has indicated, the classification and naming of physiologic races of *P. graminis tritici* from a limited geographic area by testing the 12 Stakman differentials is of limited value. However, the use of these genotypes (or the resistance genes they possess) in conjunction with additional genotypes (or genes), especially commercial cultivars and those potentially important in breeding, has done much to aid current understanding of variability in *P. graminis tritici*, and its importance in breeding for resistance. Browder (personal communication) has emphasized the value of living type cultures to characterize avirulence and virulence as permanent reference points for individual host genes. While we concur in this, we suggest that the identification and preservation of additional pairs of such cultures differing in additional genes could provide a valuable laboratory aid. Those interested in all aspects of rust genetics and in breeding for resistance should have access to a vast collection of cultures having combinations of genes for virulence and avirulence which could expedite the identification of corresponding host genes and permit the combining of them into appropriate resistant genotypes.

The derivation of the gene-for-gene concept was based on simultaneous studies of pathogens and hosts and the concept continues to be tested in order to identify possible exceptions which may provide important leads in understanding host:pathogen relationships. With modern concepts of gene action it should not be surprising that details and interpretations of a model first suggested as early as 1942 (Flor, 1942) should change. Clearly the interaction must involve a gene product:gene product interaction but because of so few exceptions (to be discussed in the following sections) from one:oneness genetically, it is possible that the interacting gene products are primary gene products, messenger RNA's and the polypeptides they specify (see Chapter 5).

IV. Genetics of Reaction and Pathogenicity

Table II summarizes some genetic results for rust systems where a gene-for-gene relationship has been established or assumed. The following discussion considers instances of deviations from one:oneness and reviews of studies having some bearing on genetic fine structure or on gene action.

A. Host Resistance

The 29 host resistance genes involved in flax rust are clustered in five closely linked or allelic groups, *K, L, M, N* and *P,* consisting of 1, 13, 7, 3 and 5

Table II. Numbers of resistance genes reported for some rust diseases.

Disease host:pathogen	No. of resistance genes	No. of loci if different	No. of loci with multiple alleles	References
Linum :M. lini	29	5	4	Lawrence (1978)
Phaseolus vulgaris :Uromyces appendiculatus	13–15	—	1	Ballantyne (1978)
Coffea arabica :Hemileia vastatrix	5	—	—	Russell (1978)
Helianthus :P. helianthi	at least 3	—	—	Miah and Sackston (1970)
Avena :P. graminis avenae	14	10	3	Simons et al. (1978)
:P. coronata	68	—	6	Simons et al. (1978)
Hordeum vulgare :P. hordei	9	—	—	Tan (1977)
Secale :P. graminis	11	10	1	Tan (1973)
Triticum :P. graminis	36	30	2	McIntosh (1973, 1978)
:P. recondita	31	24	3	McIntosh (1973, 1978)
:P. striiformia	12	10	2	McIntosh (1973, 1978)
Zea mays :P. sorghi	21	5	2	Saxena and Hooker (1968)

resistance genes, respectively (Lawrence *et al.*, 1980). For the investigation of allelism or close linkage, Shepherd and Mayo (1972) proposed a modified cis-trans test. They considered that truly codominant allelic genes might confer a different phenotype when present in the same chromosome strand $\frac{A^1A^2}{+\ +}$ compared to their presence in different strands $\frac{A^1\ +}{+\ A^2}$. In terms of disease reaction the possible phenotypes emerging from testcrosses of the latter (i.e. $\frac{A^1\ +}{+\ A^2} \times \frac{+\ +}{+\ +}$) are defined by the pathogenicity phenotypes A^1, A^2, $A^1 + A^2$ and $+ +$ (susceptible). Hence in addition to parental products, A^1A^2 and $+ +$ recombinants may be expected in equal proportions. However, if the properties of the gene product of A^1A^2 are different from those produced by A^1 or A^2 then the A^1A^2 recombinant might be phenotypically distinguishable from the combined properties of A^1 and A^2, but might not be distinguishable from $+ +$. Distinctness of the A^1A^2 gene product is more likely when both genes lie within the same contiguously transcribed DNA region and when the critical sites are sufficiently close so that the tertiary molecular structure required to confer one phenotype is not independent of the tertiary structure required for the other. In studies involving recombination at the M locus, Shepherd and Mayo obtained equal numbers of $+ +$ and MM^3 recombinants indicating that these genes are closely linked and produce different gene products. Alternatively, if a single gene product is involved then the critical sites are physically independent. On the other hand these workers obtained only $+ +$ recombinants in studies involving genes in the L group. They surmised further that if their model was correct and if the L genes were truly allelic then from 50% of the recombinants $\frac{(L^2L^{10}}{(+ +}$ and $\frac{+ +)}{+ +)}$ individuals with the original L^2 and L^{10} pathogenicity phenotypes should be recoverable. After self-pollination of six recombinants, five plants with the L^{10} phenotype and none with the L^2 phenotype were obtained. Further studies continued to produce low frequencies of L^{10} but no L^2 plants. Hence the results did not completely fulfil the requirements of the test. Although there was no explanation for the lack of L^2 individuals, it was possible to confirm that re-emergence of L^{10} from susceptible recombinants involved recombination within the L gene itself rather than other mechanisms.

Clearly, further studies are needed to increase knowledge of the genetic fine structure of resistance genes and to establish whether the above results are unique. Unfortunately studies of this type will be limited to a few host species because of difficulties in producing large testcross populations. In polyploids where occasional aneuploid and possibly susceptible individuals may occur, such studies will be possible only if suitable flanking marker genes can be identified and exploited.

In *Zea mays* genes determining reaction to *P. sorghi* are located at five loci, three of which, *Rp5*, *Rp1* and *Rp6*, are closely linked in the short arm of chromosome 10. Fourteen resistance alleles occur at *Rp1*. Saxena and Hooker (1968) noted that in some cases the pathogenicity phenotype conferred by one allele was similar to the sum of those conferred by two other alleles. In certain crosses involving stocks carrying different alleles, double recessive recombinant progenies were recovered. In five crosses recombination frequencies varied from 0·1 to 0·37% while in two crosses the frequencies of double recessives were not distinguishable from background mutation rates. In three crosses where appropriate screening procedures were carried out double resistant recombinants were recovered in frequencies similar to those for double recessives. It was suggested that $Rp1^d$ which conferred resistance to all isolates of *P. sorghi* might be a natural combination of $Rp1^a$ and $Rp1^k$, for these alleles recombined with each other but neither recombined with $Rp1^d$. Hence the *Rp1* region in maize appeared to behave similarly to the *M* region in flax.

Six resistance alleles have been allocated to the *Sr9* region in wheat. Apart from the fact that each produces a similar infection type with cultures of *P. graminis tritici* possessing the appropriate corresponding gene for avirulence they appear to have no special similarities. On the other hand, there appear to be definite similarities and perhaps increasing orders of complexity among resistance alleles in the *Lr2* and *Lr3* regions in wheat (Dyck and Samborski, 1968; Haggag and Dyck, 1973).

The genetic background and the environment have significant effects on host response even when resistance is based on a single gene. Dyck and Samborski (1974) studied wheat lines in which alleles of *Lr2* were transferred to three genetic backgrounds. The genes were most effective when transferred to Thatcher and least effective in the Red Bobs background. Luig and Rajaram (1972) noted decreasing levels of expression of genes *Sr5* and *Sr9b*, as well as reduced dominance in heterozygotes, as these genes were transferred to highly susceptible wheat backgrounds. Kerber and Dyck (1979) and The (1973) found that wheat genes transferred from diploid to tetraploid and hexaploid species confer decreasing levels of resistance as measured by infection type. Obviously genes in the additional genomes act as modifiers but the nature and numbers of modifiers are unknown. Law and Johnson (1967) showed that infection types conferred by *Lr14a* in wheat chromosome 7BL could be significantly changed by variation in the numbers of the homoeologous 7A and 7D chromosomes as well as by deletion of the opposite chromosome arm, 7BS.

Although many reports of non-allelic gene interactions in the determination of host resistance could be cited, observed results were adequately confirmed in few instances. Most of these were based on inferences of genetic behaviour from F2 phenotypic ratios without adequate confirmation of normal gametic transmission rates or of the reliability of F2 phenotypes as indicators of genotype. Baker (1966), however, presented evidence for complementary genes conferring resistance to *P. coronata* in *Avena sativa* cv. Bond when he isolated susceptible F4 lines which, when intercrossed, produced a pathogenicity phenotype characteristic of Bond.

The effects of high temperature on wheat gene *Sr6* have been exploited in genetic and physiologic studies. However, several other genes in wheat are affected by environment especially light and temperature. Wheat genes *Sr15* and *Sr17* are effective only at temperatures below about 23°C while genes such as *Sr13* and *Sr14* are more easily manipulated under high light and temperature conditions. In the wheat leaf rust system *Lr20* is inefffective at temperatures in excess of 26°C. On the other hand seedling resistance in many Frontana derivatives is more effective at higher temperatures. Sharp and co-workers (see Röbbelen and Sharp, 1978) have elegantly demonstrated the interactions between additive genes with small effects and temperature in the wheat stripe rust system.

Host genes for rust resistance have been associated with reaction to a second disease in at least two instances. In Victoria oats the presence of *Pc-2* involved in reaction to *P. coronata* is always associated with susceptibility to *Helminthosporium carbonum*, a toxin-producing facultative pathogen, and mutations for loss of *Pc-2* specificity, or for toxin insensitivity, were always associated with a reciprocal change with respect to the second pathogen. Presumably the product of *Pc-2* interacts with the product of A_{Pc-2} resulting in resistance to *P. coronata* cultures possessing A_{Pc-2} and acts as a receptor of *H. carbonum* toxin resulting in susceptibility to Victoria blight (Wallace *et al.*, 1967). In wheat the genes *Pm1*, *Sr15* and *Lr20* for resistance to certain cultures of *Erysiphe graminis tritici*, *P. graminis tritici* and *P. recondita* always occur in coupling and have never been separated by recombination. McIntosh (1977) treated appropriately marked stocks possessing these genes with the chemical mutagen ethyl methane sulphonate. Mutations of *Pm1* occurred independently of those for *Sr15* and *Lr20*. However, mutations of the two latter were always asssociated even though the magnitudes of change as measured by infection type did not appear to be identical. Many of the mutants of *Sr15* and *Lr20* produced infection types that were distinguishable from those characteristic of the original stocks and from full compatibility. Like the parental stocks, mutants with intermediate infection types were fully susceptible to cultures virulent for *Sr15* and *Lr20*. Apparently the gene products of the host mutant genes did not interact as effectively as the gene products of the original genes when exposed to the products of the corresponding avirulence genes. Such mutations could be considered "missense" mutations if the original genes were considerd to be "sense". Other mutations resulting in full compatibility could be either "missense" types in which the gene product no longer could be recognized by the product of the pathogen gene or "nonsense" types in which there was no gene product due to deletion or premature termination of DNA translation. Genetic analyses of the mutants showed that mutation always involved these loci rather than the genetic background; that is, the mutant genes conferring intermediate phenotypes behaved as alleles of the original gene (McIntosh, unpublished). McIntosh (1977) speculated that further mutagenic treatment of the mutant derivatives might restore the original phenotypes by "error correction". If these models are true then this technique might be applicable to the enhancement of resistance provided by some naturally occurring genes which confer only small effects on disease reaction and which are considered inadequate for exploitation in resistance breeding.

Several studies have suggested the production of resistance genes by mutation procedures. Simons (1979) concluded "There is no question that artificial mutagens can be used to induce mutations to specific resistance . . ." Several of these studies related to rust diseases but in no case has a gene determining resistance in an alleged mutant been named and catalogued. In certain instances we (unpublished) have shown alleged wheat mutants to possess known genes and consideration of the host materials and circumstances makes it unlikely that the genes present were generated by a mutational process. Flor (1971) reached similar conclusions after considering the results of mutation experiments in flax.

Rare genotypes of hexaploid wheat that are susceptible to cultures of *P. graminis* f. sp. *secalis*, as well as f. sp. *tritici*, can be found among stock collections or occur as transgressive segregates in certain crosses. Following the addition to such genotypes of single genes for resistance to either form species it can be shown that *secalis* sometimes lacks the corresponding gene for avirulence even when the added gene is derived from wheat. Similarly, *tritici* may attack selected rye lines with resistance to some *secalis* cultures because it lacks the corresponding genes for avirulence. Hence pathogenicity in *P. graminis secalis* or *P. graminis tritici* corresponding to certain resistance genes in wheat or rye, respectively, can be considered polymorphic and because such genes would not normally be exposed to selective forces, unless their presence *per se* is related to reproductive potential, they could be useful as genetic markers in evolutionary and competition studies. Overlapping in the occurrence of virulence genes corresponding to resistance genes in specialized pathogen species undoubtedly has greater significance in natural populations where related host species may be growing as mixed populations. Eshed (1978) experienced extreme difficulty in distinguishing between form species of *P. coronata* in Israel when collections from single hosts were found to be virulent on testers chosen to distinguish the forms. Different races of *P. coronata* f. sp. *avenae* as distinguished on a set of oat testers were no more uniform than random isolates from wild host species when tested on the host species testers.

Resistance can be expressed at various growth stages. Adult plant resistances to the cereal rusts are not uncommon and inheritance studies have shown that such resistances are often specific and simply inherited. Ellingboe (1978) pointed out that in host surveys the types of resistance found reflect the methods used to screen for resistance. If genes producing small effects on disease reaction at specified growth stages are desired then appropriate screening procedures must be used. In classical studies of host :pathogen relationships infection type has been the main screening criterion but many workers now emphasize the importance of epidemiological characters such as number of pustules, incubation period, number of spores produced per pustule and longevity of sporulation. These are host parameters often considered important components of non-specific resistance but Ellingboe (1975) concluded that "non-specific resistance is that resistance which hasn't yet been shown to be specific." Obviously as phenotypic differences become smaller it becomes increasingly difficult to characterize the contributions of single genes and to distinguish genetic and environmental effects.

B. Pathogenicity

Among the rusts, inheritance of pathogenicity in *M. lini* has been the most intensively studied. Generally inheritance studies in *M. lini*, and other rust pathogens, have shown avirulence to be controlled by genetically independent dominant, or incompletely dominant, genes. In one study of a cross between cultures of *M. lini* races 22 and 24, both virulent on flax cultivars Williston Brown and Williston Golden possessing M^1, Flor (1946) obtained avirulent F2 segregates indicating an apparent dominance of virulence, but the ratio of avirulent to virulent segregates deviated from that reasonably expected for segregation at a single locus. In a second study, Flor (1965) selfed a culture avirulent for M^1 and obtained a normal F2 ratio with dominance of avirulence. In a later critical study, Lawrence *et al.* (1981) obtained evidence for a dominant inhibitor of avirulence for M^1 and pointed out that Flor's first result conformed with a ratio of 13 virulent:three avirulent. Moreover, these authors noted that virulent cultures with genotype $I_{M^1} - A_{m^1}$—produced slightly less compatible infection types on M^1 hosts than virulent cultures with genotype $i_{M^1} i_{M^1} a_{M^1} a_{M^1}$. Further inhibitors of avirulence genes A_{L^1}, A_{L^7} and $A_{L^{10}}$ were reported by Lawrence *et al.* All four inhibitors were completely linked although I_{M^1} was genetically distinguishable from the others. Hence I_{L^1}, I_{L^7} and $L_{L^{10}}$ could be a single gene.

Lawrence (1977) confirmed Flor's studies indicating that genes A_p, A_{p^1}, A_{p^2} and A_{p^3} in *M. lini* are closely linked whereas A_{p^4} is genetically independent. He screened for recombinant progeny in crosses involving the linked group of genes using both dikaryon and monokaryon screening techniques after showing that basidiospores with the avirulence allele were truly avirulent on the host with the corresponding resistance allele, and that those with the gene for virulence were virulent.* Recombinant basidiospores from the heterozygote $\dfrac{A_p \ a_{p^1} \ a_{p^2} \ a_p}{a_p \ A_{p^1} \ A_{p^2} \ A_{p^3}}$ were selected on F1 plants of constitution PP^1, PP^2 and PP^3 each of which could resist the non-recombinant parental types as well as 50% of recombinants. Ten recombinants were recovered, three from dikaryon screening and seven from monokaryon screening. One was virulent for P and produced an intermediate infection type on the host with P^3, one was virulent for P, P^1 and P^3 and eight were virulent on all four host testers. It was considered that meiotic recombination, rather than mutation or some other process, was involved and that the probable gene order was $A_p \ A_{p^2} \ A_{p^1} \ A_{p^3}$. Despite the close linkage, Lawrence concluded that the region probably represented a series of cistrons since a single DNA strand could express multiple specificity. However, as Flor (1958) had shown that induced mutations led to loss of function of more than a single specificity, transcription of the multiple cistron region might be contiguous.

Inheritance studies in *M. lini* have had the advantage that certain inconsistent results have been studied in detail by well-designed follow-up experiments. In other pathogens, interesting results which can only be regarded as preliminary have been established in the literature and few detailed follow-up studies have been undertaken. In some instances this has been due to technical difficulties in producing the required materials.

From a selfing study of a rare isolate of *P. recondita* producing an intermediate infection type on wheat plants with *Lr9*, Samborski (1963) showed that the isolate was heterozygous for a corresponding gene for avirulence.

Samborski and Dyck (1968) found that the expression of virulence on host lines possessing *Lr2* from different sources was in some instances genetically

---

*This information was essential since Flor (1959) had shown that basidiospores from a dikaryon avirulent for Bombay flax could attack Bombay. In this respect Bombay acted as an alternate host to the particular culture. Statler and Gold (1980) reported further instances in which the pathogenicity of basidiospores differed from that of the uredospore cultures from which they were derived.

modified. Subsequently Dyck and Samborski (1974) reported that cultures virulent on plants with *Lr2a*, *Lr2b* and *Lr2c* have genotype *p2p2*, cultures avirulent for all three host genes are either *P2P2* or *P2p2* whereas cultures producing an intermediate infection type on plants with *Lr2a* but virulent for *Lr2c* are *P2p2* and heterozygous for additional gene(s) affecting the expression of *P2p2* and perhaps *P2P2* genotypes. Haggag *et al.* (1973) studied the inheritance of pathogenicity with respect to host genes at or near the wheat *Lr3* locus. Although Sinvalocho appeared to carry the same host allele as Democrat, there was evidence for an inhibitor or modifier of the *p3p3* pathogen genotype when testing was conducted on the backcross line Sinvalocho/6*Prelude. This indicated specificity not only for the *Lr3* host allele but also for its modifiers. With respect to Bage/8*Thatcher two pathogen genes segregated, one corresponding to *P3* and producing infection type 0; and a second producing infection type 2. Since there was evidence for only a single gene in Bage/8*Thatcher, the results suggested either that this host stock possessed a complex locus including *Lr3* or that there was a 2 gene:1 gene relationship between pathogen and host.

Virulence for *Lr3* in Mediterranean and Democrat was produced by the same gene. However, in a population derived by selfing a race 1 culture, virulence was recessive, whereas in a selfed culture of race 161 virulence behaved as a dominant character (Samborski and Dyck, 1968).

Linkage of genes for virulence in *P. recondita* was reported by Samborski and Dyck (1976). Gene *p3ka* showed linkage of $3·7 \pm 1·9\%$ with *p30* ($=pT$) while *p3* and *p14a* were linked with $3·9 \pm 1·6\%$ recombination.

Statler (1977) obtained evidence for duplicate recessive genes determining virulence on a host possessing *Lr1*. However, in a second study he obtained evidence for a single gene (Statler, 1979). Both Statler (1979) and Samborski and Dyck (1976) reported recessive genes determining virulence for *Lr30* ($=LrT$). Statler (1979) found linkage of $11·0 \pm 4·3\%$ between *p30* and *p3* whereas Samborski and Dyck (1976) found linkage of $3·7 \pm 1·9\%$ between *p30* and *p3KA*. Since the latter had reported that *p3* and *p3KA* were genetically independent it appears that more than one gene in the pathogen might overcome *Lr30*.

Both Statler (1977) and Samborski and Dyck (1976) commented upon the extent of heterozygosity in North American *P. recondita* cultures chosen for genetic studies. However, it seems difficult to decide whether such frequencies of heterozygosity differ significantly from those encountered in outcrossing species of higher plants.

Several studies have involved the inheritance of pathogenicity in *P. graminis tritici*. Loegering and Powers (1962) identified eight genes in the F2 of an intercross between cultures of race 111 and race 36. The corresponding host genes were distributed among seven stocks. Kao and Knott (1969) extended this work by the inclusion of Loegering and Powers' F1 culture as one parent in a further cross. They found that recessive alleles determined virulence with respect to host genes *Sr5, 6, 8, 9a, 14* and *18*. Their results in general agreed with those obtained in earlier studies by Green (1964, 1966). Kao and Knott reported that virulence for *Sr9d* (*=Sr1*) was controlled by dominant complementary genes, however this finding should be reviewed because Green and Dyck (1979) found that one *Sr9d* stock carries a second gene derived from its H-44 parent.

Green (1966) and Luig and Watson (1961) have shown that virulence on the *T. turgidum* cultivars Mindum and Arnautka is dominant. Since it is generally accepted that these stocks carry *Sr9d* it seems likely that it is the same as one of the complementary genes identified by Kao and Knott. However, avirulence for *Sr9d* has never been documented in Australian field strains although avirulence for a second temperature sensitive gene in Mindum, Arnautka and Spelmar was characteristic of the predominant strain from 1926 to 1954 but on selfing, this strain produced others avirulent at both high and low temperatures (Waterhouse, unpublished).

Green (1965b) found that pathogenicity in *P. graminis avenae* with respect to the *Avena sativa* gene *Pg3* was maternally inherited. In the same pathogen populations, virulences for genes *Pg-1, Pg-2, Pg-4* and *Pg-8* were each inherited as single recessive alleles (Johnson *et al.*, 1967).

Various early attempts were made to intercross different form species of *P. graminis*. Johnson and Newton (1933) crossed f. sp. *tritici* with *avenae* only when the former was used as the nectar-receiving parent. While the hybrid showed an extended host range to attack genotypes in host species attacked by the parent cultures, the virulence range within host species was reduced. Form species show varying degrees of inter-crossability ranging from fertile to sterile (Johnson, 1949).

Green (1971a) studied F2 populations derived from four crosses between *P. graminis tritici* and *P. graminis secalis*. F2 cultures generally resembled their F1 parents and were less virulent on wheat and rye than their respective parent cultures. Possibly the parent cultures differed by genes at many pathogenicity loci such that transgressive segregates with sufficient virulence genes to attack the tester genotypes of either host species were not obtained. On the other hand, Watson and Luig (1962) were able to identify genes for virulence on certain wheat host lines in *P. graminis secalis* following selfing.

Various abnormalities have been observed in the study of sexual hybrids of *P. graminis*. Johnson and Green (1954) noted that some infections of wheat with aeciospores from one culture resulted in uredia while others produced abortive pycnia. On the other hand, Newton and Johnson (1937) observed the production of uredospores and teleutospores on barberry in cultures with only a partial ability to produce aecia. Anikster and Wahl

(1979) consider these kinds of behaviour to be important in the evolution of macrocyclic forms of rust fungi to shorter cycled types.

Green and Johnson (1958) reported the inability of basidiospores from some cultures of *P. graminis tritici* race 15B to infect barberry. Earlier, Waterhouse (1929) had found that basidiospores from an Australian race 43 culture could not attack this host.

Miah (1968) considered the problems of collection, analysis and interpretation of data from self-fertilization of rust fungi. Generally two methods have been employed. In the reciprocal or pairing method spermatia from one pycnium are transferred to a second pycnium on a reciprocal basis. Union occurs if pycnia are of different mating types. This method provides data that are easiest to interpret. The pooled nectar method involves the transfer of bulked spermatia from several pycnia to those from which the nectar was collected. This method has been commonly used in the cereal rust fungi but the data obtained may be more subject to sampling errors. Samborski and Dyck (1968) refer to these methods as discriminate selfing and indiscriminate selfing, respectively. Miah (1968) advocated the use of a backcross method where interpretation of multiple gene interactions is easier.

V. Variation in Rust Fungi

A. Sexual

Sexual recombination is the origin of the most spectacular changes that occur in rust fungi and for heteroecious rusts, the removal of alternate host populations undoubtedly contributes to reduced variability and often to reduced disease. The removal of *Berberis vulgaris* and *Rhamnus catharticus* in the cereal-growing areas of the U.S.A. provide examples of contributions to the control of disease by these methods (Browning, 1973).

B. Asexual

1. Uredospore Migration

In areas where the uredial stage cannot survive and where the alternate host does not occur or has been largely eradicated, such as with *P. graminis tritici* in major wheat-growing areas in the U.S.A., inoculum for the initiation of epidemics must migrate from areas where the uredial stage can survive the winter. Movement of cereal rust pathogens from southern U.S.A. and, probably less frequently Mexico, to northern U.S.A. and Canada is well established. In Australia with its milder climate, cereal rust pathogens are capable of surviving through the summer on self-sown host species and grasses in areas and years where rainfall is adequate to ensure continued growth. There is also evidence that *P. graminis tritici*, and probably other species, originating in other countries may enter Australia as rare events (de Sousa, 1975). Variants of *P. graminis tritici* differing at several pathogenicity loci from localized types were detected in Australia in 1925 (126–5, 6, 7, 11), 1954 (21–0) and 1968 (326–1, 2, 3, 5, 6). In each instance these were capable of competing with the existing strains and after 1925 and 1954, at least, the latter were largely replaced by the new types.

2. Mutation

Detailed annual pathogenicity surveys of *P. graminis tritici* in Australia have shown many instances of single gene changes that were undoubtedly the result of mutation. Some of these mutant forms were selected because the matching resistance genes were present in host cultivars. Other mutants involved genes unrelated to those present in the host cultivars and their frequencies showed a wide range of variability. For example, Watson and Luig (1966) described the gradual but widespread incorporation of aviru-lence for *Sr15* into Australian *P. graminis tritici* populations as a consequ-ence of mutation and somatic hybridization despite the insignificant occurr-ence of *Sr15* in the host population. However, Luig and Watson (1970) also demonstrated that as a sequence of host resistance genes were used in cultivars the pathogen gradually accumulated the matching genes for viru-lence. Similar stepwise additions of virulence genes have been achieved in the laboratory through recurrent selection procedures following treatment of uredospores with mutagenic agents (Luig, 1979).

From both field surveys and in induced mutation experiments, mutation rates differ widely. Mutations from avirulence to virulence with respect to host genes *Sr5*, *Sr21* and *Sr9e* are common both in the field and in laboratory experiments; mutation to virulence for host gene *SrTt1* has been rare in the field and has not been produced by induced mutation; while mutations to virulence for the *Agropyron elongatum*-derived *Sr26* have never been found (Luig, 1979). Few data are available in regard to mutations from virulence to avirulence since efficient screening procedures have not been developed. Obviously, observed mutation rates are affected by dominance or recessive-ness of virulence, and homozygosity or heterozygosity of the dikaryons when virulence is recessive.

3. Somatic Hybridization

Despite the absence of sexual recombination in various situations, rust fungi are known to undergo recombination resulting from hyphal fusions and nuclear exchange (Nelson *et al.*, 1955; Nelson, 1956), or from parasexual processes. In *P. graminis,* Watson (1957) and Watson and Luig (1958) showed that the multiplicity of variants arising from mixtures could result only from a process which they designated somatic recombination. At least 20 new types have arisen over a period of 15 years from mixing the same two parents. Other rust pathogens where asexual variation has been established include *M. lini* (Flor, 1964), *P. striiformis* (Little and Manners, 1969; Goddard, 1976) and *P. coronata* (Bartos *et al.*, 1969).

The importance of somatic hybridization in the evolution of new strains in nature is more difficult to establish. However, one post-1954 group of *P. graminis tritici* strains in Australia appeared to combine certain attributes of the pre-1954 standard race 126 with other attributes of standard race 21 first found in 1954 (Luig and Watson, 1977). Watson and Luig (1968a) referred to the widespread occurrence of putative somatic hybrids of *P. graminis tritici* and *P. graminis secalis* occurring on *Agropyron scrabrum* in Queensland. Although these hybrids are avirulent on commercial wheat and rye cultivars they now appear to cause some damage to commercial barley (Luig, personal communication) and may serve as a source of virulence genes for certain wheat host genes following further introgressive crosses with *P. graminis tritici* (Luig, 1979).

In a recent review of work covering a period of 60 years Watson (1981) examined the role that uredospore migration, mutation and somatic hybridization have played on variation in *P. graminis tritici* in Australia. This is a relatively isolated continent and, because the sexual host is rare, observed variation can be attributed almost entirely to asexual processes.

4 Progressive Increases in Virulence

Watson and Luig (1968b) selected variants of *P. graminis tritici* showing clearly distinctive incompatible infection types when interacting with host genes $Sr6$ and $Sr11$. Obviously, homozygosity and heterozygosity of a corresponding pathogenicity gene could allow for two distinctive incompatible infection types but more than two levels of incompatability were observed for each of the above host genes. Further studies have indicated that such progressive increases in virulence occur with respect to many of the genes for resistance to stem rust and leaf rust in wheat and the actual levels of incompatibility serve as additional markers when considering the relationships between strains. The genetic nature of such variation is unknown but it would be most rewarding to establish if it is caused by variation within pathogenicity loci or at modifying loci. In many respects progressive increases in virulence appear to involve variation similar to that produced by mutation of the host gene(s) *Sr15/Lr20* where the host variation represented by distinctive incompatible infection types involved variation within the *Sr15/ Lr20* locus (loci) (McIntosh, 1977). Green (1971b) referred to the eroded effectiveness of host gene *Sr7a*. We interpret this as an example of progressive increases in virulence.

C. The Relationship between Virulence, Reproductive Potential and Competitive Ability

Van der Plank (1968) emphasized the possible significance of various observations that the frequencies of genes for virulence in pathogen populations declined following reductions in the frequencies of the corresponding host resistance genes. He reasoned that in some instances at least, there was a negative association between the presence of genes for virulence and reproductive potential in pathogen genotypes possessing them. Assuming the relationship, he emphasized the strategies which could be used to take advantage of this "stabilizing selection" and considered that specific resistance was useful as a resistance breeding strategy only in those instances where stabilizing selection occurred. Other workers have not been so enthusiastic in regard to the significance of the observations. Watson (1970), Nelson (1973) and Crill (1977) cited works supporting and not supporting the negative association. It seems that virulence and reproductive potential are independent genetic characters and the long-term survival of mutants with newly-derived genes for virulence depends on genetic backgrounds providing an ability to compete and survive. Hence early indications of stabilizing selection may not continue in the long term.

On the other hand, the precise nature of mutation events that result in the change from avirulence to virulence may influence reproductive potential. For example, some of the more extreme stages in progressive increases in virulence are more difficult to distinguish from fully virulent counterparts than they are from their fully avirulent counterparts, and in surveys where only two classes, avirulent and virulent, are considered they will often be included in the virulent class. As already discussed virulent mutants which genetically suppress the avirulence gene A_{M_1} in *M. lini* have reduced reproductive potential compared with $a_{M_1}a_{M_1}$ strains. Van der Plank (1973) inferred that avirulence genes have pleiotropic roles in fungal metabolism and argued that virulent mutations would have detrimental effects on relative fitness. Any effect on reproductive potential would depend on the precise action of the gene products and the nature of the mutation events. Obviously, mutation events leading to the absence of gene products could result in reduced fitness, whereas other virulent mutations could result from single base changes in the DNA sequence without affecting other functions of the gene product.

VI. Breeding for Resistance

Successful programs of breeding rust resistant field crop cultivars are well known and studies have been in progress since 1907 when Biffen showed that resistance in wheat to *P. striiformis* was controlled by a single gene (Russell, 1978). At that time the importance of genetic variation in pathogens was unknown. Indeed, it was only following the work of Stakman and colleagues from about 1920, of Flor from about 1940, and of Johnson and Newton in the 1930's, that simultaneous studies were undertaken on the nature of resistance in the host and on the extent of variation in the pathogen. The knowledge accumulated from such studies permitted resistance breeding procedures to be placed on a sound scientific basis.

Some of the breeding methods and techniques in selection of disease resistant cultivars have been described recently (McIntosh, 1978; Russell, 1978). Rigid strategies are impossible to outline because they will vary between locations, objectives of individual workers and the diseases involved.

A. *Measurement of Variation in the Pathogen*

In order to determine the extent of variation within the various rust pathogens the usual procedure is to undertake annual physiologic race surveys (also called pathogenicity surveys or virulence surveys). These involve the collection of rusted samples from a nominated survey area and the determination of the variation in pathogenicity in and between samples with respect to groups of host tester lines chosen for differentiating ability, or because of potential as sources of resistance. Worldwide acceptance of host testers has not been achieved because pathogen populations differ widely between geographical areas. With *P. graminis tritici* the Australian approach has been an attempt to assess the total variability within the area, and for this purpose testers with single genes for resistance provide the greatest resolution (Person, 1959). In other areas the emphasis has been on testers possessing only named genes, or on testers of relevance only to current activities of breeders. Much variability surveying has been conducted using seedlings in the greenhouse but following the recent emphasis on post-seedling and non-hypersensitive forms of resistance, especially with wheat stripe rust and barley leaf rust, a greater emphasis will need to be placed on both greenhouse and field methods suitable for assessing pathogen variability with respect to appropriate growth stages of the host. Undoubtedly these will be more time consuming and expensive and will need to be justified by improved durability of the resistances to be exploited. At the other extreme Hooker (1973) stressed that resistance to maize rust had been very successfully achieved without the need to resort to pathogenicity surveys and concluded that race determination "seems academic and of little practical value". Hooker described resistance in maize as either specific or non-specific and said that the basis of non-specific resistance had not been determined but stressed that it was known to be inherited as a polygenic character and could be easily selected in a breeding program.

B. Assessment of Resistance

Concurrent with studies of variation in pathogen populations the breeder needs to evaluate the range of variation in reaction that is available in the host. Russell (1978) discussed many of the terms which have been used in the description of resistance. The simplest classification that has been used places the mechanisms of resistance into two categories, viz. specific and non-specific. Specific resistance is often referred to as race-specific but would be more accurately called gene-specific because its expression is dependent on the concurrent presence of the corresponding gene(s) for avirulence. This type of resistance operates in seedlings and often in post-seedling stages as well. In the absence of close repulsion linkages specific resistance genes can be combined in one host plant to form broadly based resistances. In our view such genes, which can be isolated, mapped and studied in detail, have been the basis for successful control of some rust diseases in the small grain cereals. Rajaram (1968) showed that several wheats with resistance to stem rust on a global basis possessed various combinations of well-known genes. Successful control of wheat stem rust in the more rust-prone areas of North America and Australia over the last thirty years has been due to the use of a diversity of genes for resistance and the use of these genes in broadly based combinations.

Genes for specific resistance may control characters other than hypersensitivity. Genes *Sr8, Sr9b, Sr30* and others when operative against appropriate strains of *P. graminis tritici* reduce pustule size and retard the rate of disease development. Undoubtedly specificity of genes controlling other characters concerned in reducing the rate of infection will be demonstrated in the future. Variation in some characters that may be relevant to wheat stripe rust was mentioned by Russell (1978).

In addition to sources of specific resistance breeders are anxious to exploit other genetic systems concerned in rust resistance. The ultimate objective is durable resistance, a term Johnson and Law (1975) used to cover those situations where protection against disease has been maintained for considerable periods of time. They cited the wheat cultivar Hybride de Bersee as an example of this type of resistance to stripe rust in the United Kingdom.

There are difficulties in working with resistances that cannot be readily analysed genetically. The horizontal resistance of Van der Plank (1968) has not been characterized to the extent that it can be exploited by breeders. Johnson (1979) stated that "absolute proof of horizontal resistance is impossible". We are not familiar with resistances to wheat stem and leaf rusts that are not associated with recognizable specific genes irrespective of whether the material is slow rusting or is resistant against a broad spectrum of strains. Many of the genotypes chosen for disease progress and genetic studies have a broad base for specific resistance. Studies by Rees *et al.* (1979a, 1979b) on slow rusting were confounded by the presence of genes for specific resistance. Various wheat lines selected for adult plant resistance to stem rust by Knott (1977) in Canada were found to possess combinations of genes for specific resistance when tested in Australia (Luig and McIntosh, unpublished). The situation in oats appeared to be similar. Kochman and Brown (1975) found cultivar Garry to be slow rusting but it has genes for specific resistance. Heagle and Moore (1970) isolated oat lines with moderate resistance to crown rust, and presumably the resistance was effective against many *P. coronata* strains, but Watson (unpublished) found the lines to carry genes for specific resistance. In barley Parlevliet (1976) suggested a polygenic system to control resistance to *P. hordei* but this resistance must await exposure to a wider range of pathogenic variability and to wider variations in environmental conditions. In fact in this and other crops, studies designed to investigate the genetic basis of durable resistance must disentangle the effects of genes for specific resistance from those concerned in the expression of other characters related to rust development such as infection efficiency, latent period, and sporulation rate and duration.

The procedure of Bingham (1975) aimed at the elimination of plants possessing genes for specific resistance at the seedling stage appears to be unique. However, time will tell whether this approach simply isolates lines with post-seedling specific resistance.

C. Deployment of Resistance

Since single gene resistances to rusts in field crops have generally proved transient (Russell, 1978), breeders must also look for other approaches. Experience with wheat stem rust in North America and Australia have shown that considerable success in disease control using such genes can be achieved if breeding procedures are based on the maintenance of genetic diversity and on the generation of resistance gene combinations. Diversity can be achieved with genetic variation between plants within crops, between fields or between agricultural regions, and each procedure has been recommended and tried. Provided yields can be maintained, multilines and varietal mixtures show some promise. Multiline components so far produced for control of some wheat rusts are based on well-known genes for specific resistance (Watson, 1979). There seems no doubt that multilines and mixtures afford considerable protection in the short term but argument continues as to the types of evolutionary changes that will occur in the long term, and therefore, whether a given set of resistance genes deployed in mixed host populations will afford greater usefulness in the long term than the same set of genes deployed in alternative ways.

Our approach in the control of wheat stem rust has been the use of genetic diversity in combinations of specific resistance genes with, where possible, suspected durable resistance sources such as *Sr2* present in Hope and H44 and many modern derivatives (Hare and McIntosh, 1979). In areas where current cultivars are resistant, breeding efforts should be aimed at ensuring resistance to potentially new strains that will attack those cultivars. By necessity the anticipated strains will possess virulence genes corresponding to genes already deployed. As already stated determination of the presence of additional effective genes in host breeding populations is dependent upon the availability of comprehensive culture collections with critical combinations of genes for virulence and avirulence. Co-operative testing in laboratories in other geographical areas may be necessary. Procedures in breeding for resistance to anticipated strains were outlined by Watson (1979).

At the other extreme, some workers recommend the generation of resistances based on multiple genes of small effect following recurrent selection procedures (Robinson, 1976). They assume that such genes are common and widespread in most species. Pope (1968) and Sharp and co-workers (see Röbbelen and Sharp, 1978) showed that transgressive segregations in reaction to *P. striiformis* are not uncommon in wheat crosses but we are not aware of similar observations in relation to wheat leaf rust and wheat stem rust. However, Ballantyne (1978) obtained excellent evidence for transgressive segregations in reaction to *Uromyces appendiculatus* in crosses of French beans (*Phaseolus vulgaris*). We suggest that in situations where adequate levels of resistance can be produced by transgressive segregation, breeders will enthusiastically attempt to exploit such segregates in further crosses to commercial types.

VII. Concluding Remarks

Considering the very large number of rust diseases and their importance to mankind, relatively little is known in regard to their genetics. Genetic analyses have been restricted mainly to those rusts of greatest economic significance. Despite the fact that host reaction and pathogen pathogenicity represent suitable characters for study by geneticists, they have tended to neglect the area of host pathogen interactions in the past. However, increased contributions in this area from population, ecological and molecular geneticists can be anticipated in future. Such studies should benefit those working in the more applied areas of plant pathology and plant breeding.

References

Anikster, Y. and Wahl, I. (1979). *Ann. Rev. Phytopathol.* **17**, 367–403.
Baker, E. P. (1966). *Euphytica* **15**, 313–318.
Ballantyne, B. J. (1978). Ph.D. Thesis; University of Sydney.
Bartos, P., Fleischmann, G., Samborski, D. J. and Shipton, W. A. (1969). *Can. J. Bot.* **47**, 1383–1387.
Bingham, J. (1975). *J. Roy. Agric. Soc. England* **136**, 65–77.
Browder, L. E. and Eversmeyer, M. G. (1976). *Plant Dis. Reptr.* **60**, 143–147.
Browder, L. E. and Young, H. C. (1975). *Plant Dis. Reptr.* **59**, 964–965.
Browning, J. A. (1973). *In* "Breeding Plants for Disease Resistance" (R. R. Nelson ed.), pp. 155–180. Pennsylvania State University Press, University Park and London.
Craigie, J. H. (1927). *Nature* **120**, 765–767.
Crill, P. (1977). *Ann. Rev. Phytopathol.* **15**, 185–202.
Day, P. R. (1974). "Genetics of Host-Parasite Interaction." W. H. Freeman, San Francisco.
de Sousa, C. N. A. (1975). M. Agr. Thesis. University of Sydney.
Dinoor, A. and Peleg, N. (1972). Proceedings European and Mediterranean Cereal Rusts Conference Prague, pp. 115–120.
Dyck, P. L. and Samborski, D. J. (1968). *Can. J. Genet. Cytol.* **10**, 613–619.
Dyck, P. L. and Samborski, D. J. (1974). *Can. J. Genet. Cytol.* **16**, 323–332.
Ellingboe, A. H. (1975). *Aust. Plant Pathol. Newsl.* **4**, 44–46.
Ellingboe, A. H. (1976). *In* "Physiological Plant Pathology" (R. Heitefuss and P. H. Williams, eds), Encyclopedia of Plant Physiology, New Series 4, pp. 761–778. Springer-Verlag, Heidelberg.
Ellingboe, A. H. (1978). *In* "The Powdery Mildews" (D. M. Spencer, ed.), pp. 160–181. Academic Press, London.

Eshed, N. (1978). Ph.D. Thesis, Hebrew University of Jerusalem.
Flor, H. H. (1942). *Phytopathology* **32**, 653–669.
Flor, H. H. (1947). *J. Agric. Res.* **73**, 335–337.
Flor, H. H. (1956). *Adv. Genet.* **8**, 29–54.
Flor, H. H. (1958). *Phytopathology* **48**, 297–301.
Flor, H. H. (1959). *Phytopathology* **49**, 794–795.
Flor, H. H. (1964). *Phytopathology* **54**, 823–826.
Flor, H. H. (1965). *Phytopathology* **55**, 724–727.
Flor, H. H. (1971). *Ann. Rev. Phytopathol.* **9**, 275–296.
Goddard, M. V. (1976). *Trans. Br. Mycol. Soc.* **67**, 395–398.
Green, G. J. (1964). *Can. J. Bot.* **42**, 1653–1664.
Green, G. J. (1965a). *Can. Plant Dis. Surv.* **45**, 23–29.
Green, G. J. (1965b). *Can. J. Genet. Cytol.* **7**, 641–650.
Green, G. J. (1966). *Can. J. Bot.* **44**, 1255–1260.
Green, G. J. (1971a). *Can. J. Bot.* **49**, 2089–2095.
Green, G. J. (1971b). *Can. J. Bot.* **49**, 1575–1588.
Green, G. J. and Dyck, P. L. (1979). *Phytopathology* **69**, 672–675.
Green, G. J. and Johnson, T. (1958). *Can. J. Bot.* **36**, 351–355.
Haggag, M. E. A. and Dyck, P. L. (1973). *Can. J. Genet. Cytol.* **15**, 127–134.
Haggag, M. E. A., Samborski, D. J. and Dyck, P. L. (1973). *Can. J. Genet. Cytol.* **15**, 73–82.
Hare, R. A. and McIntosh, R. A. (1979). *Z. Pflanzenzüchtg.* **83**, 350–367.
Heagle, A. S. and Moore, M. B. (1970). *Phytopathology* **60**, 461–466.
Hooker, A. L. (1973). *In* "Breeding Plants for Disease Resistance" (R. R. Nelson ed.), pp. 132–154. Pennsylvania State University Press, University Park and London.
Johnson, R. (1979). *Phytopathology* **69**, 198–199.
Johnson, R. and Law, C. N. (1975). *Ann. Appl. Biol.* **81**, 385–391.
Johnson, T. (1949). *Can. J. Res. C.* **27**, 45–65.
Johnson, T. and Green, G. J. (1954). *Can. J. Agric. Sci.* **34**, 313–315.
Johnson, T. and Newton, M. (1933). Proc. World's Grain Exhibition and Conference, Canada, Vol. II, 220–223.
Johnson, T., Green, G. J. and Samborski, D. J. (1967). *Ann. Rev. Phytopathol.* **5**, 183–200.
Kao, K. N. and Knott, D. R. (1969). *Can. J. Genet. Cytol.* **11**, 266–274.
Kerber, E. R. and Dyck, P. L. (1979). *In* "Proceedings 5th International Wheat Genetics Symposium" (S. Ramanujam ed.), pp. 358–364. Indian Soc. Genetics and Plant Breeding, New Delhi.
Knott, D. R. (1977). *In* "Induced Mutations Against Plant Diseases," pp. 81–88, I.A.E.A., Vienna.
Kochman, J. K. and Brown, J. F. (1975). *Ann. Appl. Biol.* **81**, 33–41.
Law, C. N. and Johnson, R. (1967). *Can. J. Genet. Cytol.* **9**, 805–822.
Lawrence, G. J. (1977). Ph.D. Thesis, University of Adelaide.
Lawrence, G. J., Mayo, G. M. E. and Shepherd, K. W. (1981). *Phytopathology* **71**, 12–19.
Little, R. and Manners, J. G. (1969). *Trans. Br. Mycol. Soc.* **53**, 251–258.
Loegering, W. Q. (1966). Proc. 2nd Int. Wheat Genetics Symposium, Lund, 1963. *Heredites Suppl.* **2**, 167–177.
Loegering, W. Q. (1972). Proc. NATO-IUFRO Advanced Study Institute, 1969, U.S.D.A. Forest Serv. Misc. Publ. **1221**, 29–37.

Loegering, W. Q. (1978). *Ann. Rev. Phytopathol.* **16**, 309–320.
Loegering, W. Q. and Powers, H. R. (1962). *Phytopathology* **52**, 547–554.
Loegering, W. Q., McIntosh, R. A. and Burton, C. H. (1971). *Can. J. Genet. Cytol.* **13**, 742–748.
Luig, N. H. (1979). *In* "Proc. 5th Int. Wheat Genetics Symposium" (S. Ramanujam, ed.), pp. 533–539. Indian Society Genetics and Plant Breeding, New Delhi.
Luig, N. H. and Rajaram, S. (1972). *Phytopathology* **62**, 1171–1174.
Luig, N. H. and Tan, B. H. (1978). *Aust. J. Biol. Sci.* **31**, 545–551.
Luig, N. H. and Watson, I. A. (1961). *Proc. Linn. Soc. N.S. Wales* **86**, 217–229.
Luig, N. H. and Watson, I. A. (1970). *Proc. Linn. Soc. N.S. Wales* **95**, 22–45.
Luig, N. H. and Watson, I. A. (1977). *Proc. Linn. Soc. N.S. Wales* **101**, 65–76.
McAlpine, D. (1906). "The Rusts of Australia". Department of Agriculture, Victoria.
McIntosh, R. A. (1973). *In* "Proceedings 4th International Wheat Genetics Symposium" (E. R. Sears and L. M. S. Sears, eds), pp. 893–937. University of Missouri, Columbia.
McIntosh, R. A. (1977). *In* "Induced Mutations against Plant Diseases", pp. 551–565. I.A.E.A., Vienna.
McIntosh, R. A. (1978). *In* "The Powdery Mildews" (D. M. Spencer, ed.), pp. 237–257. Academic Press, London.
McIntosh, R. A. (1979). *In* "Proceedings 5th International Wheat Genetics Symposium Vol. II" (S. Ramanujam, ed.), pp. 1299–1325. Indian Society Genetics and Plant Breeding, New Delhi.
Miah, M. A. J. (1968). *Can. J. Genet. Cytol.* **10**, 613–619.
Miah, M. A. J. and Sackston, W. E. (1970). *Phytoprotection* **51**, 17–35.
Nelson, R. R. (1956). *Phytopathology* **46**, 538–540.
Nelson, R. R. (1973). *In* "Breeding Plants for Disease Resistance" (R. R. Nelson, ed.), pp. 40–66. Pennsylvania State University Press, University Park and London.
Nelson, R. R., Wilcoxson, R. D. and Christensen, J. J. (1955). *Phytopathology* **45**, 639–643.
Newton, M. and Johnson, T. (1937). *Nature, Lond.* **139**, 800.
Parlevliet, J. E. (1976). *Euphytica* **25**, 241–248.
Person, C. (1959). *Can. J. Bot.* **37**, 1101–1130.
Pope, W. K. (1968). *In* "Proceeding 3rd International Wheat Genetics Symposium (K. W. Finlay and K. W. Shepherd, eds), pp. 251–257. Aust. Acad. Sci., Canberra.
Rajaram (1968). Ph.D. Thesis, University of Sydney.
Rees, R. G., Thompson, J. P. and Mayer, R. J. (1979a). *Aust. J. Agric. Res.* **30**, 403–419.
Rees, R. G., Thompson, J. P. and Goward, E. A. (1979b). *Aust. J. Agric. Res.* **30**, 421–432.
Röbbelen, G. and Sharp, E. L. (1978). Fortschritte der Pflanzenz. 9. Paul Parey, Berlin and Hamburg.
Robinson, R. A. (1976). "Plant Pathosystems." Springer-Verlag, Berlin, Heidelberg, New York.
Rowell, J. B., Loegering, W. Q. and Powers, H. R. (1963). *Phytopathology* **53**, 932–937.
Russell, G. E. (1978). "Plant Breeding for Pest and Disease Control." Butterworths, London, Boston.

Samborski, D. J. (1963). *Can. J. Bot.* **41**, 475–479.
Samborski, D. J. and Dyck, P. L. (1968). *Can. J. Genet. Cytol.* **10**, 24–32.
Samborski, D. J. and Dyck, P. L. (1976). *Can. J. Bot.* **54**, 1666–1667.
Saxena, K. M. S. and Hooker, A. L. (1968). *Proc. Natl. Acad. Sci.* **61**, 1300–1305.
Shepherd, K. W. and Mayo, G. M. E. (1972). *Science* **175**, 375–380.
Simons, M. D. (1979). *Ann. Rev. Phytopathol.* **17**, 75–96.
Simons, M. D., Martens, J. W., McKenzie, R. I. H., Nishiyama, I., Sadanaga, K., Sebesta, J. and Thomas, H. (1978). Oats: a standardized system of nomenclature for genes and chromosomes and catalogue of genes governing characters. U.S.D.A. Agriculture Handbook No. 509. 40 pp.
Stakman, E. C., Stewart, D. M. and Loegering, W. Q. (1962). U.S. Dept. Agric. Res. Serv. Publ. E617, 53 pp.
Statler, G. (1977). *Phytopathology* **67**, 906–908.
Statler, G. (1979). *Phytopathology* **69**, 661–663.
Statler, G. D. and Gold, R. E. (1980). *Phytopathology* **70**, 555–557.
Tan, B. H. (1973). Ph.D. Thesis, University of Sydney.
Tan, B. H. (1977). *Cereal Rusts Bull.* **5**, Part 2, 39–43.
The, T. T. (1973). Ph.D. Thesis, University of Sydney.
Van der Plank, J. E. (1968). "Plant Diseases: Epidemics and Control." Academic Press, New York.
Van der Plank, J. E. (1975). "Principles of Plant Infection." Academic Press, New York, San Francisco and London.
Wallace, A. T., Singh, R. M. and Browning, R. M. (1967). *In* "Induced Mutations and their Utilization" Edwin-Baur-Gedächtnisvorlesungen IV, abh. d. Deutsch. Akad. d. Wiss. sch. zu Berlin, Klasse für Medizin, Jahrg. 47–57.
Waterhouse, W. L. (1929). *Proc. Linn. Soc. N.S. Wales* **54**, 615–680.
Watson, I. A. (1957). *Phytopathology* **47**, 510–512.
Watson, I. A. (1970). *Ann Rev. Phytopathol.* **8**, 209–230.
Watson, I. A. (1979). *Indian J. Genetics and Plant Breed* **39**, 50–59.
Watson, I. A. (1981). *In* "Wheat Science: Today and Tomorrow" (W. J. Peacock and L. T. Evans, eds), pp. 129–147. Cambridge University Press, Cambridge.
Watson, I. A. and Luig, N. H. (1958). *Proc. Linn. Soc. N.S. Wales* **83**, 190–195.
Watson, I. A. and Luig, N. H. (1962). *Proc. Linn. Soc. N.S. Wales* **87**, 39–44.
Watson, I. A. and Luig, N. H. (1966). *Euphytica* **15**, 239–250.
Watson, I. A. and Luig, N. H. (1968a). *In* "Proceedings 3rd International Wheat Genetics Symposium" (K. W. Finlay and K. W. Shepherd, eds), pp. 227–238. Aust. Acad. Sci., Canberra.
Watson, I. A. and Luig, N. H. (1968b). *Phytopathology* **58**, 70–73.
Wilson, M. and Henderson, D. M. (1966). "British Rust Fungi." Cambridge University Press, Cambridge.

4. Physiology and Biochemistry of Spore Germination

G. WOLF

Institute of Plant Pathology and Plant Protection,
George August University, West Germany

CONTENTS

I. Introduction

Among the biotrophic fungi, the rusts have been particularly well studied with regard to the physiology and biochemistry of their spore germination. This is probably due not only to their economic importance as plant parasites, but also to the fact that the germination of uredospores occurs readily on aqueous media, independent of the host. The main purpose of these studies always was, and still is, to gain a better understanding of the biotrophic habit of these fungi. For this reason, interest has centered on determining whether the sporelings have a normal metabolic potential, or whether there are defects accounting for their dependence on the living host plant.

At first, spore germination *per se* was the only ontological process that could be studied independent of the host plant. This allowed a direct analysis of the metabolic processes of these fungi, in contrast to studies of the host-parasite-complex, where it is practically impossible to differentiate between the metabolism of the parasite and that of the host. Only after the first successful axenic culture of wheat stem rust (Williams *et al.*, 1966) was it possible to study *in toto* the growth of a fungus which, up to this time, had been considered an obligate biotrophic parasite. Since then, however, other rusts have been successfully grown in axenic culture.

Spore germination is a decisive phase in the propagation and survival of biotrophic fungi which, under normal conditions, depend on their hosts. It is therefore hardly surprising that biotrophic fungi have developed specific regulatory mechanisms which are highly adapted to environmental conditions and allow an effective colonization of the host. Characteristic for many representatives of this group of fungi are the specific infection structures, such as appressoria, infection peg, substomatal vesicles, hyphae and haustoria, which have developed as means of infecting the host via stomata.

The relatively late direct contact with the actual substrate by means of haustoria is a further characteristic of biotrophic fungi. Thus, an efficient exploitation of the endogenous reserve substances of the spore is necessary until contact with the exogenous substrate is established. It is therefore understandable that studies of the physiology and biochemistry of sporelings have concentrated particularly on the phenomenon of regulation of germination and on the metabolic potential of spores.

Primarily because of their epidemiological importance and availability, uredospores have been used for most of these studies. There is little information to date on the germination of teliospores, aeciospores or basidiospores.

Four stages of uredospore germination can be determined: (a) Hydration and swelling; (b) Dissolution of the cell-wall pore plug and appearance of the germ tube; (c) Growth of the germ tube; (d) Differentiation of the infection structures, e.g. appressorium, infection peg, substomatal vesicle, infection hyphae and haustorium.

Only the first three stages are usually thought of as being part of spore germination, although differentiation up to the formation of infection hyphae occurs in response to certain, often specific, stimuli (e.g. heat shock or membranes), and their formation depends on endogenous reserves of the spore. Differentiation of infection structures should therefore be included in reviews of the germination of uredospores.

The physiological and biochemical aspects of rust spore germination have been dealt with in a number of reviews and articles concerned with fungal spore germination in general, or with specific aspects of rust spore germination (Shaw, 1963; Shaw, 1964; Allen, 1965; Staples and Wynn, 1965; Allen, 1976; Gottlieb, 1976; Macko *et al.*, 1976; Reisener, 1976; Staples and Yaniv, 1976; Staples and Macko, 1980).

II. Control of Spore Germination

A. *Germination Inhibitors and their Identification*

It has long been known that fungal spores in dense populations, either in pustules or in suspension, do not germinate at all or do so only at a reduced rate. In some cases, this effect is the result of self-inhibition by substances present in or produced by the spores themselves. Self-inhibition of uredospore germination is undoubtedly advantageous to the rust since it prevents premature germination of spores in the pustule and thus contributes to efficient spore dispersal. Reports on the self-inhibition of fungal spore germination have been reviewed by Allen and Dunkle (1971), Macko and Staples (1973), Macko *et al.* (1976), Allen (1976) and Staples and Yaniv (1976).

Evidence for an inhibiting principle in water extracts of uredospores of *P. graminis* f. sp. *tritici* was first reported by Allen in 1955, but it took another 15 years to identify the compound. Self-inhibitors of rusts occur in very low concentrations. Characterizing the active compounds was therefore difficult and time-consuming. Self-inhibitors were extracted by stirring spores in water and then partitioning from water to ether, followed by a further purification with silica gel chromatography and gas chromatography, thus yielding a pure compound indentifiable by spectrometric analysis.

Self-inhibitors of the uredospores of seven rust species have been identified with this technique (Macko *et al.*, 1970, 1971a, 1971b, 1972; Foudin and Macko, 1974; Macko *et al.*, 1977). The self-inhibitor of *U. phaseoli*, *P. helianthi*, *P. antirrhini*, *P. arachidis*, *P. sorghi* and *P. striiformis* was identified as methyl-*cis*-3,4 dimethoxycinnamate (MDC), while that of *P. graminis* f. sp. *tritici* was found to be methyl-*cis*-ferulate (MF).

The natural inhibitors are the *cis*-isomers of these cinnamic acid esters. They can be converted by u.v. irradiation to the biologically inactive *trans*-isomers (Allen, 1972; Macko *et al.*, 1972). This points to a high degree of structural specificity. Modification of the molecule causes a dramatic change in biological activity. By testing the effect of different cinnamic acid derivatives on the germination of *P. graminis* f. sp. *tritici* uredospores, it was found that the 3-methoxy groups and the double bond of the side chain are essential. In addition, the ester bond appears to be important for biological activity (Macko *et al.*, 1972; Allen, 1976). Self-inhibitors of rust germination are amongst the most potent biologically active compounds found anywhere. They inhibit in concentrations from a few ng/ml (10^{-8} M) to a few pg/ml (10^{-11} M) (see Table I).

Cross-reactivity was observed in some instances. MDC from bean rust inhibits wheat stem rust uredospores with an ED_{50} of 3·7 ng/ml, whereas coffee rust and flax rust are insensitive to MDC (Macko *et al.*, 1972; Musumeci *et al.*, 1974). It has also been shown that no other fungi so far tested are affected by self-inhibitors from rust uredospores (Allen, 1976).

The presence of as yet unidentified self-inhibitors in uredospores of other rust fungi has been reported for *P. graminis* f. sp. *agrostis*, *P. polysora* and *M. methae* (Pritchard and Bell, 1967); *P. polygonii* and *P. carthami* (Wilson, 1958); *P. tritiania* and *P. dispersa* (Hoyer, 1962); *P. coronata* (Naito *et al.*, 1959); *U. fabae* (Marte, 1971); *U. striatus* (Pritchard and Bell, 1967); *Hemileia vastatrix* (Musumeci *et al.*, 1974); and in aeciospores of *P. sorghi* (Le Roux and Dickson, 1957) and *Cronartium comandrae* (Eppstein and Tainter, 1976, 1979). A study on the inhibition of coffee rust uredospore germination using 24 different cinnamic acid derivatives showed that the natural inhibitor may be a methoxycinnamic acid (Stahmann *et al.*, 1976).

The inhibitors may be bound to the cell wall, since they are released very

Table I. Self-inhibitors of spore germination and their potency (adapted from Macko and Staples, 1973, and Foudin and Macko, 1974).

Species	Self-inhibitor	ED_{50}(ng/ml)
Uromyces phaseoli		2·72
Puccinia helianthi, race 1		1·65
Puccinia antirrhini ATCC 3/66	methyl *cis*-3, 4-dimethoxycinnamate	0·23
Puccinia sorghi, race 1-20		0·51
Puccinia arachids		0·008
Puccinia graminis tritici race 56	methyl *cis*-ferulate	0·18

readily, Because cinnamate esters are preferably soluble in nonpolar solvents, the ease with which the inhibitor is released into water suggests that it may be bound *in vivo* to a yet unknown carrier.

As shown by Allen and Dunkle (1971), the inhibitor is already present in infected plants. It could be detected in susceptible bean plants at early stages of infection, before the onset of spore formation of *U. phaseoli* (three to four days after inoculation) (Elnaghy and Heitefuss, 1976). This indicates that the production of the inhibitor is initiated in the mycelium.

B. Mode of Action of Self-Inhibitors

The most detailed knowledge on the mode of action of self-inhibitors is available for *cis*-ferulate, the germination inhibitor found in *P. graminis* f. sp. *tritici*. Germination of rust uredospores begins with hydration and swelling, followed by the formation of a germ-tube, which then penetrates the spore wall through one of the preformed pores. The action of the inhibitor is restricted to the first 10–20 minutes, when the germ-tube extrudes through the cell wall. Once outside, the germ-tube is no longer sensitive.

An electron-microscopic study of the early stages of germination showed that, after removal of the inhibitor, the germ pore plug dissolved within 10 minutes, after which the germ-tube appeared. If the inhibitor is added when the pore plug is only partially removed, germination stops. These observations (Hess *et al.*, 1975) point to a block in the removal of the germ pore plug.

The hydration of the uredospores is apparently not affected by the self-inhibitor, since swelling and respiration are normal when the inhibitor is present (Maheshwari and Sussman, 1970). Thus, the only event of spore germination sensitive to MF is the pore-plug digestion, which occurs within 30 minutes of the removal of the inhibitor.

Protein synthesis does not seem to be a prerequisite for wall digestion (Hess *et al.*, 1976). Cycloheximide, which inhibits incorporation of $[^{14}C]$-L-leucine into proteins of the wheat stem rust uredospores, does not prevent pore-plug dissolution. Methyl-*cis*-ferulate, on the other hand, prevents cell-wall digestion, but has no effect on L-leucine incorporation. This indicates that pore-plug dissolution does not depend on the synthesis of new protein. Similar effects of the bean rust inhibitor MDC on pore-plug dissolution of *U. phaseoli* were observed during an electron-microscopic study carried out by Staples and Yaniv (1976). It can be speculated that hydrolytic enzymes, in the region of the germ pore, involved in the degradation of the cell-wall material have already been synthesized in the dormant spores and are specifically blocked by the inhibitor.

C. Stimulation of Germination

Regulation of rust uredospore germination is apparently dependent not only on self-inhibitors but also on endogenous stimulators. One of the first endogenous compounds reported as stimulating fungal spore germination was *n*-nonanal, the stimulator obtained from *P. graminis* f. sp. *tritici* (French and Weintraub, 1957). More recently, nonanal was identified as a major component among volatiles collected directly from uredospores of seven different rust species (see Table II; Rines *et al.*, 1974).

Furthermore, 6-methyl-5-hepten-2-one was identified as an additional endogenous stimulator of *P. graminis* f. sp. *tritici* and *P. striiformis* (Rines *et al.*, 1974). These endogenous, volatile stimulators acted on all species in nearly the same optimal dose of less than 1 p.p.m. (French *et al.*, 1975b). The effect of the identified endogenous stimulators is not species specific. Thus, nonanal stimulated the germination of other rusts and Ustilago species. The same was observed for methylheptenone, the second stimulator of *P. graminis* f. sp. *tritici* (French *et al.*, 1975a).

On the other hand, it is surprising that nonanal, although detected in relatively high amounts in *U. phaseoli*, does not stimulate its germination (French *et al.*, 1975a).

Nonvolatile endogenous compounds such as coumarin, phenols and other substances were also shown to have a stimulatory effect on the germination of uredospores of *P. graminis* f. sp *tritici* (van Sumere *et al.*, 1957; French and Weintraub, 1957).

Besides these endogenous stimulators, over 60 different exogenously supplied compounds have been shown to stimulate the uredospore germination of the wheat stem rust. They range in length from 5–12 carbon atoms

Table II. Nonanal recovered in distillates of uredospores of various rusts (adapted from Rines *et al.*, 1974).

Species	1 of nonanal/g of spores
Puccinia graminis f. sp. *tritici*, race 56	0·046
Puccinia graminis f. sp. *tritici*, race 15B	0·050
Puccinia striiformis	0·048
Puccinia reconditia	0·055
Uromyces phaseoli	0·034
Puccinia helianthi	0·026
Puccinia coronata	+
Puccinia sorghi	+

+ Nonanal present as predominant distillate component but amount not determined.

and include alcohols, aldehydes, ketones, amines, cyclic and noncyclic derivates, and sulfur derivates (French, 1961; French and Gallimore, 1971; French *et al.*, 1975b). Germination of uredospores of *U. phaseoli* was stimulated by linear, branched, saturated and unsaturated methyl ketones with 6–9 carbons, and by cyclic unsaturated ketones, particularly by β-ionone. Unlike the spores of *P. graminis* f. sp. *tritici*, the spores of *U. phaseoli* were not stimulated by linear or cyclic aldehydes or by alcohols (French *et al.*, 1975a).

Although a great deal of information is available on the chemical properties of the stimulators of rust uredospore germination, knowledge of their mode of action is limited. Spores of *U. phaseoli*, *P. graminis* f. sp. *tritici* and *P. coronata* exposed for only 20 seconds to vapors of β-ionone or nonanal, respectively, germinated with up to 95% efficiency. This points to a sensitive chemical switch (French *et al.*, 1977). It is possible that stimulators facilitate the release of inhibitors from the spore, thus allowing spore germination.

Is is noteworthy that compounds active as germination stimulators have been found in many natural products, particularly in essential oils and perfumes. In addition to the endogenous stimulators, it is possible that similar but yet unknown substances from host plants may contribute to the regulation of germination of biotrophic fungi on plant surfaces.

III. Induction of Infection Structures

The successful entry of rusts into the host requires the development of special structures of the germ-tube. These infection structures (appressoria, infection peg, substomatal vesicle and infection hypha) are characteristic for each rust species. Their formation requires special stimuli. *In vivo*, physical stimuli caused by contact with the leaf surface are evidently involved in the induction process. This field has recently been reviewed by Staples and Macko (1980). In an elegant study, Wynn (1976) was able to show that up to 92% of the *U. phaseoli* spores formed appressoria above the stomata of bean leaves or on plastic replicas of the leaf, whereas on surfaces or replicas of nonhost leaves, only 5% of the germ-tubes formed appressoria. To exclude substances of host origin adhering to the replica surface, a series of replicas was used. Even on the last of at least 12 replicas of bean leaf surfaces, the number of appressoria formed was the same as on the first.

As shown by scanning electron microscopy, the stomatal lip is apparently responsible for the induction process (see M. Heath Chapter 6).

These observations agree with earlier results of Dickinson (1970, 1971, 1972), who found that all types of infection structures of *P. coronata* can be induced on plastic or cellulose nitrate membranes with or without paraffin oil. Collodion membranes plus oil have proved to be very effective for the induction of infection structures of bean rust germ-tubes (Maheshwari *et al.*, 1967).

Another type of induction was reported for wheat stem rust. Two hours after germination commenced at 22°C, uredosporelings of *P. graminis* f. sp. *tritici* were, by exposure to a temperature of 30–37°C (heat shock) for 75–90 minutes followed by a return to the lower temperature, induced to form infection structures comparable to those formed *in vivo* (Maheshwari *et al.*, 1967). Dense spore suspensions (5 mg/ml) do not form infection structures unless the medium has been replaced before heat shock (Allen and Dunkle, 1971). Since the self-inhibitor has no effect on differentiation, a diffusible substance may be responsible for this inhibition.

In addition to the triggering effect of physical stimuli, chemical induction of infection structure formation has also been found. Allen (1957) showed that a volatile compound present in spore eluates of *P. graminis* f. sp. *tritici* was able to induce infection structures in the germinating spore. In a recent study by Macko *et al.* (1978), acrolein (2-propenal) was identified as the active component. The natural compound as well as synthetic acrolein were active. In a standard assay, 50–80% of the germinating uredospores formed infection structures at 0·1–0·4 nmole of acrolein per 6 ml gas diffusion vessel.

A survey of over 60 compounds structurally similar to acrolein indicated that short-chain aliphatic carbonyl compounds with a conjugated double bond are able to induce infection structures whereas saturated aliphatic aldehydes and ketones, saturated and unsaturated aliphatic alcohols, and phenylacrolein were morphogenetically inactive.

It should be mentioned that substances from the host plant may also play a role in the regulation of differentiation processes. Grambow and Riedel (1977) were able to achieve a high rate of infection structure formation in uredosporelings of *P. graminis* f. sp. *tritici* when these were germinated in the presence of a complex volatile fraction from wheat leaves and nonvolatile components from an epicuticular extract. It was also reported that hexenols (especially *cis*-3-hexenol), well-known as volatile plant constituents, have a striking induction effect, but only, however, in the presence of the nonvolatile fraction (Grambow and Grambow, 1979).

Thus, infection structure formation also appears to be subject to a regulatory mechanism involving inhibitors and stimulators, in principle similar to that regulating the onset of germination.

IV. Carbohydrate Metabolism

The biological function of carbohydrates in spores and mycelium of the rust fungi are generally two-fold: they form important structural elements of the cell wall and they play an important role as energy storage substances especially for germination. It is thus surprising that, comparatively speaking, little work has been done on carbohydrates in these structures. Historically, polysaccharides of fungal cell walls were first used as a taxonomic criterion with considerable success (Bartnicky-Garcia, 1968). More recently polysaccharides or other carbohydrate-containing molecules have attracted increased attention because of their possible role in cell-cell recognition phenomena, including those between hosts and their fungal parasites (Callow, 1978).

A. Soluble Carbohydrates

With respect to the content of soluble carbohydrates in rust uredospores, the literature contains only a few reports (see Shaw, 1964), in part with conflicting results. Perhaps the most reliable results were obtained with uniformly [^{14}C]-labelled uredospores of *P. graminis* f. sp. *tritici* (Daly *et al.*, 1967). The metabolites soluble in 80% ethanol contain approximately 20% of the total carbon of the uredospores. Of this, the sugar alcohols mannitol and arabitol are the predominant compounds (28% and 18% respectively). Another important compound in such extracts is trehalose (11%), while glucose and fructose occur only in very small amounts. During germination trehalose, mannitol and arabitol are metabolized relatively quickly (about 30% of the total in seven hours).

In this connection it is worth mentioning that approximately 7% of the total uredospore carbon is lost during the same period, most of it during the first ten minutes of germination.

B. Pathways of Carbohydrate Utilization

This subject has been reviewed previously by Shaw (1964) and Gottlieb (1976).

The Emden-Mayerhof-Parnas (EMP) pathway and the Hexosemonophosphate (HMP) pathway have been demonstrated in rust uredospores (Shu and Ledingham, 1956; Shaw and Samborski, 1957) by determination of the C_6/C_1 ratio in respired [$^{14}CO_2$] after application of 1-[^{14}C]-glucose and 6-[^{14}C]-glucose. Most of the EMP enzymes were also shown to be present in *P. graminis* f. sp.*tritici* and *U. phaseoli* (Shu and Ledingham, 1956; Caltrider and Gottlieb, 1963), together with all enzymes of the HMP shunt. The operation of the citric acid cycle (CAC) in *P. graminis* f. sp. *tritici* is also suggested by [^{14}C]-labelling studies with acetate, propionate and valerate. Labelling of intermediates in each case points to the operation of the cycle (Reisener *et al.*, 1961, 1963a, 1963b; Reisener and Jäger, 1967).

As shown by White and Ledingham (1961), most of the CAC intermediates are fully oxidized by uredospores of *P. graminis* f. sp. *tritici*. Non-keto acids are also present in resting uredospores of this fungus: they disappear, however, during germination, except citrate, which increases (Staples, 1957).

Several important enzymes of the cycle (aconitase, isocitric dehydrogenase, succinic dehydrogenase, fumarase and malic dehydrogenase) have been demonstrated in *P. graminis* f. sp. *tritici* and *U. phaseoli* (Caltrider and Gottlieb, 1963).

C. Carbohydrate Content of Spore Walls and Germ-Tube Walls

Shu *et al.* (1954) reported that uredospores of *P. graminis* f. sp. *tritici* contain 29% carbohydrate with respect to their fresh weight, including 15·2% mannan.

Detailed investigations of Joppien *et al.* (1972) and Joppien (1975) revealed that the spore wall consists of about 68% of bound neutral sugar, of which mannose was the main constituent (64%). The difference (4%) consisted of approximately equal amounts of glucose and galactose. However, the spore wall contains relatively small amounts of bound glucosamine (7·2%) and only traces of N-acetylglucosamine (Table III).

The monomers in mannan are predominately β (1-->4) linked, but some β(1-->3) bonds do occur, presumably at branching points. The non-reducing ends are probably occupied by β (1-->6) linked galactose or glucose.

Table III. Carbohydrate composition of spore walls and germ-tube walls of *P. graminis* f. sp. *tritici* (adapted from Joppien, 1975).

Carbohydrate	Spore wall (%)	Germ-tube wall (%)
Neutral sugars	68·4	55
Mannose	63·8	24
Glucose	2·5	26
Galactose	2·2	5
N-acetylglucosamine	+	14·2
Glucosamine	7·2	2

The carbohydrate monomers in the germ-tube wall occur in three different types of polysaccharides and probably in glycopeptides. *N*-acetylglucosamine occurs predominantly in chitin, while mannose, glucose and galactose make up a complex galacto-glucomannan. In addition, glucose is bound as glucan in one of the main polysaccharides in the germ-tube wall.

Most mannose and most glucose monomers are β (1-->3) linked in these polysaccharides. In addition, galactose is β (1-->3) and β (1-->6) linked; glucose and/or mannose is located as branch points. Non-reducing ends consist of glucose and/or mannose. Degradation of the polysaccharides yields oligosaccharide units consisting exclusively of either mannose or glucose. This indicates that the polysaccharides are made up of mannan and glucan sequences.

V. Respiration

The process of germ-tube growth requires energy production which depends on the oxidation of endogenous reserve substances. The process of respiration was studied in the early phase of work on the physiology of rusts (Shaw, 1964). Earlier studies used manometric methods which are not ideal for this type of material. According to the more detailed work of Maheshwari and Sussman (1970), who used the polarographic method for O_2 measurement, respiration of germinating uredospores of *P. graminis* f. sp. *tritici* is characterized by four distinct phases. During Phase 1 (0–30 minutes), respiratory activity begins immediately after vapor-phase hydration and reaches a maximum during this interval and before the appearance of the germ-tube. Phase 2 (30–90 minutes) is characterized by a respiratory decline coinciding with the appearance of the germ-tube. Phase 3 (90–300 minutes), which corresponds to the maximum of germ-tube growth, is marked by a second and greater maximum in respiratory rate. Thereafter, a gradual decline is observed after the germ-tubes have reached more than half of their maximal length (Phase 4).

The addition of a partially purified self-inhibitor, which inhibited germination, did not affect the initial rise in respiratory rate.

The respiratory quotient (RQ), a rough guide to the general type of endogenous substrate being oxidized, declined from an initial high value of 0·88 to low values of about 0·33 to 0·59, indicating that at first carbohydrates and later lipids, were being used as substrates.

With indirect methods using the respective inhibitors for determination of enzyme activities, it was possible to show that the terminal electron transport system is present in rusts tested thus far (White and Ledingham, 1961;

Caltrider and Gottlieb, 1963; Staples, 1957). NADH-cytochrome c̠ oxidase and cytochrome c̠ oxidase were found in uredospores of *P. graminis* f. sp. *tritici* and *U. phaseoli*, succinic oxidase in *P. rubigo-vera*.

VI. Lipid Metabolism

Uredospores of rusts are known to contain large oil droplets. Early analytical data on the gross composition had shown a relatively high lipid content (18–25%). This led to the concept that lipids may be the chief storage metabolities. This was supported by a low content of carbohydrates (0·01–0·12%; Caltrider *et al.*, 1963) and a low RQ of respiration during germination. However it should be noted that the analytical techniques used by the workers would not have detected trehalose and sugar alcohols. An indication of the importance of lipids as reserves for germination was first found by Shu *et al.* (1954), who were able to show that lipids were depleted during germination.

A. Fatty Acid Content and Metabolism

In a series of studies using gas chromatography, Tulloch and co-workers analysed the fatty acid composition of different spore types of several rust species and other fungi (Tulloch *et al.*, 1959; Tulloch, 1960, 1964; Tulloch and Ledingham, 1960, 1962, 1964). A number of saturated acids (myristic, palmitic and stearic acids) and unsaturated acids (oleic, olenic and linolenic acids) were detected in all species. These are known to occur commonly in the plant kingdom.

A very unusual fatty acid, *cis*-9,10-epoxyoctadecanoic acid, was present in high amounts as a glyceride in the oil of most rust species, excepting the spores of two species of Gymnosporangium and one species each of Puccinia and Uromyces. According to present evidence, this fatty acid occurs only in rusts.

Palmitic (5–49% of the total fatty acids), oleic (2–62%), linoleic (2–64%), linolenic (1–73%) and *cis*-9, 10-epoxyoctadecanoic acid (1–52%) were the most common fatty acids in most species. Stearic, myristic, palmitoleic, pentadecanoic, heptadecanoic, arachidic, liconenoic and behemic acid were found in lower amounts.

The relative composition of the oils in the same spore type of different rusts appeared to be species specific, whereas races of any one species were very similar with respect to their fatty acid composition. In addition, uredospores and teliospores had a similar fatty acid composition within each species, but that of aeciospores was often quite different. As could be expected, the host plant seemed to have no influence on the oil composition of the uredospores.

In spite of the apparent importance of the lipid metabolism, few experimental data are available on the characteristics of the enzymes involved, except from the work by Knoche and Horner (1970) on the possible involvement of lipases. They showed that the majority of the lipolytic activity of disrupted uredospores was associated with a lipid-containing particulate fraction. With triolein as substrate, both 1,3- and 1,2-diglycerides were formed.

To elucidate the lipid metabolism of rust uredospores, a series of studies was undertaken using exogenous substrates. By measuring respiratory rate and uptake of radioactive precursors, it was found that spores of wheat stem rust take up short-chain fatty acids readily, depending on their chain length, in the order valerate > butyrate > propionate > acetate (Farkas and Ledingham, 1959; Reisener et al., 1961, 1963a). A considerable amount of the radiocarbon residing in the precursors is released as CO_2. It is also incorporated in relatively high amounts into the soluble fraction of spore material such as free amino acids and carbohydrates, and less readily into insoluble carbohydrates and proteins.

Valerate is utilized by β-oxidation to yield acetate which is metabolized via the citric acid cycle. This suggestion is supported by experiments using 2-[^{14}C]-valerate. The label recovered in glutamic acid resided mainly in C-2 to C-4.

With propionate labelled with [^{14}C] in the 1, 2 or 3 position, it was shown that C-1 is released as CO_2 and that C-2 and C-3 are utilized in the same way as acetate, with C-2 of propionate being used like the carbonyl carbon of acetate (Reisener et al., 1963b).

Although there is no doubt that lipids serve as respiratory substrates during spore germination, it is not certain what proportion of CO_2 released through respiration is derived from lipids. It can be speculated that lipids are not only involved in catabolic processes, but are also used for the synthesis of other compounds. This is illustrated by experiments with [^{14}C]-acetate, [^{14}C]-valerate and [^{14}C]-propionate (Staples, 1962; Reisener et al., 1963a, 1963b). In these studies considerable radioactivity was recovered in amino acids, organic acids and carbohydrates, mainly in polymeric carbohydrates of the cell wall. This would point to the possibility that the glyoxalate cycle is capable of coupling fatty acid degradation and sugar synthesis.

Indeed, the key enzymes of the glyoxalate pathway have been demonstrated in uredospores of *Melampsora lini* (Frear and Johnson, 1961), *Uromyces phaseoli* and *Puccinia graminis* (Gottlieb and Caltrider, 1962; Caltrider and Gottlieb, 1963).

B. Phospholipids

Phospholipids, which commonly occur as structural components of cell membranes, have been studied in some detail. As found by Langenbach and Knoche (1971a), dormant uredospores contain phosphatidylcholine and phosphatidylethanolamine as the main phospholipids; diphosphatidylglycerol, phosphatidylinositol and another phosphatidylinositide were present as minor components. Germ-tube wall preparations were found to contain phosphatidylcholine and phosphatidylethanolamine in about the same concentrations as was observed in the resting spore, while the amount of diphosphatidylglycerol was about three times higher.

Experiments with [^{32}P]-labelled uredospores of *U. phaseoli* showed that the phospholipid content dropped to 40% of the pregermination level within the first 20 minutes of germination. Between two and three hours after germination, phospholipid levels increased to approximately 80% of the pregermination level, and after five hours, germination catabolism had reduced the amount of [^{32}P]-lipids to the level peresent prior to the anabolic phase. Synthesis occurred again between five and ten hours with a maximum at ten hours (Langenbach and Knoche, 1971b). These changes probably reflect a degradation and synthesis of cell membranes during germinaton.

For the synthesis of phospholipids [methyl-^{14}C]-L-methionine, [^{14}C]-D, L,serine, [^{14}C]-choline and [^{14}C]-ethanolamine were utilized by germinating uredospores. The active one-carbon units for methionine and serine appear to be involved in the methylation of phosphatidylethanolamine to form phosphatidylcholine.

C. Hydrocarbons and Related Compounds

The surface lipids of rust uredospores have been examined only in *P. striiformis* (Jackson *et al.*, 1973). The major components are β-diketones, *n*-alcohols (80% octacosanol) and hydrocarbons, especially C_{27}, C_{29} and C_{31} alkanes. Additionally, free fatty acids, triglycerides, wax esters and sterol esters were found.

D. Carotenoids

Carotenoids are pigments contributing to the colour of uredospores of rusts. Uredospores of wheat stem rust, wheat leaf rust, oat crown rust and flax rust contain varying amounts of β-carotene, γ-carotene and lycopene (Irvine *et al.*, 1954). In a detailed analysis of the unsaponifiable fraction, Hougen *et al.*

(1958) found in wheat stem rust uredospores γ-carotene, β-carotene, *cis-γ-*carotene, phytoene, lycopene and *cis-β-*carotene, in decreasing amounts in the order mentioned.

E. Sterols

Sterols are typical constituents of membranes of eucaryotic cells. Many fungi require sterols in the medium for growth and/or reproduction. Extensive review articles on fungal sterols were published by Weete (1973), Weete and Laseter (1974) and Brennan and Lösel (1978).

Some of the earlier work on rusts dealt with sterols because it was assumed that these compounds, possibly derived from the host, are of special importance for these biotrophic fungi.

Rust uredospores are known to contain a number of sterols (see Table IV).

Many of these compounds occur in several of the investigated species, with stigmast-Δ^7-enol being the most abundant in most species.

Exogenously supplied acetate and mevalonate were shown to be precursors for the synthesis of sterols in uredospores (Lin *et al.*, 1972). In uredospores of *U. phaseoli* uniformly labelled with [^{14}C]-sterol, synthesis was also observed in the absence of exogenous substrates (Lin and Knoche, 1974).

Table IV. Sterols from rust uredospores.

Sterol	Fungus	Reference
$\Delta^{7,24(28)}$-Stigmasten-3β-ol	*Uromyces phaseoli*	Lin and Knoche (1974)
	Uromyces phaseoli	Hoppe and Heitefuss (1975b)
	Melampsora lini	Jackson and Frear (1968)
Δ^7-Stigmasten-3β-ol	*Uromyces phaseoli*	Hoppe and Heitefuss (1975b)
	Melampsora lini	Jackson and Frear (1968)
	Puccinia graminis	Nowak *et al.* (1972)
	Puccinia striiformis	Weete and Laseter (1974)
$\Delta^{5,7}$-Stigmastadienol	*Melampsora lini*	Jackson and Frear (1968)
Stigmasterol	*Uromyces phaseoli*	Lin *et al.* (1972)
28-Isofucosterol	*Uromyces phaseoli*	Lin and Knoche (1974)
β-Sitosterol	*Uromyces phaseoli*	Lin *et al.* (1972)
Campesterol	*Uromyces phaseoli*	Lin *et al.* (1972)
	Puccinia graminis	Nowak *et al.* (1972)
Cholesterol	*Puccinia graminis*	Nowak *et al.* (1972)
22-Dihydroergosterol	*Puccinia graminis*	Miller *et al.* (1967)
Ergost-Δ^7-enol	*Puccinia graminis*	Hougen *et al.* (1958)
	Puccinia striiformis	Weete and Laseter (1974)

The quantity of sterols increased during the first six hours of a 12 hour period of germination and decreased to a level that was still 1·5 times greater than the initial level.

These results on sterol synthesis are supported by later work (Lin and Knocke, 1974; Bansal and Knoche, 1981), which showed that an enzyme complex isolated from this material can perform a two-step successive methylation of Δ^{24}-sterols with s-adenosylmethionine as donor, similar to the reactions found in algae and higher plants. Though this pathway has not been investigated extensively, it is probable that rusts can synthesize sterols independent of their host.

VII. Nucleic Acid and Protein Metabolism

Much attention has been focused on the synthesis of nucleic acids and proteins during spore germination with particular reference as to whether synthesis of DNA, RNA and protein is essential for germ-tube growth and differentiation.

A. DNA Metabolism

The approach taken to test the requirement of macro-molecular biosynthesis during germination involved the use of antibiotics or analogues which block DNA synthesis. Using actinomycin D, hydroxyurea and cordycepin, Staples et al. (1975) studied DNA synthesis and nuclear division in connection with germ-tube growth and the formation of infection structures of bean rust uredospores on collodion membranes. They found that none of the antimetabolities influenced spore germination, although actinomycin D and hydroxyurea did inhibit DNA synthesis, nuclear division and infection structure formation, whereas cordycepin inhibited nuclear division, DNA synthesis, formation of appressoria or initiation of the vesicle. They concluded that the formation of appressoria and vesicles, but not germ-tube growth, requires DNA synthesis, but that none of these processes depend on nuclear division.

In a study of DNA synthesis during germination and infection-structure development of uredospores of *U. phaseoli*, Staples (1974) showed that only mitochondrial DNA (mt DNA) is synthesized during germ-tube growth. However, when the appressoria were formed, the synthesis of nuclear DNA was detected, and synthesis of mitochondrial DNA ceased. In later work DNA polymerase was isolated and characterized from non-germinated bean rust uredospores (Staples and Yaniv, 1978).

Tückhardt (1976) was able to detect DNA synthesis in *U. phaseoli* within only 60 minutes after the onset of germination and demonstrated that it continued up to 12 hours. This DNA was not further characterized, but, according to Staples (1974), it may be mt DNA.

B. RNA Metabolism

The effect of RNA synthesis inhibitors such as actinomycin D (5 μg/ml) and 5-fluorouracil (100 μg/ml) on the *in vitro* germination and formation of infection structures of *P. graminis* f. sp *tritici* was studied by Dunkle and Allen (1969). Neither inhibitor interfered with germ-tube growth, but both prevented later differentiation if present during the heat-shock treatment. These results suggest that germ-tube growth is not dependent on RNA synthesis. Conversely, initiation of infection structure formation seems to be accompanied by RNA synthesis.

The conclusion that germ-tube growth is independent of RNA synthesis is corroborated by a direct analysis of RNA. Trocha and Daly (1970) were able to show, by quantitative estimation of nucleosides after hydrolysis of uniformly [^{14}C]-labelled spores, that RNA does not increase during the early stages of germination, but rather decreases after 16 hours of germination. No variation of base composition was detected. For *P. graminis* f. sp. *tritici*, Kim and Rohringer (1974) found that the amount of extractable RNA decreased as germination progressed. The decrease in extractable RNA (up to 40%) occurred in both nondifferentiated and differentiated sporelings. A more detailed study of RNA metabolism was done by Chakravorty and Shaw (1972). After [^{32}P]-labelling of flax spores during six hours of hydration, they found that 28S, 18S and 5S ribosomal RNA was synthesized in the absence of the host. The newly transcribed rRNA was associated with mature ribosomal subunits and with monomeric ribosomes. Similar results were reported for *U. phaseoli* by Stallknecht and Mirocha (1971). Synthesis of ribosomal as well as soluble RNA after incorporation of [^{14}C]O$_2$ was first observed one and a half hours after the onset of germination and was highest at six hours.

In a gel electrophoretic study of RNA from germinating flax rust uredospores, Quick (1973) observed a progressive breakdown of rRNA and the formation of increasing amounts of low molecular weight RNA migrating between 5S and rRNA. It appeared to be a naturally occurring degradation product accompanying germination and not an extraction artefact.

Unexpectedly, only 5S rRNA and tRNA synthesis were observed within two to four hours of germination in *P. graminis* f. sp. *tritici* (Kim and Rohringer, 1974).

By using polyacrylamide electrophoresis. Tückhardt (1976) found in a systematic study of different stages of germination of *U. phaseoli* on water that rRNA is synthesized during all stages of germination up to 12 hours.

A degradation of 25S rRNA, however, was observed beginning at four hours of germination. This coincided with the appearance of a $1 \cdot 1 \times 10^6$ Dalton fraction migrating between the 25S and 18S rRNA. As shown by pulse and pulse-chase experiments with [^{32}P], this newly detectable fraction was not radioactive. It is evidently a degradation product of a pre-existing rRNA of the resting spore and not of the newly synthesized rRNA. Auto-

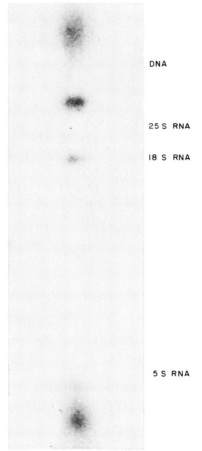

Fig. 1. Autoradiogram of ^{32}P-labelled nucleic acids separated in a polyacrylamide gel by electrophoresis. The nucleic acids were isolated from uredospores of U. phaseoli 60 minutes after commencement of germination and 30 minutes after application of ^{32}P to the medium.

radiograms showed that 5S rRNA and tRNA were also synthesized during the entire process of germination (Fig. 1).

Electrophoretic patterns of soluble RNA revealed a much larger number of tRNA bands during the first two hours of germination as compared with later stages. In spores which had been germinated for two hours or longer, RNA possessing high specific activity, possibly mRNA, was detected between the 5S and the high molecular weight rRNA.

The presence of polyribosomes in bean rust uredospores (Staples *et al.*, 1968, 1971) indicates that mRNA also occurs in this material. RNA fractions exhibiting properties typical of mRNA (MAK column chromatography and gel electrophoresis, AMP content) have been extracted and partially characterized (Ramakrishnan and Staples, 1970; Staples *et al.*, 1971). This RNA sedimented between 4S and 19S and stimulated the incorporation of amino acids into polypeptides in a cell-free protein-synthesizing-system from *E. coli*.

Since germinating rust spores are able to synthesize RNA, they should also contain the appropriate enzymes, i.e. DNA-dependent RNA polymerases. Indeed, Manocha (1973) was able to isolate by column chromatography an enzyme which was inhibited by rifampicin, an inhibitor of bacterial and mitochondrial polymerases.

Tückhardt (1976) isolated two DNA-dependent RNA polymerases from *U. phaseoli* which were insensitive to rifampicin and could be characterized, using α-amantin, as RNA polymerases I and II. Both enzymes were present in resting spores and increased slightly in activity up to nine hours of germination.

In summary, although there is no net synthesis of RNA and DNA, these two types of macromolecules are synthesized during germination and are evidently turned over at comparable rates. The synthesis of all types of nucleic acids begins very early in uredospores of *U. phaseoli* after the onset of germination (Fig. 1), similar to the synthesis of these macromolecules in other fungi (Brambl *et al.*, 1978).

Whether the observed degradation of rRNA and the absence of detectable nuclear DNA synthesis are peculiar to the species tested, or these are properties typical of biotrophic fungi in general are questions requiring further research. The results of experiments using antimetabolites to study the occurrence of DNA and RNA synthesis during sporeling development have to be interpreted with caution. Such inhibitors can have additional effects on other biosynthetic pathways or may not penetrate into the cell.

C. Protein Metabolism

Since the earlier studies of Shu *et al.* (1954), Staples *et al.* (1962), and

Reisener and Jäger (1967), who claimed that rust uredospores germinate without a net synthesis of proteins, much work has been centered on protein metabolism. A study of the effect of antimetabolites on germination and differentiating uredospores of *P. graminis* f. sp. *tritici* was carried out by Dunkle and Allen (1969). Inhibitors of protein synthesis, such as puromycin (100 μg/ml) and *p*-fluoro-phenylalanine (100 μg/ml), had no influence on spore germination, but did prevent the formation of infection structures if present after the inductive phase (heat treatment). Cycloheximide (10 μg/ml), however, inhibited germination as well as differentiation. Hess *et al.* (1976) showed that cycloheximide (50 μg/ml) not only inhibited germination of *P. graminis* f. sp. *tritici* completely, but also prevented the incorporation of [^{14}C]-leucine into acid-insoluble protein. In the absence of an inhibitor, [^{14}C]-leucine incorporation into protein could be detected within the first ten minutes of incubation. Based on the results obtained with cycloheximide, which seems to inhibit protein synthesis more efficiently than inhibitors such as puromycin (see also Brambl *et al.*, 1978), it became apparent that protein synthesis is necessary for spore germination.

The above mentioned lack of net synthesis in germinating rust uredospores was reinvestigated in detail by Trocha and Daly (1970). Working with uniformly [^{14}C]-labelled spores of *U. phaseoli*, they first removed soluble components and then determined the content of labelled amino acids after proteolytic hydrolysis. By these means they were able to show that protein synthesis occurred and that protein content increased 20–25% during the first three to five hours of germination, but decreased after 16 to 24 hours. Although their results clearly show that rust spores have a functional protein-synthesizing system, the rate of synthesis seems low compared to that found in saprophytic fungi such as Botryodiplodia and Fusarium, whose protein content increased five to tenfold during the first few hours of germination (van Etten, 1968; Cochrane *et al.*, 1971). To find an explanation for the low synthetic rate occurring in rusts, Staples and co-workers have studied the synthetic capacity in cell-free systems of the bean rust. Although the ribosome content did not decline during germination (Staples *et al.*, 1971), ribosomes from germ-tubes of *U. phaseoli* rapidly lost the capacity to incorporate amino acids after about six hours (Staples *et al.*, 1971) when dependent on endogenous messengers. Furthermore, spores which had germinated for 20 hours incorporated much less [^{14}C]-leucine into ribosomal proteins compared to freshly hydrated spores (by a ratio of 1:25) (Staples *et al.*, 1972).

One of the most obvious biochemical changes observed in germinating bean rust uredospores was the decline in ribosomal transferase (EF-1) activity (Staples *et al.*, 1971). An essay of ribosomes for polyuridylic acid-directed binding and EF-1-dependent binding of phenylalanyl-tRNA showed that ribosomes from germinated spores had one-fifth the activity of those from hydrated spores.

The above-mentioned authors further demonstrated through reassociation experiments with ribosomal subunits that reassociated ribosomes of germinating spores (20 hours) were 6% more active in the phenylalanyl-tRNA assay than ribosomes of hydrated spores. Hybrid ribosomes were 26% as active as ribosomes from hydrated spores. A similar decline in activity was observed for transferase (EF-1). This indicated that the ribosomes were damaged during the ageing process. Ribosomal ageing could be shown directly by estimating the ratio of ribosomal protein to RNA, which declined from 0·48 for hydrated to 0·22 for germinated (20 hours) spores. A direct examination of ribosomal protein by electrophoresis corroborated these results; specific proteins were lost after spores were germinated, while one band from each subunit appeared to increase.

That ribosomal ageing is not typical for other rust species was shown by Staples and Yaniv (1973). A comparison of the uredospores of *U. phaseoli*, *P. helianthi*, *P. graminis* f. sp. *tritici* and *M. lini*, and the aeciospores of *C. fusiforme* showed that the sedimentation constants and buoyant densities of monoribosomes were the same for all species. But it was also found that the amount of monosomes of *U. phaseoli* and *P. helianthi* recoverable after gradient centrifugation declined drastically during germination, whereas those of *P. graminis* f. sp. *tritici*, *M. lini* and *C. fusiforme* were fully recoverable after 20 hours of germination.

Similar differences were observed for EF-1-dependent binding of the phenylalanyl-tRNA to polyuridylic acid *in vitro*: the binding activity in extracts of bean rust and sunflower rust declined during a 20-hour germination period but extracts of sporelings of the other species exhibited constant activity. Whether these properties can be related to differences in the capability of axenic growth will have to be investigated.

Earlier studies of Staples and co-workers emphasized changes in the activity of ribosomes from uredospores germinated on water (Staples and Yaniv, 1973; Staples *et al.*, 1968, 1972). In more recent work on changes occurring after the induction of infection structures (Yaniv and Staples, 1974), it was shown that spores germinated on collodion membranes with oil always contained a larger amount (22%) of polysomes than spores germinated on water (7%). Moreover, spores germinated on membranes incorporated, in the absence of polyuridylic acid (i.e. utilizing their endogenous messenger), three times more [^{14}C]-phenylalanyl-tRNA than ribosomes from spores germinating on water. This suggested that the formation of infection structures is accompanied by a stimulation of protein synthesis.

In a recent study, Backhaus (1977) analysed the synthesis and turnover of individual proteins (*in vivo*) using two-dimensional separation techniques

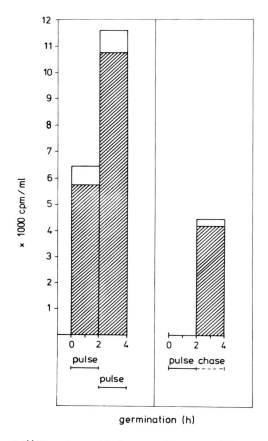

Fig. 2. Uptake of [^{14}C]-amino acids into uredospores of *P. graminis* f. sp. *tritici* during germination. Column height indicates the activity in crude extracts. The non-hatched portions indicate the amount of radioactivity incorporated into polypeptides larger than 10 000 Daltons.

(isoelectric focusing and polyacrylamide electrophoresis) after pulse and pulse-chase labelling with a [^{14}C]-amino acid mixture. Although the number of separated proteins as well as the amount of soluble acidic proteins of *P. graminis* f. sp. *tritici* (70 spots) remained unchanged up to 12 hours of germination, incorporation of radioactivity into these proteins showed pronounced differences. Incorporation began within the first two hours of germination and continued up to 12 hours. The number of labelled proteins after pulse-labelling increased up to seven hours and then declined. At the maximum point, 50% of the total number of proteins were labelled, whereas after 12 hours, only 25% were radioactive. These results agree well with those of Trocha and Daly (1970).

A pulse-chase experiment demonstrated that there is a considerable turn-over of newly synthesized proteins. Within a two-hour chase period, the number of labelled proteins declined by half, indicating that degradation had occurred (Fig. 2). At no time did synthesis appear to be limited by a lack of amino-acid precursors. Ninety to ninety-five per cent of the radioactivity taken up by the spore or germ-tube was found in the fraction containing molecules with a molecular weight of less than 10 000. The pulse-chase experiment also showed that most of the low-molecular weight fraction is lost during the chase period, probably by leaching (Fig. 2).

These results are in agreement with earlier findings of a low rate of protein synthesis in rusts, and they show that proteolysis may play an important role during germ-tube growth.

Despite the differing experimental approaches used in studies of various rust species, present knowledge of nucleic acid and protein metabolism can be tentatively summarized as follows:

(1) Germinating uredospores of rust fungi are able to synthesize both nucleic acids and proteins.

(2) The onset of synthesis occurs after hydration and after the removal of germination inhibitors.

(3) In contrast to saprophytic fungi, there is little or no net synthesis of nucleic acids or proteins.

(4) The minimal net synthesis could reflect a low synthetic rate, or it could reflect the occurrence of degradative processes, i.e. a high turnover rate. Whether or not the inhibition of polypeptide synthesis found in *in vitro* systems also occurs in spores germinating on the host surface is a question requiring further detailed study.

(5) The formation of infection structures (characteristically with nuclear division) appears to be accompanied by a higher synthetic activity. The formation of infection structures and DNA synthesis associated with nuclear division is more sensitive to antimetabolites than they are during germ-tube growth. This suggests that the formation of infection structures is a more complex process requiring the synthesis of additional macromolecules.

VIII. Conclusions

Although there are many studies on composition and metabolism of dor-mant and germinating uredospores, it is difficult to give a comprehensive review of this field because of the use of a wide range of rusts and different methods used in these studies. Nevertheless, it is possible to attempt a critical evaluation which might be useful in suggesting further worthwhile research in this field. Generally, it can be said that dormant and germinating uredospores of rusts probably have the same metabolic potential as spores and sporelings of saprophytic fungi, except for a much lower metabolic rate. This quantitative difference could be explained by assuming that the germinating rust spore utilizes the endogenous reserve substances at a slower rate to allow for the time interval between germination on the surface of the host plant and the first opportunity to take up exogenous substrates inside the host.

A second characteristic feature of rust uredospores and uredosporelings is the ease with which they lose substances of low and high molecular weight into the germination medium. Since the same phenomenon was also observed with axenic rust cultures (see the chapter of Maclean), one could suggest that the cell membrane of the rust has different properties from those of saprophytic fungi. At any rate, such an hypothesis could set up interesting experimental approaches using modern methods of cell biology. In addition to research on membrane lipids, this suggests that membrane-bound proteins should receive more attention. Furthermore, membrane properties may differ in sporelings depending on their stage of differentiation into appressorium and fungal structures inside the host. In recent years methods have been developed to induce at will differentiation of infection structures of several rusts. This provides an opportunity for further detailed studies of mechanisms involved in differentiation. However, it should be kept in mind that results obtained *in vitro* do not necessarily reflect the situation of the differentiating sporeling on the surface of the host, especially if one considers that specific physical and chemical stimuli originating from the host have been shown to control differentiation of the fungus. Earlier research was largely limited to a static approach in the analysis of fungal metabolism, to the determination of metabolite content, synthetic capacity and metabolic pathways. This is the basis for further work in which the dynamic interaction between synthesis and degradation and between environment and differentiation can be studied more fully.

References

Allen, P. J. (1955). *Phytopathology* **45**, 259–266.
Allen, P. J. (1957). *Pl. Physiol.* **32**, 385–389.
Allen, P. J. (1965). *Ann. Rev. Phytopath.* **3**, 313–342.
Allen, P. J. (1972). *Proc. Natl. Acad. Sci. U.S.* **69**, 3497–3500.

Allen, P. J. (1976). *In* "Encyclopedia of Plant Physiology" new series, 4, (R. Heitefuss and P. H. Williams, eds), pp. 51–85. Springer-Verlag, Berlin.

Allen, P. J. and Dunkle, L. D. (1971). *In* "Morphological and Biochemical Events in Plant-parasite Interaction" (S. Akai and S. Ouchi, eds), pp. 23–58. The Phytopathol. Soc. of Japan, Tokyo.

Backhaus (1977). Dissertation, University Göttingen.

Bansal, S. K. and Knoche, H. W. (1981). *Phytochemistry* **20**, 1269–1277.

Bartnicky-Garcia, S. (1968). *Ann. Rev. Microbiol.* **22**, 87–108.

Brambl, R., Dunkle, L. D. and van Etten, J. L. (1978). *In* "The Filamentous Fungi" 3, "Developmental Mycology" (J. E. Smith and D. R. Berry, eds), pp. 94–118. Edward Arnold, London.

Brennan, P. J. and Lösel, D. M. (1978). *Advances in Microbiol. Physiol.* **17**, 48–179.

Callow, J. A. (1978). *Adv. Bot. Res.* **4**, 1–49.

Caltrider, P. G. and Gottlieb, D. (1963). *Phytopathology* **53**, 1021–1030.

Caltrider, P. G., Ramanchandran, S. and Gottlieb, D. (1963). *Phytopathology* **53**, 86–92.

Chakravorty, A. K. and Shaw, M. (1972). *Physiol. Pl. Path.* **2**, 1–6.

Cochrane, J. C., Rado, T. A. and Cochrane, V. W. (1971). *J. Gen. Microbiol.* **65**, 45–55.

Daly, J. M., Knoche, H. W. and Weise, M. V. (1967). *Pl. Physiol.* **42**, 1633–1642.

Dickinson, S. (1970). *Phytopath. Z.* **69**, 115–124.

Dickinson, S. (1971). *Phytopath. Z.* **70**, 62–70.

Dickinson, S. (1972). *Phytopath. Z.* **73**, 347–358.

Dunkle, L. D. and Allen, P. J. (1969). *Science* **163**, 481–482.

Elnaghy, M. A. and Heitefuss, R. (1976). *Physiol. Pl. Path.* **8**, 253–267.

Eppstein, D. A. and Tainter, F. H. (1976). *Phytopathology* **66**, 1395–1397.

Eppstein, D. A. and Tainter, F. H. (1979). *Appl. Environm. Microbiol.* **37**, 143–147.

van Etten, J. L. (1968). *Arch. Biochem. Biophys.* **125**, 13–21.

Farkas, G. L. and Ledingham, G. A. (1959). *Can. J. Microbiol.* **5**, 141–151.

Foudin, A. S. and Macko, V. (1974). *Phytopathology* **64**, 990–993.

Frear, D. S. and Johnson, M. A. (1961). *Biochem. Biophys. Acta* **47**, 419–421.

French, R. C. (1961). *Bot. Gaz.* **122**, 194–198.

French, R. C. and Gallimore, M. D. (1971). *J. Agric. Food Chem.* **19**, 912–915.

French, R. C. and Gallimore, M. D. (1972). *Phytopathology* **62**, 116–119.

French, R. C. and Weintraub, R. L. (1957). *Arch. Biochem. Biophys.* **72**, 235–237.

French, R. C., Gale, A. W., Graham, C. L., Latterell, F. M., Schmitt, C. G., Marchetti, M. A. and Rines, H. W. (1975a). *J. Agric. Food Chem.* **23**, 766–770.

French, R. C., Gale, A. W., Graham, C. L. and Rines, H. W. (1975b). *J. Agric. Food Chem.* **23**, 4–7.

French, R. C., Graham, C. L., Gale, A. W. and Long, R. K. (1977). *J. Agric. Food Chem.* **25**, 84–88.

Gottlieb, D. (1976). *In* "The Fungal Spore: Form and Function" (D. J. Weber and W. M. Hess, eds), pp. 141–163. John Wiley, New York.

Gottlieb, D. and Caltrider, P. G. (1962). *Phytopathology* **52**, 11.

Grambow, H. J. and Grambow, G. E. (1979). *Z. Pflanzenphysiol.* **90**, 1–10.

Grambow, H. J. and Riedel, D. (1977. *Physiol. Pl. Path.* **11**, 213–224.

Hess, S. R., Allen, P. J., Nelson, D. and Lester, H. (1975). *Physiol. Pl. Path.* **5**, 107–112.

Hess, S. R., Allen, P. J. and Lester, H. (1976). *Physiol. Pl. Path.* **9**, 265–272.

Hoppe, H. K. and Heitefuss, R. (1975). *Physiol. Pl. Path.* **5**, 273.

Hougen, F. W., Craig, B. M. and Ledingham, G. A. (1958). *Can. J. Microbiol.* **4**, 521.
Hoyer, H. (1962). *Zentbl. Bakt. Parasitkde. Abstr.* II **115**, 266–296.
Irvine, G. M., Golubchuk, M. and Anderson, J. A. (1954). *Can. J. Agric. Sci.* **34**, 234–239.
Jackson, L. L. and Frear, D. S. (1968). *Phytochemistry* **7**, 651.
Jackson, L. L., Dobbs, L., Hildebrandt, A. and Yokiel, R. A. (1973). *Phytochemistry* **12**, 2233–2237.
Joppien, S. (1975). Dissertation, University Aachen.
Joppien, S., Burger, A. and Reisener, H. J. (1972). *Arch. Microbiol.* **82**, 337–352.
Kim, W. K. and Rohringer, R. (1974). *Can. J. Bot.* **52**, 1309–1317.
Knoche, H. W. and Horner, T. L. (1970). *Pl. Physiol.* **46**, 401.
Kottke, H. J. (1976). Dissertation, Technische Universität Aachen.
Langenbach, R. J. and Knoche, H. W. (1971a). *Pl. Physiol.* **48**, 728–734.
Langenbach, R. J. and Knoche, H. W. (1971b). *Pl. Physiol.* **48**, 735–739.
Le Roux, P. M. and Dickson, J. G. (1957). *Phytopathology* **47**, 101–107.
Lin, H. K. and Knoche, H. W. (1974). *Phytochemistry* **13**, 1795.
Lin, H. K., Langenbach, R. J. and Knoche, H. W. (1972). *Phytochemistry* **11**, 2319.
Macko, V. and Staples, R. C. (1973). *Bull. Torrey bot. Club* **100**, 223–229.
Macko, V., Staples, R. C., Gershon, H. and Renwick, J. A. A. (1970). *Science* **170**, 539–540.
Macko, V., Staples, R. C., Allen, P. J. and Renwick, J. A. A. (1971a). *Science* **173**, 835–836.
Macko, V., Staples, R. C. and Renwick, J. A. A. (1971b). *Phytopathology* **61**, 902.
Macko, V., Staples, R. C., Renwick, J. A. A. and Pirone, J. (1972). *Physiol. Pl. Path.* **2**, 347–355.
Macko, V., Staples, R. C., Yaniv, Z. and Granados, R. R. (1976). *In* "The Fungal Spore: Form and Function" (D. J. Weber, and W. M. Hesse, eds), pp. 73–100. John Wiley, New York.
Macko, V., Trione, E. J. and Young, S. A. (1977). *Phytopathology* **67**, 1473–1474.
Macko, V., Renwick, J. A. A. and Rissler, J. F. (1978). *Science* **199**, 442–443.
Maheshwari, R. and Sussman, A. S. (1970). *Phytopathology* **60**, 1357–1364.
Maheshwari, R., Allen, P. J. and Hildebrandt, A. C. (1967). *Phytopathology* **57**, 855–862.
Manocha, M. S. (1973). *Can. J. Microbiol.* **19**, 1175–1177.
Marte, M. (1971). *Phytopathol. Z.* **72**, 335–343.
Miller, W. L., Kalafer, M. E., Gaylor, J. L. and Delwiche, C. V. (1967). *Biochemistry*, New York **6**, 2673.
Musumeci, M. R., Moraes, W. B. C. and Staples, R. C. (1974). *Phytopathology* **64**, 71–73.
Naito, N., Yani, T. and Sato, Y. (1959). *Ann. Phytopath. Soc. Japan* **24**, 234–238.
Nowak, R., Kim, W. J. and Rohringer, R. (1972). *Can. J. Bot.* **50**, 185.
Pritchard, N. J. and Bell, A. A. (1967). *Phytopathology* **57**, 932–934.
Quick, W. A. (1973). *Physiol. Pl. Path.* **3**, 419–430.
Ramakrishnan, L. and Staples, R. C. (1970). *Contrib. Boyce Thompson Inst.* **24**, 197–202.
Reisener, H. J. (1976). *In* "The Fungal Spore: Form and Function" (D. J. Weber and W. M. Hess, eds), pp. 165–184. John Wiley, New York.
Reisener, H. J. and Jäger, K. (1967). *Planta* (Berl.) **72**, 265–283.

Reisener, H. J., McConnell, W. B. and Ledingham, G. A. (1961). *Can. J. Biochem. Physiol.* **39**, 1559–1566.

Reisener, H. J., Finlayson, A. J. and McConnell, W. B. (1963a). *Can. J. Biochem. Physiol.* **41**, 1–7.

Reisener, H. J., Finlayson, A. J. and McConnell, W. B. (1963b). *Can. J. Biochem. Physiol.* **41**, 737–743.

Rines, H. W., French, R. C. and Daasch, L. W. (1974). *J. Agric. Food Chem.* **22**, 96–100.

Shaw, M. (1963). *Ann. Rev. Phytopath.* **1**, 259–294.

Shaw, M. (1964). *Phytopath. Z.* **50**, 159–180.

Shaw, M. and Samborski, D. J. (1957). *Can. J. Bot.* **35**, 389–407.

Shu, P. and Ledingham, G. A. (1956). *Can. J. Microbiol.* **2**, 489–495.

Shu, P., Tanner, K. G. and Ledingham, G. A. (1954). *Can. J. Bot.* **32**, 16–23.

Stahmann, M. A., Musumeci, M. R. and Moraes, W. B. C. (1976). *Phytopathology* **66**, 765–769.

Stallknecht, G. F. and Mirocha, C. J. (1971). *Phytopathology* **61**, 400–405.

Staples, R. C. (1957). *Contrib. Boyce Thompson Inst.* **19**, 19–31.

Staples, R. C. (1962). *Contrib. Boyce Thompson Inst.* **21**, 487–497.

Staples, R. C. (1974). *Physiol. Pl. Path.* **4**, 415–424.

Staples, R. C. and Macko, V. (1980). *Experim. Mycol.* **4**, 2–16.

Staples, R. C. and Syamananda, R. (1962). *Phytopathology* **52**, 28–29.

Staples, R. C. and Wynn, W. K. (1965). *Bot. Rev.* **31**, 537–564.

Staples, R. C. and Yaniv, Z. (1973). *Physiol. Pl. Pathol.* **3**, 137–145.

Staples, R. C. and Yaniv, Z. (1976). *In* "Encyclopedia of Plant Physiology" new series, 4, (R. Heitefuss and P. H. Williams, eds), pp. 86–103. Springer-Verlag, Berlin.

Staples, R. C. and Yaniv, Z. (1978). *Exp. Mycology* **2**, 290–294.

Staples, R. C., Syamananda, R., Kao, V. and Block, R. J. (1962). *Contrib. Boyce Thompson Inst.* **21**, 345–362.

Staples, R. C., Bedigian, D. and Williams, P. H. (1968). *Phytopathology* **58**, 151–154.

Staples, R. C., Yaniv, V., Ramakrishnan, L. and Lipetz, J. (1971). *In* "Morphological and Biochemical Events in Plant-Parasite Interaction" (S. Akai and S. Ouchi, eds), pp. 59–90. The Phytopath. Soc. of Japan, Tokyo.

Staples, R. C., Yaniv, Z. and Bushnell, W. R. (1972). *Physiol. Pl. Path.* **2**, 27–35.

Staples, R. C., App, A. A. and Ricci, P. (1975). *Arch. Microbiol.* **104**, 123–127.

Van Sumere, C. F., Van Sumere de Preter, C., Vining, L. C. and Ledingham, G. A. (1957). *Can. J. Microbiol.* **32**, 847–862.

Trocha, P. and Daly, J. M. (1970). *Pl. Physiol.* **46**, 520–526.

Tückhardt, K. E. (1976). Dissertation, University Göttingen, pp. 84.

Tulloch, A. P. (1960). *Can. J. Chem.* **38**, 204–207.

Tulloch, A. P. (1964). *Can. J. Microbiol.* **10**, 359.

Tulloch, A. P. and Ledingham, G. A. (1960). *Can. J. Microbiol.* **6**, 425–434.

Tulloch, A. P. and Ledingham, G. A. (1962). *Can. J. Microbiol.* **8**, 379–387.

Tulloch, A. P. and Ledingham, G. A. (1964). *Can. J. Microbiol.* **10**, 351–358.

Tulloch, A. P., Craig, B. M. and Ledingham, G. A. (1959). *Can. J. Microbiol.* **5**, 485–491.

Weete, J. D. (1973). *Phytochemistry* **12**, 1843.

Weete, J. D. and Laseter, J. L. (1974). *Lipids* **9**, 575–581.

White, G. A. and Ledingham, G. A. (1961). *Can. J. Bot.* **39**, 1131–1148.

Williams, P. G. and Ledingham, G. A. (1964). *Can. J. Bot.* **42**, 497–505.
Williams, P. G., Scott, K. J. and Kuhl, J. L. (1966). *Phytopathology* **56**, 1418–1419.
Wilson, E. M. (1958). *Phytopathology* **48**, 595–600.
Wynn, W. K. (1976). *Phytopathology* **66**, 136–146.
Yaniv, Z. and Staples, R. C. (1974). *Phytopathology* **64**, 1111–1114
Yaniv, Z. and Staples, R. C. (1978). *Exp. Mycology* **2**, 279–289.

5. Biochemistry of Host Rust Interactions

A. K. CHAKRAVORTY (PART A) and K. J. SCOTT (PART B)

Department of Biochemistry, University of Queensland, St. Lucia, Australia

Part A: Primary Metabolism: Changes in the Gene Expression of Host Plants During the Early Stages of Rust Infection

CONTENTS

I. Introduction

The rust fungi, like many other obligate parasites, exhibit a striking degree of host specificity. Each species of rust fungus has several physiological races that are distinguishable by their virulence or avirulence toward different cultivars of host plants. These cultivars of host plants belong to the same species and are usually distinguishable only by their resistance or susceptibility to various races of the rust fungi.

The growth and development of the rust fungi on their hosts are controlled by genes for resistance in the host and genes for virulence in the pathogen (Flor, 1971). The concept of "gene-for-gene relationship" has been of great value in explaining the host specificity of phytopathogenic fungi at the level of classical genetics (see McIntosh and Watson, Chapter 3, this volume). However, the molecular basis of host parasite interactions that lead to either resistance or susceptibility of different cultivars of host plants to different physiological races of the rust fungi remains largely unknown.

During rust infection, there are major changes in the intermediary metabolism of the host (Scott, 1972; also see Part B of this chapter). In many instances, these metabolic transitions have been correlated with the appearance of enzymes and isozymes whose kinetic and catalytic properties are significantly different from those of the enzymes from healthy (uninoculated) plants (reviewed by Chakravorty and Shaw, 1977a). When a cultivar of host is inoculated with spores of a given race of rust fungus, there are biochemically detectable changes that indicate alteration in the patterns of gene expression in the host (Chakravorty and Shaw, 1977b). Thus, changes at the level of gene expression at a very early stage of host parasite interactions may play an important role in determining whether a host cultivar is susceptible or resistant to a race of the pathogen. While the host plant is genetically predisposed to be either susceptible or resistant to the pathogen, biochemical characterization of these changes in gene expression and of the controlling factors are extremely useful in understanding the mechanism of disease resistance at the molecular level.

Studies on biochemical changes in resistant and susceptible cultivars of host plants following inoculation with rust fungi have suggested the operation of metabolic control chains as summarized in Fig. 1. According to this scheme, host parasite interactions trigger changes in the patterns of gene expression in the host plant at a very early stage. The direction and magnitude of these changes determine whether the host will be susceptible or resistant to the pathogen. The basic tenets of this concept are that (1) the expression of disease resistance and susceptibility involves a series of metabolic events rather than a single biochemical step and (2) at the metabolic level, both resistance and susceptibility are active processes.

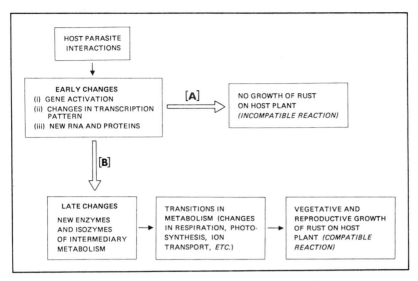

Fig. 1. Regulation of disease development through metabolic control chains. According to this scheme, during the initial stages of host rust interactions there are changes in the transcription patterns in host tissues. The direction and magnitude of these changes determine whether the outcome of these interactions will be [A] resistance or [B] susceptibility of the host to a given race of the rust fungus. Implicit in the scheme is the notion that gene-for-gene interactions elicit a particular metabolic control chain in the host plant that leads to either compatible or incompatible reaction (Chakravorty and Shaw, 1977a). The magnitude of early changes can determine the whole spectrum of intermediate reactions.

In the differentiated tissues of higher plants, primary metabolism can be operationally defined as gene expression. The major steps in gene expression and the controlling factors are depicted in Fig. 2. These steps include transcription (the synthesis of messenger RNA precursors by RNA polymerase using DNA in chromatin as template), posttranscriptional processing of mRNA, translation (protein synthesis by ribosomes using mRNA as template) and, in some cases, posttranslational modifications of newly synthesized proteins. In this chapter, we discuss changes in primary and intermediary metabolism during host rust interactions in two separate parts, for the sake of convenience. The main objective of this part is to consider recent advances in our knowledge concerning alterations in transcription patterns in host plants during the early stages of rust infection, with particular emphasis on the mechanisms that are known to control gene expression in higher organisms.

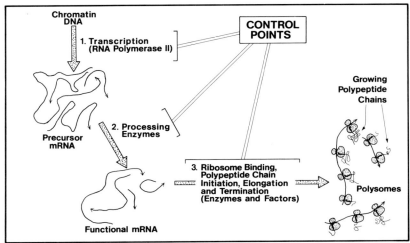

Fig. 2. Control of gene expression at the levels of transcription, posttranscriptional processing and translation. The structural genes (polypeptide specifying sequences) in chromatin are transcribed by RNA polymerase II into messenger RNA precursors. Major control elements at this step include chromosomal regulatory proteins which modulate the template activity of chromatin DNA and regulatory subunits of RNA polymerase II that determine transcriptive specificity of the enzyme. The messenger RNA precursors undergo posttranscriptional processing which includes sequential cleavage reactions and 5'PO$_4$ as well as 3'OH end modification, to become functional (translatable) mRNA. The arrow in each precursor and functional mRNA indicates the 3'OH end of the molecule. The translation of mRNA is controlled by several protein factors. The diagram shows three different size classes of mRNA in three discrete size classes of polysomes.

II. Studies on *in vivo* RNA Synthesis

In higher plants, transitions in metabolism that lead to rapid growth and differentiation are preceded and accompanied by the synthesis of new types of RNA and proteins. During the early and later stages of rust infection of plants, many major changes in metabolism occur for which new RNA and protein molecules are required. The possibility that there may be changes in RNA metabolism during the early stages of rust infection has been explored by workers in several laboratories using cytospectrophotometric, ultrastructural, autoradiographic as well as biochemical techniques (Heitefuss, 1966; Chakravorty and Shaw, 1977b). A wealth of information on changes in RNA metabolism has been obtained from these investigations. Cytospectrophotometric studies (Bhattacharya *et al.*, 1965; Shaw, 1967) revealed changes in the nuclear activities of a susceptible cultivar of wheat (*Triticum aestivum* L. var. Little Club) inoculated with race 15B of *Puccinia graminis tritici*. In the nuclei of host cells in the proximity of developing fungal mycelium, an appreciable increase in the content of RNA and nonhistone proteins has been detected at 48 hours after inoculation with rust uredo-spores. This observation has provided direct evidence for parasitically induced changes in the primary metabolism of host cells.

It has now been well documented that infection of plants by the rust fungi is accompanied by an increase in the rate of RNA synthesis (Rohringer and Heitefuss, 1961; Wolf, 1967; Chakravorty and Shaw, 1971; Tani *et al.*, 1971, 1973) and in the accumulation of RNA in the inoculated host tissues (Bhat-tacharya *et al.*, 1965; Tani *et al.*, 1971, 1973). That RNA synthesis may be a prerequisite for disease development has been suggested by a stage specific inhibition of the infection process by low concentrations of a transcription inhibitor, actinomycin D (Shaw, 1967). This antimetabolite inhibits the infection of a genetically susceptible cultivar of flax (*Linum usitatissimum*, var. Bison) by flax rust (*Melampsora line*, race 3) only if it is administered no later than five to six hours after inoculation.

Tani *et al.* (1971, 1973) used MAK (methylated albumin coated Kiesel-guhr) chromatography and gel electrophoresis to fractionate [^{32}P]-labelled RNA from susceptible as well as resistant cultivars of oat (*Avena sativa* L., vars. Victoria and Shokan 1, respectively) inoculated with crown rust (*P. coronata*, race 226). At various times after inoculation, they observed a significant increase in the incorporation of [^{32}P] into transfer RNA (tRNA), ribosomal RNA (rRNA) as well as messenger-like RNA in both susceptible and resistant cultivars of host plants.

The *in vivo* labelling pattern of RNA in a susceptible cultivar of flax (*L. usitatissimum*, var. Bison) inoculated with flax rust (*M. lini*, race 3) has revealed a two- to three-fold increase in the rate of [^{32}P] incorporation into all major classes of RNA fractionated by salt precipitation and sucrose density gradient centrifugation (Chakravorty and Shaw, 1971). This increase in the rate of RNA synthesis is accompanied by a change in the nucleotide composition of a fraction of newly synthesized RNA at 48 hours after inoculation. A comparison of the nucleotide composition of the 1M NaCl soluble RNA fraction from healthy and inoculated flax cotyledons as well as from germinating rust uredospores has revealed the preferential synthesis of one or more molecular species of RNA with a high A + U/G + C ratio at an early stage of rust infection.

Von Broembsen and Hadwiger (1972) used a highly sensitive technique to explore changes in messenger RNA (mRNA) populations associated with the expression of specific resistance genes in flax. They inoculated a number of near-isogenic lines of flax, containing different resistance genes in either the P or M locus, with race 1 of *M. lini* which carries virulence genes against

G

some and avirulence genes against other host lines. The healthy and inoculated (six, 12 and 18 hours after inoculation) seedlings were incubated with [^{3}H]-labelled and [^{14}C]-labelled L-leucine respectively. After incubation, the healthy and inoculated plants were combined and the proteins soluble in a low ionic strength buffer were extracted. These proteins were concentrated by ammonium sulphate precipitation and fractionated by Sephadex G-200 gel filtration. The [^{14}C]/[^{3}H] radioactivity ratios of these fractions were determined to analyse the relative rates of synthesis of various size classes of soluble proteins in healthy and inoculated plants. The results of these experiments have suggested the accelerated synthesis of different size classes of soluble proteins in different gene-for-gene combinations, indicating the selective translation of characteristic size classes of mRNA in each gene-for-gene combination. These changes occur earlier after inoculation and are more pronounced in the genetically resistant lines than in the susceptible near-isogenic lines.

Yamamoto *et al.* (1976) and Tani and Yamamoto (1978) reported that the expression of resistance of oat (*A. sativa*, var. Shokan 1) to an avirulent race of *P. coronata* (race 226) requires the accelerated synthesis of mRNA and proteins which is detectable at 12 hours after inoculation. In leaves inoculated with a virulent race of *P. coronata* (race 203), they found no changes in the rates of [^{3}H] uridine and [^{14}C] leucine incorporation during this period. Tani *et al.* (1973) examined the template activity of the crude RNA preparations from healthy and rust infected oat leaves in a cell-free protein synthesizing system. They found a significant increase in the template activity in the RNA from infected (susceptible) leaves at four days after inoculation. In the RNA from a resistant variety, no difference in template activity could be detected between the RNA from healthy and inoculated leaves, despite an increase in the synthesis of total RNA in the inoculated leaves.

The foregoing discussion on the *in vivo* labelling patterns of RNA and proteins during host rust interactions shows that there are changes in transcription patterns in both resistant and susceptible cultivars of host plants. The time of appearance and the magnitude of these changes are dependent upon the host rust combinations as well as the techniques used to monitor these changes.

III. Studies on *in vitro* Translation of Polysomal Messenger RNA

A. *Changes in Polysomal Messenger RNA Populations During the Early Stages of Rust Infection*

The observation that in various gene-for-gene combinations of flax and *M. lini* characteristic size classes of soluble protein are synthesized *in vivo* at an accelerated rate (von Broembsen and Hadwiger, 1972) suggested significant changes in polysomal mRNA populations at a very early stage of host rust interactions. This possibility has been explored by carrying out experiments on cell-free protein synthesis with polysomes from healthy and rust infected wheat leaves (Pure *et al.*, 1979, 1980). The basis of this method is represented diagramatically in Fig. 3. The technique involves the isolation of polysomes (containing mRNA-bound ribosomes) by sucrose density gradient centrifugation (Dyer and Scott, 1972). The number of ribosomes bound to each mRNA molecule is proportional to the length of mRNA. In Fig. 3, three discrete size classes of polysomes are shown. Each ribosome in a polysome contains a tRNA-bound polypeptide chain which has been initiated *in vivo*. These polysomes are then incubated with the soluble components of the cell-free protein synthesizing system from wheat embryos (the S176 fraction) and radioactive amino acids under optimal conditions (Musk

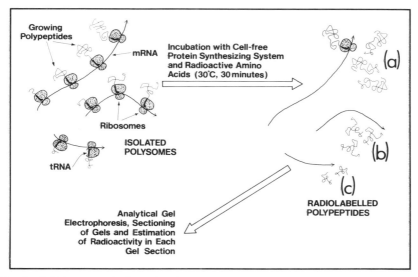

Fig. 3. Cell-free translation of polysomal messenger RNA. Polysomes containing different size classes of mRNA each bound to a varying number of ribosomes are isolated by sucrose density gradient centrifugation. These are incubated in a cell-free protein synthesizing system to allow all *in vivo* initiated polypeptide chains to be elongated to completion with radioactive amino acids (Musk *et al.*, 1979). Since each ribosome translates an mRNA only once and there is no reinitiation of polypeptide chain synthesis (Pure *et al.*, 1979), the size classes of polypeptides radiolabelled *in vitro* are proportional to the size classes of mRNA molecules that specify them. At step 2, (a), (b) and (c) represent three different size classes of polypeptides alongside the mRNA molecules specifying them. The radiolabelled polypeptides are fractionated into different size classes by analytical gel electrophoresis, as shown in Fig. 4.

et al., 1979). Under these conditions, various size classes of polypeptides are radiolabelled *in vitro*. These polypeptides, initiated *in vivo* and elongated *in vitro*, are analysed by polyacrylamide gel electrophoresis in the presence of sodiumdodecyl sulphate. The gels are then sectioned and the radioactivity in each gel slice determined in a liquid scintillation spectrometer.

Musk *et al.* (1979) characterized this cell-free protein synthesizing system reconstituted with polysomes from mature wheat leaves and the S176 fraction from wheat embryos. Their findings are summarized below. The time course of protein synthesis and the observation that the polypeptides radiolabelled *in vitro* are of large size (molecular weight 10 000 to 90 000 daltons) suggest that nearly all of the polypeptide chains initiated *in vivo* are elongated to completion. This conclusion is supported by the finding that in the cell-free protein synthesizing system, about 60% of the radiolabelled polypeptides (10 000 to 90 000 daltons) are released from the ribosomes and are found in the supernatant fraction when the ribosomes are centrifuged down after 30 min of incubation. Aurintricarboxylic acid, a potent inhibitor of polypeptide chain initiation, has little or no effect on wheat leaf polysomal mRNA translation. Considered together, these observations suggest that upon elongation of the *in vivo* initiated polypeptide chains to completion, the ribosomes reinitiate very few, if any, new polypeptide chains. Since each mRNA-bound ribosome translates the mRNA only once, there is a positive correlation between the size classes of polypeptides radiolabelled *in vitro* and the size classes of mRNA that specify them. As shown in Fig. 2, three size classes of polypeptides, marked (a), (b) and (c) respectively, are specified by three discrete size classes of mRNA molecules, the size of the polypeptides being proportional to the length of the mRNA specifying them. Thus, this technique of cell-free translation of polysomal mRNA and analytical gel electrophoresis of polypeptides radiolabelled *in vitro* represents a powerful tool for monitoring changes in polysomal mRNA populations preceding metabolic transitions that are known to occur during host rust interaction.

Pure *et al.* (1979) isolated polysomes from healthy wheat leaves of a susceptible cultivar (*T. vulgare*, var. W2691) and from those inoculated with the stem rust fungus (*P. graminis tritici*, race 126-ANZ-6, 7) at three days after inoculation. These were incubated for cell-free protein synthesis in the presence of [^{3}H]-labelled amino acids. The polypeptides radio-labelled *in vitro* with polysomes from healthy and inoculated leaves respectively were analysed by gel electrophoresis separately and the results (gel section number against radioactivity) were plotted together as shown in Fig. 4, for comparison. The radioactivity profiles in Fig. 4 reveal that polypeptides of a wide range of size classes are synthesized by polysomes from both sources. Interestingly enough, with identical concentrations of polysomes from

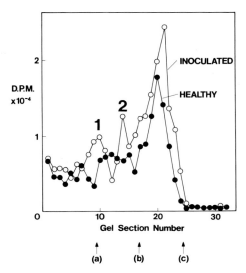

Fig. 4. Polyacrylamide gel electrophoresis of polypeptides synthesized by polysomes from healthy and inoculated wheat leaves. Polysomes from healthy and inoculated leaves (three days after inoculation) were incubated with the soluble components of cell-free protein synthesizing system from wheat embryos and [³H] L-amino acids mixture for 30 min at 30°C. The radiolabelled polypeptides were analysed by polyacrylaminde gel electrophoresis in the presence of sodiumdodecyl sulphate. The vertical arrows indicate the positions of marker proteins as follows: (a) phosphory-lase b (rabbit muscle, molecular weight 94 000), (b) bovine serum albumin (molecular weight 68 000) and (c) cytochrome c (equine heart, molecular weight 12 400). The profiles represent polypeptides synthesized with polysomes from healthy (●) and inoculated (○) leaves respectively. Reproduced from Pure *et al.* (1979).

healthy and inoculated leaves, the relative quantities of various size classes of polypeptides radiolabelled *in vitro* are quite different. The polysomes from inoculated leaves synthesize substantially greater quantities of some size classes of polypeptides than do the polysomes from healthy leaves. These include at least three size classes of radiolabelled polypeptides of approximate molecular weights:

(a) 70 000 to 80 000 (peak marked 1 in Fig. 4),
(b) 50 000 to 60 000 (peak marked 2) and
(c) 10 000 to 30 000 (the major peak).

Since each ribosome in the polysome preparations translates an mRNA only once and there is no reinitiation of new polypeptides *in vitro*, these results suggest that in the polysomes from inoculated leaves there are many size classes of mRNA molecules that outnumber those present in the polysomes from healthy leaves (Pure *et al.*, 1979).

In this cell-free protein synthesizing system, polypeptides radiolabelled with [^{35}S] methionine by polysomes from healthy and inoculated wheat leaves respectively have been electrophoresed and the gels autoradio-graphed (Pure *et al.*, 1980). A comparison of the radioactive polypeptide bands has revealed that there are at least five distinct size classes of polypeptides that are synthesized only by polysomes from inoculated leaves. With the polysomes from healthy leaves, these polypeptides are either not synthesized at all or only weakly radiolabelled, suggesting that the mRNA molecules specifying these polypeptides are either absent in the polysomes from healthy leaves or present in insignificant quantities. These results show a visually observable change in wheat leaf polysomal mRNA populations at three days after inoculation with the rust fungus.

The data presented in Fig. 4 were obtained by electrophoresing the polypeptides radiolabelled under identical conditions by polysomes from healthy and inoculated leaves respectively, in separate gels. The radioac-tivity profiles obtained with polysomes from healthy and inoculated leaves were then superimposed, for comparison. A disadvantage inherent in this method is that the accuracy of measuring differences in the radiolabelling of various size classes of polypeptides by polysomes from healthy and inoculated leaves depends largely on the assumption that the sections in the two gels are exactly corresponding. This problem has been overcome by radiolabelling the polypeptides synthesized by polysomes from healthy and inoculated leaves with different radioisotopes, mixing the polypeptides synthesized by polysomes from these two sources and electrophoresing them in the same gel.

The polysomes from healthy and inoculated leaves were incubated under identical conditions but separately in a cell-free protein synthesizing system with [^{3}H] methionine (denoted [^{3}H]-H) and [^{35}S] methionine ([^{35}S]-I) respectively. After incubation, the [^{3}H]- and [^{35}S]-labelled polypeptides were combined and subjected to polyacrylamide gel electrophoresis in the same gel. The ratios of [^{35}S]-I/[^{3}H]-H in the gel sections were then plotted against gel section number. The control profile was obtained by similarly plotting [^{35}S]-I/[^{3}H]-I ratios in different gel sections. The data (Pure *et al.*, 1980) have shown that in the control gel, the ratios of [^{35}S]-I/[^{3}H]-I in all gel sections are close to one. In the gel sections containing [^{35}S]-labelled polypeptides synthesized by polysomes from inoculated leaves and [^{3}H]-labelled polypeptides synthesized by those from healthy leaves, the corres-ponding [^{35}S]-I/[^{3}H]-H ratios are significantly different from different size classes. There are five discrete size classes of polypeptides that show far greater [^{35}S] methionine incorporation than the corresponding size classes labelled by [³H] methionine. These size classes of polypeptides include the additional bands visualized in the autoradiographs of polypeptides radiolabelled by polysomes from inoculated leaves at three days after inoculation (Pure *et al.*, 1980).

These data conclusively demonstrate significant changes in wheat leaf polysomal mRNA populations at a relatively early stage of rust infection. This is consistent with earlier observations that there are dramatic changes in the transcription pattern of higher plants during the initial stages of rust infection (Chakravorty and Shaw, 1971; Tani *et al.*, 1973).

B. Chloroplast or Cytoplasmic Messenger RNA?

The polysomes isolated from wheat leaves contain two major classes of mRNA: those of chloroplast and the others of cytoplasmic origin. The chloroplast mRNA molecules are transcribed from the chloroplast genome and are translated by 70S ribosomes (Sager and Hamilton, 1967; Svetailo *et al.*, 1967). The cytoplasmic mRNA molecules are transcribed from nuclear DNA and are translated by 80-83S ribosomes (Hoober and Blobel, 1969; Boulter *et al.*, 1972). While the chloroplast monosomes and dimers are readily distinguishable from those of cytoplasmic origin by their sucrose density gradient centrifugation patterns, it is virtually impossible to distinguish between the higher order polysomes (three or more ribosomes bound to an mRNA) by this method (Simpson *et al.*, 1979).

The experiments outlined in the preceding section have shown dramatic changes in polysomal mRNA populations in wheat leaves during rust infection. It has been of considerable interest to investigate whether these changes involve chloroplast or cytoplasmic mRNA. Musk *et al.* (1979) and Pure *et al.* (1980) studied the effect of organelle-specific inhibitors of protein synthesis on wheat leaf polysomal mRNA translation *in vitro*. It has been demonstrated previously that chloramphenicol and lincomycin specifically inhibit translation by chloroplast 70S ribosomes and have no effect on translation by cytoplasmic 80-83S ribosomes (Ellis, 1969; Ellis and Hartley, 1971; Lamb *et al.*, 1968).

In the cell-free protein synthesizing system reconstituted with wheat leaf polysomes and the S176 fraction from wheat germ, chloramphenicol preferentially inhibits the synthesis of certain size classes of polypeptides (Musk *et al.*, 1979). These include polypeptides that correspond in size to proteins known to be synthesized in isolated chloroplasts. It is known for example, that the large subunit of the enzyme ribulosediphosphate carboxylase is specified by chloroplast DNA and the mRNA coding for this polypeptide is translated by 70S ribosomes (Blair and Ellis, 1973; Gooding *et al.*, 1973).

The small subunit of this enzyme is coded by nuclear DNA and the mRNA for this polypeptide is translated by cytoplasmic 80-83S ribosomes (Gooding et al., 1973; Gray and Kekwick, 1974). In the cell-free protein synthesizing system, chloramphenicol selectively inhibits the radiolabelling of polypeptides that correspond in size to the large subunit of ribulosediophosphate carboxylase and has little or no effect on the synthesis of polypeptides that correspond in size to the small subunit of this enzyme (Pure et al., 1980). Chloramphenicol inhibits total polysomal mRNA translation by 25 to 30%, indicating that at least 25% of the polysomes isolated from wheat leaves contain chloroplast mRNA which are translated by 70S ribosomes (Musk et al., 1979).

Pure et al. (1980) carried out the cell-free translation of mRNA with polysomes from healthy and inoculated (susceptible variety, at three days after inoculation) wheat leaves in the presence of chloramphenicol and lincomycin. The electrophoretic profiles of radiolabelled polypeptides synthesized by polysomes from healthy and inoculated leaves have shown that even in the presence of chloramphenicol, some size classes of polypeptides are synthesized in far greater quantities by polysomes from inoculated leaves than by those from healthy leaves. These results have suggested that changes in wheat leaf polysomal mRNA populations during the early stages of rust infection involve, at least in part, cytoplasmic mRNA. Thus, host parasite interactions appear to affect the transcription patterns of both chloroplast and nuclear genomes.

In plant and animal cells, changes in transcription patterns are controlled at two levels: (a) changes in the template activity of DNA in chromatin and (b) changes in the transcriptive specificity of RNA polymerases. Therefore, to gain an insight into the mechanism by which rust infection elicits changes in the primary metabolism of susceptible cultivars of wheat, some properties of wheat leaf chromatin and enzymes involved in RNA metabolism have been investigated.

IV. Molecular Mechanisms Controlling Transcription Patterns in Higher Plants

A great deal is now known about the process of gene expression in plants, animals and fungi and the mechanisms that control it at the level of transcription. Until recently, it was generally assumed that transcription in higher plants is controlled by the same mechanisms as in bacteria. However, the discovery that the genomic organization in procaryote and eucaryote cells is sufficiently different led to the concept that there are fundamental differences in the mechanisms that control transcription in plants and bacteria.

Unlike the genome of procaryotes which is mostly naked DNA, the nuclear genome of plants and animals is packaged into DNA-protein complexes called chromatin. The enzymes that transcribe the genome in plants and bacteria are also different. Unlike the single form of the catalytic core of bacterial RNA polymerase which synthesizes all classes of RNA, plant and animal cells have multiple molecular forms of RNA polymerases. These polymerases can be grouped into three classes: RNA polymerases I, II and III, according to the kind of RNA molecules they synthesize. RNA polymerase I synthesizes the high molecular weight ribosomal RNA precursor, RNA polymerase II synthesizes mRNA precursors and RNA polymerase III, tRNA precursors as well as the low molecular weight rRNA precursor respectively. The transcription of DNA into RNA in plants, animals and fungi (higher and lower eucaryotes) is controlled by two major mechanisms which are not mutually exclusive: (a) template activity of chromatin as determined by the structure and arrangement of chromosomal proteins (Georgiev, 1972; Stein *et al.*, 1974) and (b) the transcriptive specificity of the multiple molecular forms of RNA polymerases, particularly of those belonging to the class RNA polymerase II (Biswas *et al.*, 1975; Duda, 1976). The mechanisms of the control of gene expression and the changes in the transcription pattern of chromatin observed during the initial stages of rust infection are outlined in the following sections.

A. *Changes in the Template Activity of Chromatin*

1. Structure of chromatin

Chromatin consists of DNA, several classes of histones (basic proteins called H1, H2A, H2B, H3 and H4), nonhistone proteins and a small amount of RNA. The histone proteins interact with themselves and with DNA to form bead-like structures called nucleosomes. The subunit structure of chromatin is diagramatically represented in Fig. 5. The nucleosomes are repeating subunits separated by chromatin fibres containing 150–200 base pairs of DNA. The nucleosomes are about 125Å in diameter and each contains about 200 base pairs of DNA. The evidence for this repeating subunit structure of chromatin is based on data from direct visualization (Oudet *et al.*, 1975), nuclease digestion patterns (Hewish and Burgoyne, 1973; Oosterof *et al.*, 1975) and studies on the spontaneous self-association of isolated histones (Kornberg, 1974).

A nucleosome is formed by two histone tetramers each containing the histones H2A, H3, H4 and H2B, around which the DNA is supercoiled (Kornberg, 1974, 1975; Oudet *et al.*, 1975). The nucleosomes are joined by

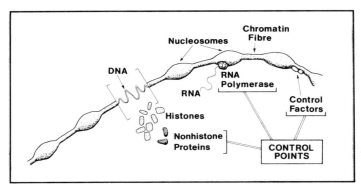

Fig. 5. Diagram of chromatin structure and function. Chromatin in the interphase nucleus consists of repeating bead-like structures called nucleosomes which are joined by fibre regions. Each nucleosome contains two histone tetramers around which the DNA is supercoiled. A nucleosome is separated into its DNA and protein components to show its composition. Several nucleosomes and fibre regions constitute a transcriptional unit (active gene) which is transcribed by RNA polymerase into a precursor RNA. The transcription of genes is controlled by regulatory (nonhistone) proteins in chromatin that determine which genes are "open" to transcription, control factors in chromatin such as hormone-receptor complexes that stimulate the transcription of specific genes and RNA polymerase associated regulatory subunits that select the genes to be transcribed.

fibre regions in which tertiary packaging of DNA is caused by the histone H1 and a heterogeneous group of nonhistone proteins. The interphase nucleus of eucaryote cells contains euchromatin (template active chromatin) and heterochromatin (template inactive). A transcriptional unit in euchromatin which is transcribed by RNA polymerase into a large precursor RNA, consists of several nucleosomes and the chromatin fibre regions separating them. It is generally believed that when a gene is derepressed, the transcriptional unit undergoes allosteric changes into the "open" form without physical displacement of histones from the nucleosomes (Gottesfeld *et al.*, 1975; Lewin, 1975).

There is substantial evidence suggesting that the template activity of DNA in chromatin is controlled by interactions between nuclear nonhistone proteins (regulatory proteins) and DNA (Stein *et al.*, 1974; Kostraba *et al.*, 1975; Reeves and Jones, 1976). The selectivity of transcription, on the other hand, is largely governed by the interactions between the catalytic subunits and the regulatory factors of RNA polymerases (Hardin *et al.*, 1972; Chuang and Chuang, 1975; Teissere *et al.*, 1975; Rizzo and Cherry, 1975).

It is conceivable that changes in transcription patterns that occur in wheat leaves during the early stages of rust infection result from a parasitically induced alteration in the structure of host chromatin. This possibility can be readily examined by comparing the *in vitro* transcription of chromatin from healthy and rust infected wheat leaves. It has been demonstrated that at least functionally, isolated chromatin is identical to that in the intact nuclei. Competitive DNA–RNA hybridization and nearest neighbour frequency analyses of *in vivo* and *in vitro* synthesized RNA have suggested that the portions of chromatin DNA which are transcribed in intact cells are the same as those transcribed *in vitro* by RNA polymerases using chromatin as template (Paul and Gilmour, 1966; Bonner *et al.*, 1968). It has also been demonstrated that the chromosomal proteins can be dissociated from chromatin isolated from plants (Bekhor *et al.*, 1969) and animals (Huang and Huang, 1969) by gentle techniques and reassociated. The native and reconstituted chromatin were found to be identical as determined by competitive hybridization between DNA and the RNA transcribed from native and reconstituted chromatin respectively. These studies have revealed the fidelity of transcription of isolated chromatin by RNA polymerases.

2. Template activity of chromatin

The chromatin preparations from healthy wheat leaves (*T. vulgare*, var. W195) and from those inoculated with the stem rust fungus (*P. graminis tritici*, race 126-ANZ-6, 7) were incubated under standard RNA polymerase assay conditions but without any added RNA polymerase. The RNA molecules synthesized by endogenous (chromatin-bound) RNA polymerases using chromatin DNA as template and [³H]-labelled nucleoside triphosphates as substrate were analysed by polyacrylamide gel electrophoresis. The results of these experiments have revealed that the electrophoretic patterns of RNA molecules synthesized by endogenous RNA polymerases in the chromatin from healthy and inoculated leaves respectively are significantly different (Flynn *et al.*, 1976; Dey *et al.*, 1980; A. K. Dey, A. K. Chakravorty and K. J. Scott, unpublished).

The chromatin isolated from a number of susceptible cultivars of wheat (both healthy and two to four days after inoculation with stem rust uredospores) was programmed with *E. coli* RNA polymerase holoenzyme (the bacterial enzyme containing all catalytic and regulatory subunits) and the RNA molecules radiolabelled *in vitro* were analysed by gel electrophoresis. The electrophoretic migration of various size classes of RNA synthesized by this enzyme using chromatin from healthy and inoculated wheat (var. W195) leaves is presented in Fig. 6. A comparison of the radioactivity profiles obtained with chromatin from healthy and inoculated plants suggests that when incubated under identical conditions and with the same enzyme (*E.*

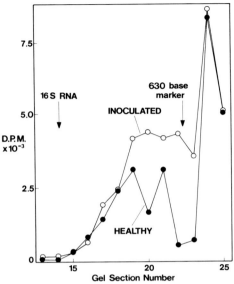

Fig. 6. Polyacrylamide gel electrophoresis of RNA synthesized by *E. coli* RNA polymerase using the chromatin from healthy and rust infected wheat leaves as template. Chromatin isolated from healthy and rust infected wheat leaves (four days after inoculation) was incubated with *E. coli* RNA polymerase under standard polymerase assay conditions (Lin *et al.*, 1974). The [³H]-labelled RNA was electrophoresed in polyacrylamide gels. The gels containing the different size classes of RNA synthesized with chromatin from healthy and inoculated leaves respectively as template were sectioned and the radioactivity in each section determined in a liquid scintillation spectrometer. The markers were: 16S ribosomal RNA and 630 nucleotides long RNA molecules.

coli RNA polymerase holoenzyme), the chromatin from inoculated leaves supports the synthesis of some size classes of RNA molecules (gel section numbers 19 to 23 in Fig. 6) in significantly greater quantities than does the chromatin from healthy leaves. These observations have suggested dramatic differences in the template activity of chromatin from healthy and inoculated leaves.

The results of electron microscopic (Shaw, 1967), cytospectrophotometric (Bhattacharya *et al.*, 1965) as well as the biochemical experiments outlined above have provided strong evidence for changes in the structure and function of host chromatin during the early stages of rust infection. It is conceivable that chromosomal regulatory proteins induced during the initial stages of host rust interactions are responsible for the observed changes in the template activity of chromatin. The induced synthesis of these proteins may provide a molecular mechanism that, at least in part, couples host parasite interactions to changes in transcription patterns (Chakravorty and Shaw, 1977a).

B. Changes in the Transcriptive Specificity of RNA Polymerases

In both higher and lower eucaryotes, mRNA precursor molecules are synthesized by a group of enzymes known as RNA polymerase II which differs from RNA polymerases I and III in its chromatographic behaviour, α-amanitin sensitivity and cation requirement. RNA polymerase II is highly sensitive to the mushroom toxin, α-amanitin. It includes several molecular forms that have some common subunits and are interconvertible (Mullinix *et al.*, 1973; Kedinger *et al.*, 1974; Chambon, 1975). It is possible that each of these molecular forms has a specific function and that different molecular forms of RNA polymerase II are responsible for the transcription of different parts of the genome.

The two major classes of RNA polymerase (RNA polymerase I and RNA polymerase II) have been isolated from healthy and rust infected wheat leaves of a cultivar (*T. vulgare*, var. W195) which is susceptible to *P. graminis*, race 126-ANZ-6, 7 (Flynn *et al.*, 1976). These have been partially purified by diethylaminoethyl (DEAE) cellulose and DEAE Sephadex column chromatography. The wheat leaf RNA polymerase I elutes from DEAE cellulose at 0·1M ammonium sulphate and is not affected by α-amanitin. RNA polymerase II elutes at 0·2M ammonium sulphate and is completely inactivated by α-amanitin (5 μg/ml), suggesting that the enzyme preparation is essentially free from other RNA polymerases. The activity of both RNA polymerases (I and II) is completely inhibited by actinomycin D and unaffected by rifampicin, a potent inhibitor of bacterial RNA polymerase. A comparison of the kinetic and catalytic properties of the two RNA polymerases from healthy and inoculated (two and four days after inoculation) wheat leaves has revealed that both enzymes have undergone significant changes with respect to V_{max}, K_m, divalent cation requirement and template specificity.

The transcriptive activity of RNA polymerase II from healthy and inoculated wheat leaves with native (double stranded) and denatured (single stranded) salmon sperm and calf thymus DNA as template, is presented in Table I. The data show a consistently higher specific activity of the RNA polymerase II from inoculated leaves than that of the corresponding enzyme from healthy leaves with all templates except for native calf thymus DNA.

In view of the biological function of RNA polymerase II, changes in catalytic properties and template preference of this enzyme have particular

Table I. Template preference of RNA[a] polymerase II from healthy and rust infected wheat leaves.

Template DNA	Specific activity ($p$mol [³H] GMP incorporated mg⁻¹ protein)		Specific activity ratio (inoculated/ healthy)
	Healthy	Inoculated	
Native salmon sperm	13·5	24·7	1·8
Denatured salmon sperm	7·5	15·8	2·1
Native calf thymus	6·5	7·0	1·1
Denatured calf thymus	3·0	6·4	2·1

[a]RNA polymerase II was isolated from healthy wheat leaves (*T. vulgare*, var. W195) and from those inoculated with the stem rust fungus (*P. graminis tritici*, race 126-ANZ-6, 7) at four days after inoculation, by DEAE cellulose chromatography (Lin *et al.*, 1974). The enzyme was incubated with either double stranded (native) or single stranded (denatured) DNA in standard reaction mixtures (Lin *et al.*, 1974) that included [³H] GTP, for 25 min at 33°C.

relevance to the alterations in the transcription patterns in wheat leaves and the resulting changes in polysomal mRNA populations during the early stages of rust infection.

The activity of RNA polymerases I and II with chromatin isolated from healthy and inoculated wheat leaves has also been investigated. The partially purified chromatin contain tightly bound RNA polymerases which are not removed during purification (endogenous RNA polymerases). The chromatin preparations have been used as template for RNA synthesis by endogenous RNA polymerases as well as by RNA polymerases I and II isolated from healthy and inoculated wheat leaves. The specific activity ($p$mol [³H] GMP incorporated mg⁻¹ protein) of endogenous RNA polymerases in the chromatin from inoculated leaves has been found to be at least two-fold greater than that in the chromatin from healthy leaves (J. Flynn, unpublished observation).

The results presented in Table II show the synthesis of RNA by RNA polymerases I and II from healthy and inoculated leaves with native chromatin isolated from these sources as template. As the radioactivity due to the synthesis of RNA by endogenous RNA polymerases has been subtracted from these values, the data represent largely the synthesis of additional RNA molecules by the RNA polymerases added. With the exception of two combinations (RNA polymerases I and II from healthy leaves each programmed with chromatin from inoculated leaves), the RNA polymerases I and II from healthy and inoculated leaves synthesize additional RNA molecules. These results suggest that the transcription initiation sites in the

Table II. The activities of RNA polymerases I and II from healthy and inoculated wheat leaves with chromatin as template.

RNA polymerase and source	Specific activity[a] ($p$mol [^{3}H] GMP incorporated mg^{-1} protein) with chromatin from:	
	Healthy	Inoculated
RNA polymerase I		
Healthy	1·7	0
Inoculated	3·6	4·6
RNA polymerase II		
Healthy	0·3	0
Inoculated	4·4	5·0

[a]The activities due to endogenous (chromatin-associated) RNA polymerases have been substracted from the activities obtained with added RNA polymerases to arrive at the data presented. Isolation of chromatin from healthy and rust infected wheat leaves (three days after inoculation) and incubations to determine RNA polymerase activities were carried out as described by Lin *et al.* (1974).

chromatin DNA are not completely saturated by endogenous RNA polymerases in the case of six out of eight combinations in Table II. A comparison of the chromatin directed RNA synthesis by RNA polymerases I and II in different enzyme-chromatin combinations shows dramatic changes in the template specificity of these two enzymes as well as in the template activity of chromatin, at three days after inoculation with the rust fungus. Thus, with the same chromatin preparation as template (for example, chromatin from healthy leaves), the RNA polymerases I and II from inoculated leaves synthesize significantly greater quantities of RNA than do the corresponding enzymes from healthy leaves. Both RNA polymerases from healthy leaves show no detectable activity with chromatin from inoculated leaves as template but do synthesize RNA when the template is chromatin from healthy leaves, suggesting changes in the template activity of chromatin in the rust infected leaves.

The RNA molecules synthesized by RNA polymerase II from healthy and inoculated leaves with chromatin as template have been analysed by polyacrylamide gel electrophoresis. There are several distinct size classes of RNA which are synthesized by the enzyme from inoculated leaves; with the enzyme from healthy leaves, some of these RNA molecules are either not synthesized or are synthesized at a very low level.

The RNA polymerases I and II have been isolated from the wheat stem rust fungus grown in axenic cultrue. These enzymes show chromatographic properties similar to polymerases from other fungi (Adman *et al.*, 1972; Buhler *et al.*, 1976) and are substantially different from the corresponding enzymes from healthy and inoculated wheat leaves. Thus, the observed quantitative and qualitative changes in wheat leaf RNA polymerases are likely to be parasitically induced, rather than due to a direct contribution of rust enzymes to the polymerases isolated from inoculated leaves.

These findings provide a basis for further work on the mechanism by which the rust fungi cause dramatic changes in the transcriptive specificity of RNA polymerases in their hosts and at an early stage of disease development. RNA polymerases in both higher and lower eucaryotes are complex, multimeric proteins containing several high and low molecular weight polypeptide subunits (Mullinix *et al.*, 1973; Smith and Bogorad, 1974; Biswas *et al.*, 1975). In addition to the catalytic subunits, the activity of RNA polymerases is controlled by regulatory polypeptide factors (Biswas *et al.*, 1975) which are capable of altering the rates of transcription initiation as well as RNA chain elongation. It is possible that changes in the properties of RNA polymerases I and II in the inoculated host are due to such regulatory factors. The operation of a similar mechanism has been demonstrated in the hormonal control of transcription in higher plants (Teissere *et al.*, 1975).

C. Ribonuclease and Rust Infection

Like the RNA polymerases, ribonucleases (RNases) play an important role in the initial steps of gene expression. The precursor RNA molecules synthesized by the three classes of RNA polymerases undergo posttranscriptional processing before these become biologically functional (Darnell, 1975). The processing of all three major classes of RNA (mRNA, rRNA and tRNA) involves several steps including sequential cleavage of the precursor RNA catalysed by nucleases with endonucleolytic and exonucleolytic modes of action. Because of their central role in RNA metabolism, RNases have been studied extensively in plants during viral and fungal infections (Diener, 1961; Rohringer *et al.*, 1961; Reddi, 1966). More recently, studies on both quantitative and qualitative changes in RNase activity during the infection of higher plants by the rust fungi have been reported (reviewed by Chakravorty and Shaw, 1977a,b).

In a number of susceptible varieties of host plants inoculated with rust uredospores, the RNase activity extractable in a low ionic strength buffer increases in two well defined phases: an early phase (early RNase) and a late phase (late RNase). In a genetically resistant cultivar of flax (*L. usitatissimum*, var. Bombay) inoculated with flax rust (*M. lini*, race 3) only the early phase of RNase increase occurs. Quantitative changes in RNase activity in a

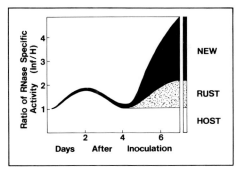

Fig. 7. Changes in the soluble ribonuclease activity of a susceptible cultivar of flax (var. Bison) at different times after inoculation with flax rust (race 3). The enzyme was extracted from healthy and inoculated flax cotyledons with a low ionic strength buffer and partially purified by acid precipitation (Scrubb *et al.*, 1972). The new enzyme represents an RNase component which is different in its properties from the host RNase and the corresponding enzyme fraction isolated from flax rust. The relative amounts of new RNase, rust RNase and host RNase are based on differences in their thermal stability, DEP sensitivity and substrate preference, as indicated in the text.

susceptible cultivar of flax (var. Bison) following inoculation with race 3 of flax rust are shown in Fig. 7. These quantitative changes are accompanied by significant changes in the properties of the enzyme including thermal stability, diethylpyrocarbonate (DEP) sensitivity and substrate preference as judged by the relative rates of hydrolysis of synthetic homoribopolymers [polyadenylate or poly(A), polycytidylate or poly(C), polyguanylate or poly(G) and polyuridylate or poly(U) respectively] by the enzymes from healthy flax cotyledons, incoculated cotyledons and the flax rust fungus grown in axenic culture (Scrubb *et al.*, 1972). The enzymes from both flax cotyledons and the rust fungus are heat stable. The enzyme from rust infected cotyledons (seven days after inoculation) loses 60% of the activity at 100°C for 10 min. The flax enzyme is inhibited by only 25% when preincubated with DEP. The RNase from rust mycelium is inhibited by 50%. The enzyme from infected plants shows much greater DEP sensitivity. The substrate preference of the RNase from infected flax cotyledons is quite different from that of the enzyme from healthy flax cotyledons as well as the substrate preference of the rust mycelial enzyme. When the RNase from infected cotyledons is heated under conditions that inactivate the heat labile component, the remaining enzymatic activity shows DEP sensitivity and substrate preference that are intermediate between those of the flax enzyme and the rust enzyme. These findings allow an approximate estimation of the relative quantities of the host RNase, rust RNase and the new enzyme at different stages of disease development, as shown in Fig. 7.

There is considerable evidence suggesting that the changes in the kinetic and catalytic properties of RNase during the early and later stages of rust infection represent a specific phenomenon which is caused by compatible host rust interactions (Scrubb *et al.*, 1972; Chakravorty *et al.*, 1974a,b,c; Harvey *et al.*, 1974). There is also a positive correlation between the growth of the rust fungus in the host tissue and changes in the amounts and properties of RNase (Scrubb *et al.*, 1972; Chakravorty *et al.*, 1974c). Thus, when a susceptible cultivar of wheat (*T. aestivum*) carrying the temperature sensitive Sr6 allele for rust resistance is grown under conditions that allow the formation of only a few surviving intercellular hyphae but prevent further growth of the fungus, it has been demonstrated that (a) there are changes in the catalytic properties of RNase within two days after inoculation, (b) the appearance of the late RNase requires and accompanies the growth of the pathogen and (c) the quantitative and qualitative changes in RNase activity are not a function of the amount of rust fungus present in the inoculated leaves (Chakravorty *et al.*, 1974c).

Mechanical injury of healthy flax cotyledons causes a substantial increase in RNase activity but the properties of the RNase from intact and mechanically injured cotyledons are identical. Mechanical injury causes an increase in the activity of a number of hydrolytic enzymes including deoxyribonuclease (DNase), phosphodiesterase, acid phosphatase and alkaline phosphatase whereas in the inoculated cotyledons only RNase activity increases (Scrubb *et al.*, 1972). The appearance of the new RNase in the inoculated cotyledons can be prevented by cycloheximide, an inhibitor of protein synthesis, suggesting that the heat labile, highly DEP sensitive RNase component includes newly synthesized protein molecules (Scrubb *et al.*, 1972).

The quantitative and qualitative changes in RNase activity have been studied in a number of host rust combinations some of which consist of two different hosts inoculated with the same rust fungus as well as others comprising the same host cultivar inoculated with different races of the rust fungus (Chakravorty *et al.*, 1974b,c; Harvey *et al.*, 1974; Chakravorty and Shaw, 1977a). These include: (i) wheat (Sr6)—wheat stem rust, races 56, 56A and 126-ANZ-6,7; (ii) wheat (sr6)—race 56; (iii) barley (*Hordeum vulgare* L., var. Montcalm)—wheat stem rust, race 56; (iv) *Ribes (Ribes nigrum* L.)—blister rust (*Cronartium ribicola*) and (v) pine (*Pinus monticola*)—blister rust. The biphasic increase in RNase activity has been observed in all these compatible host rust combinations. However, a comparison of the direction and magnitude of changes in thermal stability, DEP sensitivity and in the relative rates of hydrolysis of the substrates poly (A), poly (C) and poly (U) are significantly different from one host rust combination to another, no two combinations showing the same pattern. These findings suggest that the direction and magnitude of changes in the kinetic and catalytic properties of RNase during rust infection are determined by both the host plant and the rust fungus.

Changes in the properties of two barley leaf RNases (the pH5 insoluble RNase and soluble RNase) have been observed to occur during the early stages of infection by the powdery mildew fungus, *Erysiphe graminis*, race 3 (Chakravorty and Scott, 1979). The specificity of these enzymes for cleavage sites in the natural substrates, polysomal mRNA and the RNA of intact ribosomes, has also been investigated (Simpson *et al.*, 1979). In susceptible varieties of barley, a decrease in chloroplast polysome content is one of the earliest detectable biochemical symptoms of powdery mildew infection (Dyer and Scott, 1972). The experiments on *in vitro* polysomal mRNA hydrolysis by the pH5 insoluble RNase and the finding that the ribosomes isolated from inoculated barley leaves (susceptible cultivars) contain a significantly greater number of mRNA fragment-bound monosomes than those from healthy leaves (Simpson *et al.*, 1979) have suggested that the pH5 insoluble RNase may be responsible for chloroplast polysomal mRNA hydrolysis *in vivo*, resulting in the observed decrease in polysome content. Further experiments have revealed that at 24 and 48 hours after inoculation, there is a remarkable change in the substrate preference of the barley leaf pH5 insoluble RNase. As for the action of this enzyme on the natural substrates, the pH5 insoluble RNase from inoculated leaves hydrolyses chloroplast polysomal mRNA *in vitro* and produces a far greater number of mRNA fragment-bound monosomes than does the corresponding enzyme from healthy leaves (Chakravorty *et al.*, 1980). These results have suggested a biological role of the pH5 insoluble RNase during the early stages of host parasite interactions. Presumably, it selectively hydrolyses those polysomal mRNA molecules *in vivo* whose function is not essential for the immediate survival of either the host plant or the powdery mildew fungus. Interestingly, neither the changes in the action of this enzyme on polysomal mRNA nor the decrease in polysome content can be detected in genetically resistant varieties of barley (Dyer and Scott, 1972; Chakravorty *et al.*, 1980).

It is conceivable that in rust infected host plants, the new RNase has functions similar to those observed in powdery mildew infected barley leaves. In addition, the dramatic changes in transcription patterns during the early stages of host rust interactions (Chakravorty and Shaw, 1971; Tani *et al.*, 1971, 1973; Pure *et al.*, 1979, 1980) and the appearance of RNase activity with unique catalytic properties (Chakravorty and Shaw, 1977a,b) suggest that the new RNase may have essential roles in the processing of precursor RNA molecules. In the flax-rust system, the early phase of RNase increase coincides with an increase in the rate of synthesis of all major classes of RNA as well as the accelerated synthesis of RNA molecules with high A + U/G + C ratios (Chakravorty and Shaw, 1971). The appearance of late RNase is accompanied by a substantial increase in the RNA content of the infected host (Chakravorty and Shaw, 1977b). Thus, there is a striking parallelism between the quantitative as well as qualitative changes in RNase activity and increases in the rate of synthesis as well as accumulation of RNA during the early and intermediate stages of rust infection. These observations strengthen the view that quantitative and qualitative changes in RNase activity during rust infection of all susceptible cultivars of host plants investigated so far, represent a part of an elaborate control system at the post-transcriptional processing level of primary metabolism.

V. Concluding Remarks

The application of recently developed techniques of biochemistry and molecular biology has revealed dramatic changes in the primary metabolism of host plants during the early stages of host rust interactions. Changes in the patterns of gene expression during these stages are readily detectable at the levels of transcription *in vivo* and *in vitro*, polysomal mRNA populations and the patterns of polypeptides translated *in vivo*. It is reasonable to assume that changes in the primary metabolism of the host plant at stages where either susceptibility or resistance is determined may play a decisive role in the further course of disease development. The mechanism by which host parasite interactions govern the direction and degree of these changes, and hence, the metabolic control chains leading to either susceptible or resistance reaction is only beginning to be understood. Current investigations on the transcriptive specificity of RNA polymerases, template activity of chromatin and the action of RNases on the natural substrates in the rust and powdery mildew infected plants have indicated that these fastidious obligate parasites have evolved sophisticated mechanisms to control the host plant's primary metabolism. The ability or inability of the pathogen to elicit changes in the metabolic control chains of the host in a particular direction is very likely to form the molecular basis of host parasite specificity. Owing to advances in biochemical methodology and the availability of appropriate biological materials, it should be possible to test the validity of this model.

Acknowledgements

The original work presented in this article has been supported by grants from

Australian Research Grants Committee and the Wheat Industry Research Council of Australia. I am grateful to my present and former associates, Glenn Pure, Peter Musk, Arun Dey and John Flynn who contributed to this work. All the illustrative material has been prepared by Glenn Pure.

References

Adman, R., Schultz, L. D. and Hall, B. D. (1972). *Proc. Nat. Acad. Sci, U.S.A.* **69**, 1702–1706.

Bekhor, I., Kung, G. M. and Bonner, J. (1969). *J. Molec. Biol.* **39**, 351–364.

Bhattacharya, P. K., Naylor, J. M. and Shaw, M. (1965). *Science* **150**, 1605–1607.

Biswas, B. B., Ganguly, A. and Das, A. (1975). *Prog. Nucleic Acids Res. Mol. Biol.* **15**, 145–184.

Blair, G. E. and Ellis, R. J. (1973). *Biochim. Biophys. Acta* **319**, 223–234.

Bonner, J., Dahmas, M., Fambrough, D., Huang, R. C. C., Marushige, K. and Tuan, D. (1968). *Science* **159**, 47–56.

Boulter, D., Ellis, R. J. and Yarwood, A. (1972). *Biol. Rev.* **47**, 113–175.

Buhler, J. M., Iborra, F., Sentenac, A. and Fromageot, P. (1976). *J. Biol. Chem.* **25**, 1712–1717.

Chakravorty, A. K. and Scott, K. J. (1979). *Physiol. Plant Pathol.* **14**, 85–97.

Chakravorty, A. K. and Shaw, M. (1971). *Biochem. J.* **123**, 551–557.

Chakravorty, A. K. and Shaw, M. (1977a). *Biol. Rev.* **52**, 147–179.

Chakravorty, A. K. and Shaw, M. (1977b). *Ann. Rev. Phytopathol.* **15**, 135–151.

Chakravorty, A. K., Shaw, M. and Scrubb, L. A. (1974a). *Physiol. Plant Pathol.* **4**, 313–334.

Chakravorty, A. K., Shaw, M. and Scrubb, L. A. (1974b). *Physiol. Plant Pathol.* **4**, 335–358.

Chakravorty, A. K., Shaw, M. and Scrubb, L. A. (1974c). *Nature* **247**, 577–580.

Chakravorty, A. K., Simpson, R. S. and Scott, K. J. (1980). *Plant Cell Physiol.* **21**, 425–432.

Chambon, P. (1975). *Ann. Rev. Biochem.* **44**, 613–638.

Chuang, R. Y. and Chuang, L. F. (1975). *Proc. Nat. Acad. Sci., U.S.A.* **72**, 2935–2939.

Darnell, J. E. (1975). *In* "The Eucaryote Chromosome" (W. J. Peacock and R. D. Brock, eds), pp. 185–198. Australian National University Press, Canberra.

Dey, A. K., Chakravorty, A. K. and Scott, K. J. (1980). *Proc. Aust. Biochem. Soc.* **13**, 95.

Diener, T. O. (1961). *Virology* **14**, 177–189.

Duda, C. T. (1976). *Ann. Rev. Plant Physiol.* **27**, 119–132.

Dyer, T. A. and Scott, K. J. (1972). *Nature* **236**, 237–238.

Ellis, R. J. (1969). *Science* **163**, 477–478.

Ellis, R. J. and Hartley, M. R. (1971). *Nature New Biol.* **233**, 193–196.

Flor, H. H. (1971). *Ann. Rev. Phytopathol.* **9**, 275–296.

Flynn, J. G., Chakravorty, A. K. and Scott, K. J. (1976). *Proc. Aust. Biochem. Soc.* **9**, 44.

Georgiev, G. P. (1972). *Curr. Top. Devel. Biol.* **7**, 1–60.

Gooding, L. R., Roy, H. and Jagendorf, A. T. (1973). *Arch. Biochem. Biophys.* **159**, 324–335.

Gottesfeld, J. M., Murphy, R. F. and Bonner, J. (1975). *Proc. Nat. Acad. Sci,* *U.S.A.* **72**, 4404–4408.

Gray, J. C. and Kekwick, R. G. O. (1974). *Eur. J. Biochem.* **44**, 491–500.

Hardin, J. W., Cherry, J. H., Morre, D. J. and Lembi, C. A. (1972). *Proc. Nat. Acad. Sci., U.S.A.* **69**, 3146–3150.

Harvey, A. E., Chakravorty, A. K., Shaw, M. and Scrubb, L. A. (1974). *Physiol. Plant Pathol.* **4**, 359–371.

Heitefuss, R. (1966). *Ann. Rev. Phytopathol.* **4**, 221–244.

Hewish, D. R. and Burgoyne, L. A. (1973). *Biochim. Biophys. Res. Comm.* **52**, 502–510.

Hoober, J. K. and Blobel, G. (1969). *J. Molec. Biol.* **41**, 121–138.

Huang, R. C. C. and Huang, P. C. (1969). *J. Molec. Biol.* **39**, 365–378.

Kedinger, C., Gissinger, F. and Chambon, P. (1974). *Eur. J. Biochem.* **44**, 421–436.

Kornberg, R. D. (1974). *Science* **184**, 868–871.

Kornberg, R. D. (1975). *In* "The Eucaryote Chromosome" (W. J. Peacock and R. D. Brock, eds), pp. 245–253. Australian National University Press, Canberra.

Kostraba, N. C., Montagna, R. A. and Wang, T. Y. (1975). *J. Biol. Chem.* **250**, 1548–1555.

Lamb, A. J., Clark-Walker, G. D. and Linnane, A. W. (1968). *Biochim. Biophys. Acta* **161**, 415–427.

Lewin, B. (1975). *Nature* **254**, 651–653.

Lin, C. Y., Guilfoyle, T. J., Chen, Y. M., Nagao, R. T. and Key, J. L. (1974). *Biochim. Biophys. Res. Comm.* **60**, 498–506.

Mullinix, K. P., Strain, G. C. and Bogorad, L. (1973). *Proc. Nat. Acad. Sci., U.S.A.* **70**, 2386–2390.

Musk, P., Chakravorty, A. K. and Scott, K. J. (1979). *Plant Cell Physiol.* **20**, 1359–1369.

Oosterof, D. K., Hozier, J. C. and Rill, R. L. (1975). *Proc. Nat. Acad. Sci., U.S.A.* **72**, 633–637.

Oudet, P., Gross-Bellard, M. and Chambon, P. (1975). *Cell* **4**, 281–300.

Paul, J. and Gilmour, J. S. (1966). *J. Molec. Biol.* **16**, 242–244.

Pure, G. A., Chakravorty, A. K. and Scott, K. J. (1979). *Physiol. Plant Pathol.* **15**, 201–209.

Pure, G. A., Chakravorty, A. K. and Scott, K. J. (1980). *Plant Physiol.* **66**, 520–524.

Reddi, K. K. (1966). *Proc. Nat. Acad. Sci., U.S.A.* **56**, 1207–1214.

Reeves, R. and Jones, A. (1976). *Nature* **260**, 495–500.

Rizzo, P. J. and Cherry, J. H. (1975). *Plant Physiol.* **55**, 574–577.

Rohringer, R. and Heitefuss, R. (1961). *Can. J. Bot.* **39**, 263–267.

Rohringer, R., Samborski, D. J. and Person, C. (1961). *Can. J. Bot.* **39**, 775–784.

Sager, R. and Hamilton, M. G. (1967). *Science* **157**, 709–711.

Scott, K. J. (1972). *Biol. Rev.* **47**, 537–572.

Scrubb, L. A., Chakravorty, A. K. and Shaw, M. (1972). *Plant Physiol.* **50**, 73–79.

Shaw, M. (1967). *Can. J. Bot.* **45**, 1205–1220.

Simpson, R. S., Chakravorty, A. K. and Scott, K. J. (1979). *Physiol. Plant Pathol.* **14**, 245–258.

Smith, H. J. and Bogorad, L. (1974). *Proc. Nat. Acad. Sci., U.S.A.* **71**, 4839–4842.

Stein, G. S., Spelsberg, T. C. and Kleinsmith, L. J. (1974). *Science* **183**, 817–825.

Svetailo, E. N., Philippovich, I. I. and Sissakian, N. M. (1967). *J. Molec. Biol.* **24**, 405–415.

Tani, T. and Yamamoto, H. (1978). *Physiol. Plant Pathol.* **12**, 113–121.

Tani, T., Yoshikawa, M. and Naito, N. (1971). *Ann. Phytopathol. Soc. Japan* **37**, 43–51.

Tani, T., Yoshikawa, M. and Naito, N. (1973). *Ann. Phytopathol. Soc. Japan* **39**, 7–13.

Teissere, M., Penon, P., Van Huystee, R. B., Azou, Y. and Ricard, J. (1975). *Biochim. Biophys. Acta* **402**, 391–402.

von Broembsen, S. L. and Hadwiger, L. A. (1972). *Physiol. Plant Pathol.* **2**, 207–215.

Wolf, G. (1967). *Phytopathol. Z.* **59**, 101–194.

Yamamoto, H., Tani, T. and Hokin, H. (1976). *Ann. Phytopathol. Soc. Japan* **42**, 583–590.

Part B: Intermediary Metabolism

CONTENTS

I. Introduction

Invasion of plant tissues by a pathogen inevitably leads to a loss of some metabolic control in those host tissues and consequently to alterations in metabolic pathways. The extent of this alteration will depend on the nature of the host/parasite interaction. For example, in the case of interaction between host and biotrophic parasites such as rust fungi, the interaction is a very delicately balanced one at least until the invading pathogen has sporulated. Hence changes in host metabolism are likely to be subtle. We have seen from Part A of this chapter that important subtle changes do occur both in transcription and translation in host tissue after invasion by rust fungi. In the remainder of this chapter we will be concerned with changes which occur in intermediary metabolism after infection of hosts with rust fungi.

II. Carbohydrate Metabolism

Until the early 1970's mechanisms of synthesis and degradation of carbohydrates (photosynthesis and respiration) in plant cells infected with rust fungi had probably been more fully investigated than any other aspect of their metabolism. Early attempts to explain resistance and susceptibility centered round carbohydrate metabolism and subsequent workers were concerned with extending these early observations.

A. Supply of Carbohydrates for Pathogen Development

It has been known for more than 50 years that development of rust fungi on susceptible hosts is dependent upon an adequate supply of carbohydrates.

Fromme (1913) demonstrated that light exclusion retarded development of *Puccinia coronata* on oats. The retardation was related to the length of time the plants were kept in the dark rather than when the dark period was given. Thus retardation occurred if the dark period was given at time of inoculation or after the pathogen was established. He suggested that rust development was dependent upon transitory products of photosynthesis and proposed that this was the explanation for the inability of rust fungi to develop on any form of artificial medium. Mains (1917) extended the observations of Fromme by showing that it was not the presence of chlorophyll which was necessary for the development of *P. sorghi* on maize or *P. coronata* on oats but rather it was the supply of carbohydrates. These could be either formed by photosynthesis or given directly to seedlings or to leaf segments in the dark. Mains postulated that the obligate nature of rust fungi could be explained by their requirement for some transient or nascent organic products related to carbohydrates, which they secured from living host tissue.

B. Respiration

After the importance of carbohydrates in the development of rust fungi had been established, workers turned their attention to pathways of utilization of carbohydrates in infected leaves. It quickly became apparent that the rate of respiration in rust-infected tissue was higher than in non-infected controls and it is now generally accepted that increased respiration results from infection by a wide range of pathogens including all pathogenic fungi and bacteria.

Four questions have been asked concerning this increase in respiration.

(a) What are the relative contributions of the host and pathogen to the increase?

(b) Are there changes in the pathways of respiration of host tissue after infection?

(c) What is the cause of the increase in respiration?

(d) What is the significance of this increase in the successful establishment of the host-parasite complex, i.e. does the respiratory patttern after infection have any role in determining resistance and susceptibility?

The respective contributions of the host and pathogen to the rise in respiration has been the subject of conjecture over many years. Underlying this controversy is the inherent difficulty of working with a two-membered system where it is almost impossible to physically separate one member from the other. However, it is now generally accepted that whilst the growing fungus, especially at the time of sporulation, contributes significantly to the measured rate, host tissue respiration in the neighbourhood of fungal colonies is stimulated approximately two-fold.

When considering the pathway of the increase in respiration, two possibilities need to be borne in mind. Rates of carbohydrate flow through existing pathways may be enhanced or alternatively a qualitative change in existing pathways may occur. Several lines of evidence favour the second possibility (Daly et al., 1957). Aerobic CO_2 production by rust infected hypocotyls of safflower is much less affected by NaF than that of non-infected control hypocotyls. The RQ was approximately one suggesting that β-oxidation of fatty acids is not the source of two carbon fragments required for the enhanced respiration. Also the inhibition of anaerobic CO_2 production was the same in both types of tissues indicating that extra carbon for respiration was not being produced via glycolysis. Daly et al. (1957) also observed that the C_6/C_1 ratio was appreciably reduced in rust infected safflower, bean or wheat. Non-infected controls showed a ratio between 0·5 and 0·8 whereas in infected tissues it fell below 0·25.

The above observations are consistent with an enhanced operation of the pentose phosphate pathway. However, attempts to establish changes in respiratory pathways on the basis of C_6/C_1 ratios and inhibitor studies are far from convincing and supportive evidence should be sought before definitive conclusions can be drawn. Such supportive evidence can be found from studies concerned with changes in enzymatic activities. The activities of glucose-6-phosphate dehydrogenase and 6-phosphogluconate dehydrogenase, the first two enzymes of the pentose phosphate pathway, are enhanced in mildewed and rusted cereal leaves (Lunderstädt et al., 1962; Scott, 1965; Scott and Smillie, 1963, 1966). The pattern of these increases is coincident with the pattern of increase in respiration and there is no corresponding change in the activities of enzymes of the glycolytic pathway or the TCA cycle.

It remains to relate the enhanced activity in the pentose phosphate pathway to the observed increase in oxygen consumption. Except where degenerative and necrotic changes have set in, all available evidence suggests that only two terminal oxidases play a significant role in oxygen consumption in plant tissues. These are the cytochrome oxidase system and glycolate oxidase. Enhanced utilization of glucose through the pentose phosphate pathway is not directly coupled to re-oxidation of co-enzymes via either of these terminal oxidase systems. Therefore, it seems possible either that this pathway is not linked to increased oxygen consumption or a mechanism for shuttling electrons from NADPH outside the mitochondria to NAD^+ inside the mitochondria has to be evoked. NADPH formed as a result of action of the two dehydrogenases of the pentose phosphate pathway could be readily utilized in rust infected tissue for synthesis of lipids and sugar alcohols especially round the period of sporulation. Reoxidation of NADPH via a shuttle system involving the electron transport system of the mitochondria is an attractive idea in terms of energy synthesizing systems but there is little evidence to support it. One possibility which has been raised is a shuttle system involving the malic enzyme (Scott, 1972). Enhanced activities of this enzyme have been reported in rusted bean leaves (Mirocha and Rick, 1967) and in barley leaves infected with the powdery mildew fungus (Ryrie and Scott, 1968). Consistent with the view that increased oxygen consumption occurs via the electron transport system of the mitochondrion is the observation that the concentration of NAD^+ increases in infected leaves in a pattern striking similar to that of oxygen consumption.

The cause of the rise in respiration in infected host tissue still remains a matter of speculation. Regulatory mechanisms controlling leaf respiration could be directly affected by toxins or other substances that are released after infection. Indeed, Allen (1953) proposed that the increased respiration in diseases such as rusts and mildews was caused by the production of a diffusible toxin which uncoupled the rate limiting ATP formation from oxygen consumption in mitochondrial electron transport. Although this paper of Allen stimulated much research in this area, it now seems unlikely that the increase in respiration is attributable to an uncoupling factor.

Respiration like most other metabolic pathways is regulated from one second to the next, depending upon the cell's energy requirements and is effected basically by four different mechanisms. The first involves a regulation of the basic parameters affecting the activities of individual enzymes. These include intracellular pH, concentrations of enzymes and intermediates, essential metal ions and co-factors. The second mechanism involves regulatory enzymes through allosteric effects; the third is concerned with genetic control of the rate of enzyme synthesis; and the fourth mechanism is control exerted through the influence of hormones.

Upsets in concentrations of substrates, metal ion activators and co-enzymes can be effected most drastically in the cell by a breakdown in organelle compartmentation. Special metabolic functions are easily recognized as being uniquely compartmentalized within specific organelles. For example, the chloroplast is not only the site of the photosynthetic apparatus but many other important reactions such as sucrose and starch synthesis. The TCA cycle, electron transport and oxidative phosphorylation are localized in the mitochondrion. The glycolytic and pentose phosphate pathway occur in the cystosol. Organelles provide centres for localizing not only enzymes but also specific metabolities and co-factors. The outer membrane of the chloroplast is semi-permeable with respect to the movement of metabolites between the chloroplast and the cytoplasm. The pyridine nucleotides as well as pentose and hexose diphosphates and hexose monophosphates do not move freely across the chloroplast membrane (Heber and Santarius, 1965; Latzko and Gibbs, 1969). There is strong evidence that chloroplast degeneration occurs in cereals after infection with rust fungi resulting in a breakdown in compartmentation. It is well established that the operation of the pentose phosphate pathway in the cytosol is limited by the availability of the $NADP^+$ and the possibility is raised as to whether enhanced activity of this pathway results from increased availability of $NADP^+$ by release from the chloroplast after membrane degeneration.

Most experiments which indicate that the rise in respiration may be coupled to photosynthesis and the chloroplast have been performed on cereals infected with powdery mildew rather than rust fungi. Millerd and Scott (1963) observed that respiration was not enhanced after inoculation of susceptible etiolated barley leaves with the powdery mildew fungus. This observation was extended by Scott and Smillie (1963, 1966) who showed that the onset of the increase in respiration and the decrease in photosynthesis (see below) occurred about the same time commencing about two days after inoculation. Also no increase was observed in respiration nor in activities of the pentose phosphate pathway dehydrogenases in infected leaves lacking functional chloroplasts but supplied with organic carbon.

This question was further studied by Ryrie and Scott (1968) by studying changes in the intracellular compartmentation of $NADP^+$ after infection of susceptible barley leaves with the powdery mildew fungus. The rationale behind these experiments was that most of the $NADP^+$ in the leaf is localized in the chloroplast and its movement across the chloroplast membrane is strictly limited. It was believed that the NADP-requiring-dehydrogenases of the pentose phosphate pathway are located in the cytosol and that the activities of these enzymes are limited by $NADP^+$ availability and in turn the overall activity of the pathway is limited by the activity of the dehydrogenases. Thus any increase in the cytosol $NADP^+$ concentration would result in enhanced pentose phosphate pathway activity. In non-infected barley leaves, 98% of the $NADP^+$ occurred in the chloroplast. In

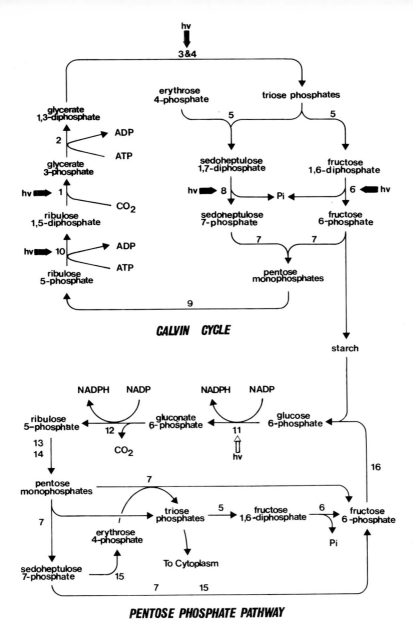

Fig. 1. Oxidative and reductive (Calvin cycle) pentose phosphate pathways. This scheme shows the flow of carbon and the interconversion of carbohydrates in chloroplasts. All possible interconversions are not shown but the basic flow of carbon is indicated. Enzyme control points affected by light are shown ➡ light activated; ⇨ light inhibited. Individual enzymes are numbered: 1. Ribulose 1, 5-diphosphate carboxylase; 2. Phosphoglycerate kinase; 3. Phosphoglyceraldehyde dehydrogenase; 4. Triose isomerase; 5. Aldolase; 6. Fructose 1,6 diphosphate phosphatase; 7. Transketolase; 8. Sedoheptulose 1, 7-diphosphate phosphatase; 9. Epimerase; 10. Phosphoribulokinase; 11. Glucose 6-phosphate dehydrogenase; 12. gluconate 6-phosphate dehydrogenase; 13. Pentose epimerase; 14. Pentose isomerase; 15. Transaldolase; 16. Hexose isomerase.

age comparable leaves six days after inoculation, only 40% of the $NADP^+$ was localized in the chloroplast. On the basis of these results it was suggested that the enhanced pentose phosphate pathway activity results from a break-down of compartmentation of $NADP^+$ in the chloroplast (Ryrie and Scott, 1968).

More recent experiments (Anderson et al., 1974) showing that the pentose phosphate pathway dehydrogenases also occur in the chloroplast necessitates a re-investigation of the experiments of Ryrie and Scott (1968). The chief storage form of carbohydrate in the chloroplast is starch. However, for export of carbohydrate from the chloroplast the starch is degraded to triose phosphate and exported primarily as such. Two pathways can account for this breakdown namely the glycolytic sequence and the oxidative pentose phosphate pathway and both occur in chloroplasts of a range of higher plants. The two dehydrogenases occur in the stroma along with those other enzymes of the pentose phosphate pathway with overlap with the reductive pentose cycle (Fig. 1). The ratio of chloroplast to cyto-plasmic glucose-6-P dehydrogenase activity ranges from 1:3 to 1:1·5 (Ander-son et al., 1974). Both sources of enzymes are inactivated by light and dithiothreitol as well as a low $NADPH/NADP^+$ ratio which as far as the chloroplast enzyme is concerned is accentuated by ribulose bis-phosphate. It is conceivable that with the fall off in photosynthesis, control mechanisms are released on the two chloroplast dehydrogenases resulting in enhanced activity. An indication of the enhancement of the pathway can be gained by an examination of the NADPH mole fraction after infection. This is defined as the ratio $\dfrac{NADPH}{NADP^+ + NADPH}$. From the results of Ryrie and Scott (1968), in barley leaves six days after inoculation with powdery mildew the mole fraction is 0·302 compared with 0·196 in non-infected controls.

C. Photosynthesis

Except in regions of "green islands" it appears that the overall rate of photosynthesis declines as rust infection progresses commencing about three days after inoculation. Expressed on a fresh weight basis, light reaction activities such as the Hill reaction and photoreduction of $NADP^+$, chloro-phyll-protein complexes and activities of enzymes of the Calvin cycle all decline in a very similar manner. However, expressed on a chlorophyll basis there is no difference between infected and non-infected tissues. Some specific experiments on the decline in photosynthesis are discussed below.

Montalbini and Buchanan (1974) have investigated the effect of rust infection on photophosphorylation using chloroplasts isolated from leaves of *Vicia faba* infected with *Uromyces fabae*. These workers conducted time course analyses of both cyclic and non-cyclic photophosphorylation. Cyclic photophosphorylation involves photosystem I and energy of light is converted into chemical energy of ATP according to the following equation:

$$ADP + Pi \overset{hv}{\rightarrow} ATP$$

In non-cyclic photophosphorylation both photosystems I and II are involved and radiant energy is used to transfer electrons from water to $NADP^+$ via ferredoxin as follows:

$$2H_2O + 2ADP + 2Pi \overset{hv}{\rightarrow} 2ATP + O_2 + 4H^+$$
$$+ 4 \text{ Ferredoxin}_{(oxid.)} \qquad + 4 \text{ Ferredoxin}_{(red.)}$$

$$4 \text{ Ferredoxin}_{(red.)} \xrightarrow[\text{reductase}]{\text{Fd.NADP}} 2NADPH + 2H^+$$
$$+ 2NADP^+ + 4H^+ \qquad\qquad + 4 \text{ Ferredoxin}_{(oxid.)}$$

These workers showed that while cyclic photophosphorylation was unaffected by rust infection, non-cyclic photophosphorylation was decreased in chloroplasts from infected leaves. The onset of the decrease occurred at two days after inoculation and appeared to be due to an impairment of electron transport rather than an uncoupling of photosynthetic phosphorylation. It was suggested that the bean rust fungus produces a compound which is functionally similar to 3(3,4-dichloro-phenyl)-1,1-dimethyl urea, a urea derivative which is the basis of an important group of herbicides and which intercepts electron flow between photosystem II and I. Similar results have been obtained with chloroplasts isolated from sugar beet (*Beta vulgaris*) leaves infected with powdery mildew (*Erysiphe polygoni*) (Magyarosy et al., 1976).

In further work using the sugar beet system it has been reported (Magyarosy and Malkin, 1978) that 30 days after inoculation, cytochrome f, cytochrome b559 and cytochrome b6 were all decreased about 30% relative to chlorophyll. Preliminary experiments indicate that a decrease in electron transport cytochrome f relative to chlorophyll occurs in thylakoids four days after infection of barley leaves with the powdery mildew fungus (Holloway and Scott, unpublished observation). The chloroplast electron transport scheme as presently understood is indicated in Fig. 2.

The effect of infection on the rate of photosynthesis during the early stages of infection is less clearly understood. Several reports suggest that photosynthesis may be stimulated initially in rusted bean (Livne, 1964) and wheat (Doodson et al., 1965) and mildewed barley (Allen, 1942; Scott and Smillie, 1966). Although very little follow-up work has been done on this aspect of metabolism in rust-infected tissues, Edwards (1970) has re-examined this

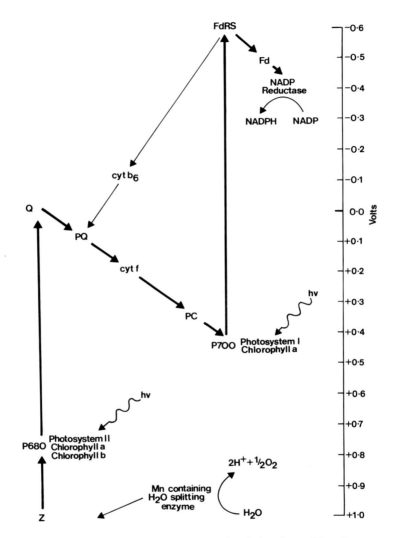

Fig. 2. Photosynthetic electron transport in cholroplasts. The electron transport scheme as presently understood is presented together with an indication of values of redox potential. Abbreviations are as follows: z: intermediate electron acceptor between H_2O splitting enzyme and photosystem II. P-680: photosystem II "core" chlorophyll. Q: photosystem II acceptor. PQ: plastoquinone. PC: plastocyanin. P-700: photosystem I "core" chlorophyll. x: photosystem I electron acceptor. FdRS: ferredoxin reducing substance. Fd: ferredoxin.

apparent stimulation of photosynthesis in mildewed barley. Interestingly enough, he has shown that the concentration of CO_2 used apparently determines whether a stimulation in photosynthesis occurs. Previous workers (Allen, 1942; Scott and Smillie, 1966) had used 1% CO_2 in their experiments and when Edwards used this concentration a stimulation was observed. However, using 0·04% CO_2, he reported no stimulation but a biphasic inhibition of photosynthesis. This raises the important question of what the pathogen is doing to host metabolism so that stimulation of photosynthesis occurs at high concentrations and inhibition at low CO_2 concentrations. One possible explanation is that photorespiration is affected by infection.

Many plants exhibit photorespiration upon illumination and these plants are characterized by having a C-3 pathway of photosynthesis. The process of photorespiration is outlined in Fig. 3. Glycolic acid, the initial substrate of photorespiration, is the major product of photosynthesis under conditions of low CO_2 concentration. During photorespiration CO_2 is released and this could result in an apparent decrease in photosynthesis. The process of photorespiration involves three organelles namely the chloroplast, the peroxisome and the mitochondrion. Release of CO_2 results from a complex mitochondrial reaction in which glycine is converted to equimolar amounts of CO_2, NH_3 and the C-1 group of N^5, N^{10}-methylene-tetrahydrofolate. Thus a consideration of photorespiration in rusted leaves would be important not only for CO_2 release but also for NH_3 release and re-assimilation. Photorespiratory CO_2 production can be as high as 80 μmol/hr/g fresh weight and it is therefore important that leaves must be capable of efficient re-assimilation of NH_3 under photorespiratory conditions.

Two mechanisms have been proposed for ammonia re-assimilation in leaves. First, NH_3 could be refixed directly into glutamate by mitochondrial glutamate dehydrogenase utilizing the NADH generated during glycine decarboxylation. A strong argument against this proposed mechanism is the high K_m of glutamate dehydrogenase for NH_3. The second proposed mechanism for re-assimilation of photo-respired NH_3 is by the sequential action of cytoplasmic glutamine synthetase and the chloroplastic enzyme glutamate synthase, depicted as follows.

Reaction I

$$NH_3 + Glutamate + ATP \xrightarrow[\text{Synthetase}]{\text{Glutamine}} Glutamine + ADP + Pi.$$

Reaction II

Glutamine + α-Ketoglutarate + Ferredoxin$_{(red.)}$

$$\xrightarrow[\text{Synthase}]{\text{Glutamine}} 2\ Glutamate + Ferredoxin_{(oxid.)} + H_2O.$$

H

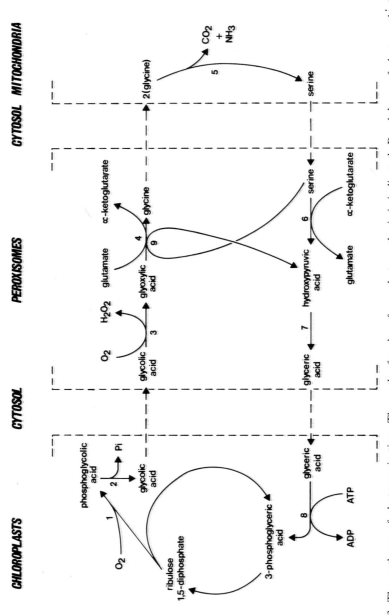

Fig. 3. The pathway of photorespiration. The path of carbon from photosynthetic products to release CO_2 and NH_3 occurs in the mitochondria. Breakdown of photosynthetic products to release CO_2 and NH_3 occurs in the mitochondria. Enzymes are numbered as follows (the enzymes of the C_3 pathway in the chloroplast are not shown): 1. Ribulose, 1,5-diphosphate oxygenase; 2. Phosphoglycolic acid phosphatase; 3. Glycolate oxidase; 4. Glutamate-glyoxylate aminotransferase; 5. Serine hydroxymethyl transferase; 6. Glutamate-hydroxypyruvate aminotransferase; 7. Glyoxylate reductase; 8. Glycerate kinase; 9. Serine-glyoxylate aminotransferase. Reactions are shown by solid lines, movement of substrate across membranes by broken lines.

Sum

$$NH_3 + ATP + \alpha\text{-Ketoglutarate} + Ferredoxin_{(red.)} \rightarrow$$
$$glutamate + ADP + Pi + Ferredoxin_{(oxid.)} + H_2O.$$

Preliminary experiments on barley leaves infected with powdery mildew indicate that this second mechanism of NH_3 assimilation is enhanced over non-infected controls (Maynard and Scott, unpublished observations). Investigations of this area of metabolism would appear well warranted in rust infected plants.

III. Green Islands

"Green islands" are areas of green leaf tissues round regions of individual infection sites where the remainder of the leaf is chlorotic (Cornu, 1881). There is disagreement in the literature on the pattern of "green island" development. Early workers (Rice, 1927; Allen, 1942) thought there was re-greening of chlorotic tissues of leaves infected with rusts and mildews. More recent work suggests that "green islands" result from retention of chlorophyll rather than breakdown and re-greening (Wang, 1961; Bushnell, 1967; Harding et al., 1968). Bennett (1971) has shown that "green islands" round rust pustules contain more Fraction I protein, chlorophyll and chloroplast ribosomes than either non-infected leaves or areas of infected leaves 1 cm or more from the infection court. He also showed that "green islands" are active in the synthesis of Fraction I protein and apparently carry out normal photosynthesis. Wang (1961) demonstrated that most of the fixed carbon dioxide in rusted bean was incorporated into newly synthesized starch. In fact, an interesting aspect of "green island" formation is the large scale deposition of starch.

MacDonald and Strobel (1970) and Bennett (1971) have investigated mechanisms regulating starch deposition in wheat leaves infected with *P. striiformis* and *P. graminis tritici* respectively. Their work was based on the observations of Ghosh and Preiss (1966) that regulation of starch biosynthesis in leaves of spinach is mediated through ADP-glucose pyrophosphorylase. This enzyme was found to be activated by glycolytic intermediates particularly 3-phosphoglycerate and inhibited by inorganic phosphate. MacDonald and Strobel (1970) showed that changes in starch synthesis in infected wheat leaves could be explained by changes in the concentration of orthophosphate in the host/parasite complex. These workers showed that starch synthesis was very rapid immediately prior to sporulation of the fungus, at a time when the orthophosphate concentration of the host/parasite complex had declined rapidly. The decline in orthophosphate could not be accounted for by synthesis of other ethanol soluble phosphate compounds in the infected leaf. It has been shown (Bennett and Scott, 1971) that uredospores are rich in polyphosphates. Synthesis of these polyphosphates at the time of sporulation may well account for a decrease in the concentration of orthophosphate in the host/parasite complex. This would relieve inhibition of ADP-glucose pyrophosphorylase by inorganic phosphate leading to starch synthesis. This in turn could regulate the flow of soluble carbohydrates from host to fungus. A consequence would be regulation of ATP supply for synthetic reactions thus establishing a balance between host senescence and parasite proliferation. Acid invertase may also play an important regulatory role in provision of substrate for synthesis and accumulation of starch round infection sites (Long *et al.*, 1975; Mitchell *et al.*, 1978).

IV. Mobilization of Metabolites in Diseased Leaves

The ability of rust fungi to redirect metabolites from host tissues is the basis of their economic importance. In cereals after infection both root growth and grain yield are seriously reduced (Doodson *et al.*, 1964; Martin and Hendrix, 1967). This appears to result from fungal utilization of metabolites normally translocated from mature leaves. Doodson *et al.* (1964) have shown that infection of wheat leaves with the stripe rust fungus results in a retention of photosynthetic products within the infected leaves. Nutrient retention has also been observed in bean leaves infected with *U. phaseoli* (Zaki and Durbin, 1965). It has also been observed that younger non-infected bean leaves supply photosynthate and orthophosphate to the infected leaves (Livne and Daly, 1966; Pozsar and Kiraly, 1966), a phenomenon not so far observed with rusted cereals. Sempio and Raggi (1966) found that infection of broad bean with *U. fabae* resulted in a marked flow of amino acids to the infected unifoliate leaf. The chief source of these amino acids was the opposite but non-infected unifoliate leaf. Sempio and Raggi proposed that the amino acids were derived from the breakdown of protein in the non-infected leaves. In all these diseases metabolites retained in, or translocated to, the infected leaf accumulate at the site of infection providing the parasite with an adequate supply of nutrients.

The mechanisms whereby rust fungi mobilize host nutrients is obviously of great importance in disease development. In earlier work, Shaw and Samborski (1956) found that accumulation of nutrients during rust infection was inhibited by respiratory inhibitors and anaerobiosis. They concluded that the accumulation was an active process and dependent upon the increased respiration that is characteristic of infected tissues. More recently, it has been shown that alterations to the flow of nutrients within cereals can be produced by changes in sink/source relationships between various organs (King et al., 1967; Neales and Incoll, 1968). Using subterranean clover infected with U. trifolii, Thrower (1965) claimed that the parasitic mycelium constituted a sufficiently vigorous sink to attract nutrients from other parts of the plant.

It has also been claimed that nutrient accumulation in diseased tissues is dependent upon the production of plant hormones by either the infected host or obligate parasite (Shaw and Hawkins, 1958; Pozsar and Kiraly, 1966). The application of IAA to the stem, petiole or midrib of a plant will promote the movement of assimilates to that site (Thrower, 1967; Milthorpe and Moorby, 1969). The effect was generally attributed to the growth promoting properties of the auxin. Davies and Wareing (1965), however, have reported that auxin-induced transport of ^{32}P-orthophosphate could be demonstrated within two hours; i.e. before growth processes became manifest. These workers suggested that phloem transport was accelerated by direct action of the auxin on the sieve tubes. Immature cotton leaves have a high concentration of IAA and import nutrients from other leaves, whereas fully expanded leaves contain much less auxin and export most of their photosynthate (Addicott, 1961). This is in keeping with the view that auxin is produced in tissues requiring nutrients from other parts of the plant.

There have been several reports of enhanced IAA concentrations in rust infected tissues. Shaw and Hawkins (1958) observed massive increases in IAA concentrations in rusted wheat leaves. Daly and Inman (1958) reported higher concentrations of auxin in hypocotyls of safflower infected with P. carthami than in non-infected controls. The maximum increase over the controls was about 600% and occurred at 16 days after inoculation as the first uredospores were being produced. So and Thrower (1976) using Vigna sesquipedalis infected with U. appendiculatus have reported an accumulation of ^{14}C-labelled photosynthate including starch around uredia together with enhanced concentration of auxin eight days after inoculation. Also they reported a progressive increase in the concentration of cytokinins which reached a maximum at the "green island" stage.

Although not directly implicated with nutrient mobilization, ethylene, another endogenous regulator, may play an important role in the establishment of rust fungi. Montalbini and Elstner (1977) reported four stages of ethylene production in bean leaves infected with U. phaseoli. Of particular interest was the observed burst of ethylene production at 13 hours after inoculation coinciding with penetration of the stomates by the germ tube. Daly et al. (1970) studied the effect of exogenously supplied ethylene on rust infection of wheat leaves carrying the temperature sensitive Sr6 locus. They report that leaves at 20°C showing a resistant reaction reverted to a susceptible type reaction when treated with ethylene.

V. Lipid Metabolism in Infected Host Tissues

It has been known for many years that permeability is increased in rust infected tissues (Thatcher, 1939, 1943; Hoppe and Heitefuss, 1974a). In an attempt to explain this phenomenon Hoppe and Heitefuss (1974b,c; 1975a, b) investigated changes in membrane lipid metabolism in beans infected with *U. phaseoli*. Their major findings were first that structural lipid components of the chloroplast namely the monogalactosyl diglyceride and digalactosyl diglyceride decreased in the regions of the pustule; secondly, infection resulted in increased amounts of unsaturated fatty acids; thirdly, infected tissues showed enhanced activities of phospholipase; and fourthly, after infection of susceptible bean leaves, two additional sterols were observed as well as increased amounts of sterols normally present in non-infected leaves. Lösel and Lewis (1974) and Lösel (1978) have reported accumulation of lipids in regions of leaves of *Tussilago farfara* and *Poa pratensis* infected with *Puccinia poarum*. Also they observed a loss of chloroplast glycolipids and phosphatidyl glycerol. This was accompanied by an increase in phosphatidyl ethanolamine, phosphatidyl choline and phosphatidyl serine. At least some of the observed increases in lipids were thought to be associated with the synthesis of fungal membranes.

VI. Summary

The foregoing in no way attempts a comprehensive review of all the published work in the area of intermediary metabolism of rust infected host tissue. It is however an attempt to highlight the more important aspects of the upset of host metabolism after rust infection, and hopefully to offer some direction for further investigations. The discipline of plant metabolism has made rapid strides in the past decade and it now seems relevant to re-investigate upsets in metabolism caused by rust fungi in the light of this newly gained knowledge. Areas one should re-evaluate are chloroplast metabolism, especially the reductive and oxidative pentose cycles; photo-respiration and mechanisms for the re-assimilation of ammonia; electron transport in the chloroplast; and mechanism of nutrient mobilization.

References

Addicott, F. T. (1961). In "Hanbuch der Pflanzen physiologie XIV Wachstum und Wachstoffe" (W. Ruhland, ed.), pp. 829–838. Springer-Verlag, Berlin.

Allen, P. J. (1942). Amer. J. Bot. 29, 425–435.

Allen, P. J. (1953). Phytopathology 43, 221–229.

Anderson, L. E., Toh-Chin, L. M. and Kyung-Eun, Y. P. (1974). Pl. Physiol. 53, 835–839.

Bennett, J. (1971). Ph.D. Thesis, University of Queensland, Brisbane, Australia.

Bennett, J. and Scott, K. J. (1971). Physiol. Pl. Pathol. 1, 185–195.

Bushnell, W. R. (1967). In "The Dynamic Role of Molecular Constituents in Plant-Parasite Interactions" (C. J. Mirocha and I. Uritani, eds), pp. 21–39. Bruce, St. Paul, Minnesota.

Cornu, M. (1881). C.r. heled. Seance. Acad. Sci., Paris 93, 1162–4.

Daly, J. M. and Inman, R. E. (1958). Phytopathology 48, 91–97.

Daly, J. M., Sayre, R. M. and Pazur, J. H. (1957). Plant Physiol. 32, 44–48.

Daly, J. M., Seevers, P. M. and Ludden, P. (1970). Phytopathology 60, 1648–52.

Davies, C. R. and Wareing, P. F. (1965). Planta 65, 139–56.

Doodson, J. K., Manners, J. G. and Myers, A. (1964). J. Exp. Bot. 15, 96–103.

Doodson, J. K., Manners, J. G. and Myers, A. (1965). J. Exp. Biol. 16, 304–17.

Edwards, H. H. (1970). Pl. Physiol. 45, 594–97.

Fromme, F. D. (1913). Bull. Torrey Bot. Club 40, 501–21.

Ghosh, H. P. and Preiss, J. (1966). J. Biol. Chem. 241, 4491–4504.

Harding, H., Williams, P. H. and McNabola, S. S. (1968). Can. J. Bot. 46, 1229.

Heber, U. W. and Santarius, K. A. (1965). Biochem. Biophysica. Acta 109, 390–408.

Hoppe, H. H. and Heitefuss, R. (1974a). Physiol. Pl. Pathol. 4, 5–9.

Hoppe, H. H. and Heitefuss, R. (1974b). Physiol. Pl. Pathol. 4, 11–23.

Hoppe, H. H. and Heitefuss, R. (1974c). Physiol. Pl. Pathol. 4, 25–35.

Hoppe, H. H. and Heitefuss, R. (1975a). Physiol. Pl. Pathol. 5, 263–271.

Hoppe, H. H. and Heitefuss, R. (1975b). Physiol. Pl. Pathol. 5, 273–281.

King, R. W., Wardlaw, I. F. and Evans, L. T. (1967). Planta 77, 261–76.

Latzko, E. and Gibbs, M. (1969). Plant Physiol. 44, 396–402.

Livne, A. (1964). Pl. Physiol. 39, 614–21.

Livne, A. and Daly, J. M. (1966). Phytopathology 56, 170–75.

Long, D. E., Fung, A. K., McGee, E. E., Cooke, R. C. and Lewis, D. H. (1975). New Phytologist 74, 173–182.

Lösel, D. M. (1978). New Phytologist 80, 167–174.

Lösel, D. M. and Lewis, D. H. (1974). New Phytologist 73, 1157–1169.

Lunderstädt, J., Heitefuss, R. and Fuchs, W. H. (1962). Naturwissenschaften 49, 403.

Macdonald, P. W. and Stroebel, G. A. (1970). Pl. Physiol. 46, 126–35.

Magyarosy, A. C. and Malkin, R. (1978). Physiol. Pl. Pathol. 13, 183–189.

Magyarosy, A. C., Schurmann, P. and Buchanan, B. B. (1976). Pl. Physiol. 57, 486–489.

Mains, E. B. (1917). Amer. J. Bot. 4, 179–220.

Martin, N. E. and Hendrix, J. N. (1967). Pl. Dis. Reptr. 51, 1074–6.

Millerd, A. and Scott, K. J. (1963). Aust. J. Biol. Sci. 16, 775–83.

Milthorpe, F. L. and Moorby, J. (1969). Ann. Rev. Pl. Physiol. 20, 117–38.

Mirocha, C. J. and Rick, P. D. (1967). *In* "Dynamic Role of Molecular Constituents in Plant-Parasite Interactions" (J. C. Mirocha and I. Uritani, eds), pp. 121–43. Bruce, St. Paul, Minnesota.

Mitchell, D. T., Fung, A. K. and Lewis, D. H. (1978). *New Phytologist* **80**, 381–392.

Montalbini, P. and Buchanan, B. B. (1974). *Physiol. Pl. Pathol.* **4**, 191–196.

Montalbini, P. and Elstner, E. F. (1977). *Planta* **135**, 301–306.

Neales, T. F. and Incoll, L. D. (1968). *Bot. Rev.* **34**, 107–25.

Pozsar, B. I. and Kiraly, Z. (1966). *Phytopathol. Z.* **56**, 297–306.

Rice, M. A. (1927). *Bull. Torrey Bot. Club* **54**, 63–153.

Ryrie, I. J. and Scott, K. J. (1968). *Pl. Physiol.* **43**, 687–92.

Scott, K. J. (1965). *Phytopathology* **55**, 438–41.

Scott, K. J. (1972). *Biol. Rev.* **47**, 537–572.

Scott, K. J. and Smillie, R. M. (1963). *Nature, Lond.* **197**, 1319–20.

Scott, K. J. and Smillie, R. M. (1966). *Pl. Physiol.* **41**, 289–97.

Sempio, C. and Raggi, V. (1966). *Phytopathol. Z.* **55**, 117–71.

Shaw, M. and Hawkins, A. R. (1958). *Can. J. Bot.* **36**, 1351–1372.

Shaw, M. and Samborski, D. J. (1956). *Can. J. Bot.* **34**, 389–405.

So, May Ling and Thrower, L. B. (1976). *Phytopath. Z.* **87**, 40–47.

Thatcher, F. S. (1939). *Amer. J. Bot.* **26**, 449–458.

Thatcher, F. S. (1943). *Canad. J. Res.*, Sect. C. **21**, 151–72.

Thrower, L. B. (1965). *Phytopathol. Z.* **52**, 269–94.

Thrower, S. L. (1967). *Symp. Soc. Exp. Biol.* **21**, 483–506.

Wang, D. (1961). *Can. J. Bot.* **39**, 1595–1604.

Zaki, A. I. and Durbin, R. D. (1965). *Phytopathology* **55**, 528–29.

6. Host Defense Mechanisms against Infection by Rust Fungi

M. C. HEATH

Botany Department, University of Toronto, Canada

CONTENTS

I. Introduction

Although many crop plants are regarded as host species for certain rust fungi, within each species there are usually cultivars considered to be resistant to infection (for the origin and genetical basis of this resistance, see Chapter 3). However, "resistance" merely implies "incomplete susceptibility", and in practise the term covers interactions involving widely different degrees of fungal growth and host response. Various degrees of susceptibility also may be recognized, based on such features as pustule size and presence of visible signs of necrosis (e.g. Stakman and Harrar, 1957). Thus when a wide range of infection types are found for the same rust disease, it often seems that the boundary between "resistance" and "susceptibility" is drawn at a rather arbitrary point along a continuous gradient of symptom expression. For the purpose of this chapter, therefore, plants which show any infection type differing, from what is regarded as maximum suscepti-bility, will be considered as expressing signs of resistance to infection. In consequence, a defense mechanism will be regarded as any process or feature of the host plant which results in detectable restriction or retardation of fungal growth, even if the end result is merely the production of fewer, otherwise normal, pustules or a slower rate of pustule appearance (slow-rusting).

Lest this preamble suggests to the reader that a detailed list of defense mechanisms is about to follow, it should be quickly pointed out that the nature of these mechanisms is generally either unknown or not unequivo-cally proven. Nevertheless, there is a wealth of information concerning the morphological, physiological and biochemical features or changes which accompany the expression of host resistance, and this can provide clues as to the types of defense mechanisms involved. Thus the following two sec-tions will outline this descriptive information and will discuss some of its implications; which host features may be acting as defense mechanisms will be considered primarily in Section IV. The similarities and differences between the resistance of host and nonhost species will be discussed in the last section of this chapter.

II. Morphological Aspects of Host Resistance

Histological studies have usually revealed no significant differences in spore germination on resistant and susceptible cultivars of the host plant (e.g. Gavinlertvatana and Wilcoxson, 1978; Heath, 1974; Sood and Sackston, 1970; Zimmer, 1965). However, infection from urediospore inoculations usually depends not only on spore germination, but also on the location of a stoma by the germ tube. For several species of rust fungi, this process seems to involve responses to the topography of the leaf surface which result in the "directional growth" of the germ tube towards the stoma, and the subse-quent "recognition" of the stoma so that an appressorium can form over it (Littlefield and Heath, 1979; Wynn and Staples, 1981). Recently, it has been suggested that even the apparently simpler process of direct penetration, such as commonly shown by basidiospore germ tubes, similarly involves contact stimuli from the plant surface (Wynn and Staples, 1981). Thus it is possible that considerable resistance to rust infection might be conferred by "incorrect" pre-penetration behaviour mediated by some surface feature of the plant. Reduced levels of appressorium formation have indeed been reported for *Puccinia coronata* f. sp. *avenae* on sheaths and peduncles of otherwise susceptible oats (Kochman and Brown, 1975), and may be involved in some types of adult plant resistance to *P. striiformis* (Russell, 1977), although the observations reported in the latter study could also be explained by lack of penetration after appressoria had formed. However, although different efficiencies in "finding" and "recognizing" stomata have been reported for *Uromyces phaseoli* var. *vignae* on resistant and susceptible host cultivars (Heath, 1974), they do not seem great enough to have any significant effect on infection. Few other histological studies have looked at pre-penetration behaviour in great detail, but several suggest that in many interactions, including types of adult plant resistance to *P. striiformis* (Mares and Cousen, 1977), slow-rusting by *P. recondita* (Gavinlertvatana and Wilcoxson, 1978), and the resistance to the latter fungus of peduncles and sheaths of wheat (Romig and Caldwell, 1964), resistance similarly is not related to any pre-penetration phenomenon (Hilu, 1965; Rothman, 1960; Sood and Sackston, 1970; Zimmer, 1965).

A more common situation, but again not universal (Hilu, 1965; Rothman, 1960; Sood and Sackston, 1970; Zimmer, 1965), seems to be the cessation of fungal growth during the actual process of penetration into resistant tissue (Allen, 1923a; Gavinlertvatana and Wilcoxson, 1978; Heagle and Moore, 1970; Kochman and Brown, 1975; Martin *et al.*, 1977; Miller *et al.*, 1976; Patton and Spear, 1977; Romig and Caldwell, 1964), and in some cases this has been attributed to morphological factors such as the occlusion of stomata by epicuticular wax (Patton and Spear, 1977) or the form of the stomata (Allen, 1923a). Like differences in pre-penetration fungal behaviour, there is no consistent relationship between a reduction in penetration frequency and the degree or type of resistance eventually exhibited by the plant.

Where a reduced frequency of penetration is accompanied by otherwise "normal" development of the fungus once it enters the tissue, as apparently is the case for infection by several *Puccinia* spp. of peduncles and (or) sheaths of otherwise susceptible wheat and oat plants (Kochman and Brown, 1975; Romig and Caldwell, 1964), resistance is expressed only as a reduction in the number of pustules per inoculum load. However, most examples of host resistance, including examples of adult plant resistance (e.g. Heagle and Moore, 1970; Mares, 1979) and slow-rusting (e.g. Clifford and Roderick, 1978; Gavinlertvatana and Wilcoxson, 1978; Martin *et al.*, 1977), are associated with a reduction in the rate and (or) extent of fungal growth within the tissue. The majority of studies on fungal growth in different infection types have been conducted with the dikaryotic stage of the fungus (i.e. derived from urediospore inoculations) where intercellular infection structures (the substomatal vesicle, infection hypha and haustorial mother cell) subsequently develop before the first haustorium invades a host cell. In a number of these investigations, apparently deleterious effects on the fungus occur during the growth of the infection hypha (Marryat, 1907; Rodrigues *et al.*, 1975; Rothman, 1960; Stakman, 1914; Tani *et al.*, 1975a; Zimmer, 1965), suggesting that the surrounding host cells are secreting a constitutive or inducible toxin. Presumably these toxins are absent (or the fungus is insensitive to them) in other situations where adverse effects on the fungus have not been detected by light or electron microscopy until after the first haustorium is initiated (Allen, 1923a; Heath, 1971, 1972; Hilu, 1965; Littlefield and Aronson, 1969; Mares, 1979; Mendgen, 1978; Skipp and Samborski, 1974). Whether the fungus is affected before or after haustorium initiation seems to be governed by the specific combination of pathogen race and host cultivar used since, at least for *P. graminis* f. sp. *tritici* in wheat, both situations have been reported for different interactions (e.g. Allen, 1923a; Stakman, 1914).

Regardless of whether infection sites can be found where the early stages of fungal growth are adversely affected, most infection hyphae in incompatible hosts produce at least one haustorium, irrespective of the infection type which eventually develops. The subsequent extent of fungal growth depends on the race—cultivar combination and varies from there being little or no intercellular growth after the formation of the first haustorium (e.g. Heath, 1971; 1972; Littlefield, 1973; Littlefield and Aronson, 1969), to situations where fungal growth more closely resembles that in susceptible cultivars, often to the extent of the production or attempted production of uredia (e.g. Heath, 1972; Littlefield, 1973; Mares, 1979; Mares and Cousen, 1977; Stakman, 1914). Typically, this fungal growth is accompanied by morphological host responses, although these are usually seen only after the first haustorium is initiated. The few documented examples of pre-haustorial responses include the collapse of wheat mesophyll cells next to infection hyphae of *P. graminis* f. sp. *tritici* that have not formed haustoria (Martin *et al.*, 1977), the disorganization of mesophyll cells of this same host species near unhealthy-looking infection hyphae of *P. glumarum* (= *P. striiformis*) (Marryat, 1907), and the deposition of irregular electron opaque material, possibly of Golgi origin, along the walls of mesophyll cells of Shokan 1 oat leaves next to stunted infection hyphae of *P. coronata* f. sp. *avenae* (Onoe *et al.*, 1976). In all these examples, the host responses accompany impaired fungal development, and it is possible that they may be caused by products liberated from the dying fungus. However, changes in host morphology may be seen in the absence of any detectable effect on infection structures: for example the development of hemispherical deposits of callose [a wound-response material (Aist, 1976) defined by its fluorescence in ultraviolet light after aniline blue staining] in mesophyll cells of cowpea Queen Anne adjacent to ultrastructurally normal infection hyphae of *U. phaseoli* var. *vignae* (Heath and Heath, 1971), and the guard cell necrosis which accompanies successful entry of *P. graminis* f. sp. *tritici* into Khapli Emmer wheat (Allen, 1926). Thus pre-haustorial host responses may not necessarily be the consequence of fungal necrosis, but may be stimulated by metabolites normally secreted by infection structures. However, since host responses are not seen in all interactions, many hosts must be either insensitive to these substances, or their responses do not result in any change in cell appearance (Heath, 1974).

After the first haustorium develops, the most common response of incompatible hosts, as seen with the light microscope, is the necrosis of haustorium-containing cells. The necrosis is usually recognized by the browning, collapse or dye uptake of the affected cell, although recently it has become fashionable to exploit the observation that dying cells may show autofluorescence under ultraviolet light, probably in association with phenol accumulation (Marte and Montalbini, 1972). These different methods of recognition probably are not comparable in the stage of cell death which they first detect (Marte and Montalbini, 1972), and this may be an important consideration when temporal correlations are attempted between cell death and effects on fungal growth. In addition, it is generally unclear how the necrosis detected by light microscope techniques relates to the various stages of cellular disorganization visualized under the electron microscope. Nevertheless, by whatever means of detection, necrosis of invaded cells in some interactions may begin very soon after the first haustorium is initiated (Heath, 1971; Littlefield and Aronson, 1969; Mendgen, 1978; Rohringer *et al.*, 1979; Skipp and Samborski, 1974); however, in others, including the studied examples of slow-rusting and adult plant resistance, it may not begin until several days after inoculation by which time several, apparently normal, haustoria may have been formed (Heath, 1972; Mares, 1979; Mares and Cousen, 1977; Martin *et al.*, 1977; Rohringer *et al.*, 1979). Non-invaded cells adjacent to invaded ones may also become necrotic or show ultrastructural signs of damage (Allen, 1927; Mares, 1979; Mendgen, 1978; Skipp *et al.*, 1974; Sood and Sackston, 1970; Van Dyke and Hooker, 1969), and in some incompatible interactions, the presence of a diffusable toxin is suggested by the development of often extensive areas of necrosis beyond the limits of fungal invasion (Allen, 1923b; Brown *et al.*, 1966; Jones and Deverall, 1977a; Littlefield, 1973; Mares and Cousen, 1977; Rothman, 1960). Experimental evidence for the involvement of such a toxin in host necrosis has been obtained for several host–rust interactions (Jones and Deverall, 1977b; 1978; Silverman, 1960; Swaebly, 1956), and the presence of a toxin also may account for the displacement of the necrotic area from the region of fungal mycelium in *P. graminis* f. sp. *tritici*—infected Khapli Emmer wheat after application of an electric current (Olien, 1957).

Regardless of whether necrosis at the infection site primarily involves haustorium-containing cells or those without haustoria, if tissue damage is extensive, macroscopic lesions develop on the plant. The size and appearance of these vary with different race–cultivar combinations; for example, the dry, green lesions produced by *U. phaseoli* var. *typica* in French bean cultivar 780, are strikingly different from the turgid, dark brown lesions characteristic of other cultivars (Heath, unpublished). This difference in lesion colour presumably reflects differences in the way in which the host cells have died (Heath, 1976), suggesting that cell death is not a uniform phenomenon. This conclusion is supported by the different appearance of necrotic cells seen by light microscopy (Littlefield, 1973) and by the multitude of changes in ultrastructure that have been reported in haustorium-containing cells prior to their complete disorganization. These changes differ in different race–cultivar interactions and includes increases in vacuolation, reduction in mitochondrial size and number, alterations in mitochondria and chloroplast structure, disappearance of crystal-containing microbodies, and apparent changes in abundance of profiles of endoplasmic reticulum (Littlefield and Heath, 1979). Unfortunately, such observations cannot be used as yet to accurately deduce the metabolic changes which must be taking place in the cell; nevertheless, their diversity strongly re-inforces the suggestion that there are many ways in which a cell may die.

Host cell disorganization may or may not (Fig. 1) be immediately accompanied by the disorganization of the encompassed haustorium (Littlefield and Heath, 1979). If haustorium death is caused directly or indirectly by host necrosis, then this observation suggests that the morphological diversity in necrotic cells is reflected in differences in the toxicity of the products released during cell death. However, it may be argued that some other phenomenon is responsible for haustorium death and that the simultaneous disorganization of host and haustorium in some interactions is coincidental. If this were so, one might expect to find at least some reports of disorganization of the haustorium *before* that of the host cell; these exist (Fig. 2) (Harder *et al.*, 1980; Mendgen, 1978; Prusky *et al.*, 1980), but are so rare (Littlefield and Heath, 1979) that a causal relationship between host and haustorium death still seems a viable hypothesis for most incompatible interactions. Furthermore, the ultrastructural evidence suggests that it is unlikely that host necrosis is normally caused by products liberated from the already dying haustorium, even if the disputed (Kim *et al.*, 1977) suggestion is correct that the dying fungus *can* cause such a response when death is artifically induced (Barna *et al.*, 1974; Király *et al.*, 1972). Nevertheless, it should be emphasized that all these deductions are based on the temporal relationship between morphological signs of necrosis in host and fungus, and as yet there is no clear morphological (or biochemical) definition of when a cell begins to die (Heath, 1980b); even the earliest ultrastructural signs of damage may be the expression of biochemical reactions which began much earlier. However, the conclusion that host necrosis is not normally caused by fungal death is supported in wheat carrying the Sr6 allele for resistance to *P. graminis* f. sp. *tritici* by studies which revealed that inhibition of fungal growth by ethionine or polyoxin D resulted in reduced, not increased, levels of necrosis (Kim *et al.*, 1977).

Even when host death is accompanied by haustorium death, it usually [but not always (Mares, 1979)] has no ultrastructural effect on intercellular portions of the fungus (Littlefield and Heath, 1979). This indication that the intercellular mycelium may stay alive has been confirmed in a number of moderately resistant infection types where mycelium has been demonstrated to grow from pieces of necrotic tissue into susceptible leaves (Burrows, 1960; Chakravarti, 1966; Littlefield and Aronson, 1969; Sharp and Emge, 1958). In contrast, *Melampsora lini* could not be successfully transplanted from a more resistant infection type (Littlefield and Aronson, 1969), and this may indicate that the fungus does die in more severe forms of resistance; however, it is also possible that the transplantation experiment failed because of the sparcity of mycelium at each infection site and the subsequent difficulty in getting the fungus into contact with the susceptible cells. No ultrastructural signs of death of the infection hypha have been found in a similar type of resistance shown by a cowpea cultivar to *U. phaseoli* var. *vignae* (Heath and Heath, 1971).

The fact that the fungus may remain alive after host necrosis does not mean that it continues growing, and there are several reports of host cell death being accompanied by the cessation of growth, or a reduction in its rate (Heath, 1972; Jones and Deverall, 1977a; Rohringer *et al.*, 1979). However, there are also several examples of "ordinary" cultivar resistance, adult plant resistance, slow-rusting and resistance of specific organs of the susceptible plant, where reduced fungal growth can be detected before the onset of necrosis, if the latter occurs at all (Gavinlertvatana and Wilcoxsin, 1978; Mares, 1979; Mares and Cousen, 1977; Martin *et al.*, 1977; Miller and Cowling, 1977; Rohringer *et al.*, 1979).

In spite of cell death being the most common response of resistant hosts to rust infection, it does not occur in every incompatible interaction (Miller and Cowling, 1977; Ogle and Brown, 1971; Stakman, 1914) and there may be other ways in which resistant plants respond to infection. For example, in cowpea cultivar Queen Anne, only about half the infection sites of *U. phaseoli* var. *vignae* show host necrosis; in the remainder, the host plasmalemma surrounding the first-formed haustorium becomes associated with what seems to be phospholipid material (Fig. 3) and concurrently, the haustorium shows signs of starvation (Heath and Heath, 1971). The haustorium subsequently becomes encased in callose-containing material (Fig. 4)

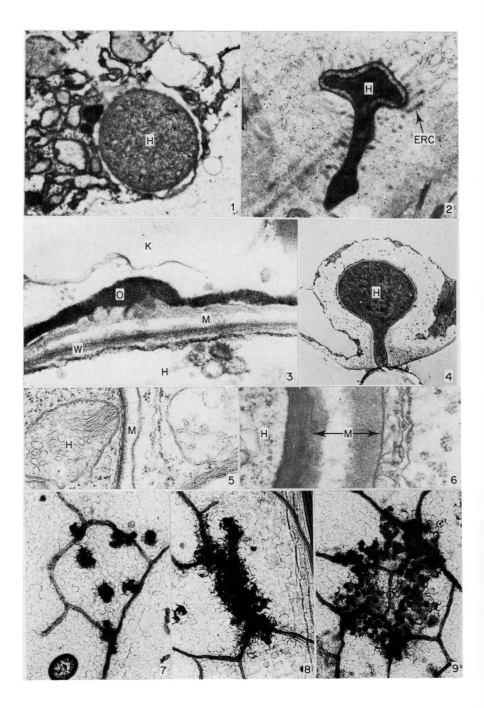

and intercellular growth is as restricted as it is in infection sites where the invaded host cell dies (Heath, 1971; Heath and Heath, 1971). Similar encasement of haustoria has been reported in light and electron microscopic studies of other incompatible (and some compatible) interactions involving rust fungi, but the fact that in each case deleterious effects on the haustorium are usually detected before encasement, suggests that this is probably a response to the already moribund or impaired fungus (Littlefield and Heath, 1979).

Several ultrastructural studies have also suggested that death of the haustorium-containing cell may be preceded by a faster accumulation of electron opaque material in the extrahaustorial matrix (the region between the

---

Figs 1 and 2. Haustoria (H) of *Puccinia graminis* f. sp. *tritici* in wheat mesophyll cells carrying the Sr6 allele for resistance. Note the ultrastructurally normal haustorium in the disorganized host cell shown in Fig. 1, and the disorganization of the haustorial cytoplasm in the normal-looking host cell shown in Fig. 2. In Fig. 2, the endoplasmic reticulum complex (ERC), which typically develops around young haustoria, is still present. Fig. 1 × 17 500; Fig. 2: × 20 400. (From Harder *et al.* (1979). Reproduced by permission of the National Research Council of Canada from *Can. J. Bot.* **57**, 2626–2634.)

Fig. 3. Electron opaque material (O) associated with the host plasmalemma surrounding the first haustorium (H) of *Uromyces phaseoli* var. *vignae* formed in cowpea cultivar Queen Anne. An incomplete callose-containing encasement (K) is present, but the haustorium already contains a large vacuole, typical of senescing fungal cells. M = extrahaustorial matrix; W = haustorium wall. × 60 500. (M. C. Heath, unpublished.)

Fig. 4. A completely encased haustorium (H) in the same host–pathogen interaction as that shown in Fig. 3. The haustorial vacuole is out of the plane of the illustrated section. × 3100. (From Heath and Heath (1971). *Physiol. Plant Pathol.* **1**, 277–287.)

Fig. 5. The extrahaustorial matrix (M) surrounding a haustorium (H) of *Melampsora lini* in a compatible flax cultivar. × 35 000. (Courtesy of M. D. Coffey, from Littlefield and Heath (1979). "Ultrastructure of Rust Fungi", Academic Press.)

Fig. 6. The extrahaustorial matrix (M) surrounding a haustorium (H) of *M. lini* in a flax cultivar containing the K gene for resistance. Note the electron opaque material in the matrix and the increased matrix width compared with that in Fig. 5 × 37 000. (From Coffey (1976). Reproduced by permission of the National Research Council of Canada from *Can. J. Bot.* **54**, 1443–1457.)

Figs 7, 8 and 9. Cleared whole leaves of flax infected with *M. lini*, race 1, illustrating the different morphological expressions of resistance conditioned by the L gene (Fig. 7), the M gene (Fig. 8), and the K gene (Fig. 9). Necrotic host cells are clearly identifiable by their uptake of the dye, acid fuchsin. All × 62. (From Littlefield (1973). *Physiol. Plant Pathol.* **3**, 241–247.)

haustorial wall and the surrounding host plasmalemma) than is seen in compatible interactions (Figs 5, 6) (Littlefield and Heath, 1979). For resistance of wheat to *P. graminis* f. sp. *tritici* conditioned by the Sr6 gene, the accumulation of this material in both resistant and susceptible cultivars seems to coincide with a reduction in ^{3}H-leucine incorporation into the haustorium and, more inexplicably, into the host cell (Manocha, 1975). A reduction in incorporation of ^{3}H-leucine into the haustorium also has been reported for one resistant cultivar of French bean infected with *U. phaseoli* var. *typica* (Mendgen, 1977), but in this case there seems to be no obvious change in appearance of the extrahaustorial matrix, nor is this reduced label incorporation characteristic of all resistant cultivars (Mendgen, personal communication).

One final point which should be regarded as significant in any discussion of rust resistance is that for any given host–pathogen combination, infection sites are rarely identical, even in a single piece of tissue. Although the majority may show similar degrees of fungal growth and host response, others are sure to be found where the fungus has grown much less or even much more than the norm, and where the usual necrosis may be reduced or absent (Heath, 1971; Mares and Cousen, 1977; Martin *et al.*, 1977; Rohringer *et al.*, 1979; Rothman, 1960; Skipp and Samborski, 1974). In some cases this variation may result in a mixture of clearly distinguishable infection types detectable by light microscopy (Ashagari and Rowell, 1975; Mares, 1979) or, in the case of mesothetic infection types, detectable both microscopically and macroscopically (e.g. Thatcher, 1943). While it could be argued that this variation merely reflects differences in the capability of each fungal individual for growth, in the author's experience this variability is much more common in resistant hosts than in susceptible ones.

III. Physiological and Biochemical Aspects of Host Resistance

The diversity in histological and ultrastructural expressions of incompatibility in rust infected plants seems likely to be a reflection of similar diversity in metabolic interactions between plant and pathogen. Thus a description of the physiological and biochemical features of a specific infection type has little meaning unless it is related to the accompanying behaviour of host and fungus. Unfortunately, the value of histological data to biochemical investigations has not always been recognized, and much early work can provide only equivocal information concerning possible defense mechanisms since it is now known that it was carried out long after the first signs of resistance were expressed. In one of the few investigations correlating histological and biochemical data, Tani and co-workers have shown that the first detected response of Shokan 1 oat leaves to infection by race 226 of *P. coronata* f. sp. *avenae* is an apparent increase in synthesis of host messenger ribonucleic acid (mRNA). This increase occurs between 12 and 16 hours after inoculation and its reality is supported by a subsequent increase, at 14–20 hours after inoculation, in protein synthesis (Tani and Yamamoto, 1979). Significantly, both events accompany the increase in Golgi activity mentioned earlier, and occur at the time when the first signs of resistance become apparent (i.e. reduced growth of the infection hypha (Tani *et al.*, 1975a). Treatment of the host with the protein synthesis inhibitor, blasticidin S, allows the infection hyphae to develop as well as they do in the compatible interaction, and also prevents the appearance of six minor protein spots which were detectable after electrophoresis of extracts from untreated, infected, incompatible tissue (Tani and Yamamoto, 1979). Increased rates of protein synthesis also have been observed during the first 18 hours after inoculation of several resistant flax cultivars with *M. lini* (Von Broembsen and Hadwiger, 1972), and thus may be a general feature of incompatible interactions involving rust fungi. Similar changes in protein synthesis were not observed in the comparable compatible interaction in either the flax or oat investigations. However, it should be pointed out that most of the results just described were deduced from comparisons of the fate of exogenously supplied radioactive precursors in inoculated and uninoculated, resistant and susceptible tissue; thus they must be interpreted with caution since the uptake and distribution of the radioactive label may not be the same in all situations (Daly, 1976).

In the oat—*P. coronata* f. sp. *avenae* interaction, increased synthesis of cytoplasmic ribosomal RNA (rRNA) was detected from 28 to at least 48 hours after inoculation, and seemed to be accompanied by increased turnover since it did not result in any net changes in total rRNA (Tani *et al.*, 1973). Although this rRNA synthesis begins at about the time of necrosis of the haustorium-invaded cell, it may, in fact, take place in the surrounding cells since many ultrastructural and biochemical studies suggest that necrotic cells, elicited by various means, induce biosynthetic activity in their neighbors (Heath, 1980b; Uritani, 1971). Initial increases in the activity of phenylalanine ammonia lyase (PAL) were observed in both compatible and incompatible interactions, but a second increase was found only in the latter at about the time of host necrosis. Negligible changes in total peroxidase activity were detected in extracts made up to 48 hours after inoculation although extracts from compatible and incompatible interactions showed slightly different changes in the relative intensities of the various electrophoretically produced bands of this enzyme (Tani and Yamamoto, 1979). In the incompatible interaction, fungal growth ceases soon after the collapse of the invaded cell (Tani *et al.*, 1975a).

Few other incompatible interactions have been examined in as much detail as the one just described. The main exception is the interaction between *P. graminis* f. sp. *tritici*, race 56, and wheat carrying the temperature sensitive Sr6 allele for resistance. Histological studies have shown that host necrosis begins soon after the formation of the first haustorium and although some colonies stop growing by about 60 hours after inoculation, about 60% continue to grow and increasing numbers of necrotic, haustorium-containing cells are formed as more haustoria develop (Skipp and Samborski, 1974). Changes in respiration rates of infected tissue are similar to those seen in near-isogenic plants carrying the sr6 allele for susceptibility (Antonelli and Daly, 1966) and necrosis is not accompanied by any detectable increase in the concentration of phenolic compounds (Seevers and Daly, 1970a); however there are slight increases in the incorporation of phenolic precursors into various aromatic compounds (Fuchs *et al.*, 1967; Rohringer *et al.*, 1967). High levels of ethylene production are not characteristic of this incompatible interaction (Daly *et al.*, 1971), but there is a rapid increase in peroxidase activity, first detectable about two days after inoculation, which is more marked than that seen in susceptible tissue (Frič and Fuchs, 1970; Seevers and Daly, 1970b). According to Daly (1972), peroxidase may act as an indoleacetic acid (IAA) decarboxylase and thus may be responsible for the subsequent increase in the latter activity in resistant leaves.

In the two systems just described, either near-isogenic resistant and susceptible host lines were used, or the incompatible and compatible interactions were produced in the same host cultivar by different races of the pathogen. However, many other studies have compared genetically unrelated resistant and susceptible plants, making it difficult to assess whether some of the apparent metabolic differences between resistance and susceptibility are merely reflections of inherent differences between the two host plants used. Nevertheless, comparative studies of stem rust–susceptible Little Club and resistant Khapli have suggested that resistance is accompanied by a more rapid increase in respiration rate (Samborski and Shaw, 1956) and accumulation of phenolic compounds (Király and Farkas, 1962), a faster increase in RNA and decrease in histone content per nucleus (Bhattacharya and Shaw, 1968), a longer sustained period of high IAA decarboxylase activity (Shaw and Hawkins, 1958), a higher sustained redox potential (Kaul and Shaw, 1960) and a decrease in leaf protein content (Quick and Shaw, 1964). Compared with Little Club, Khapli also contains higher pre-inoculation levels of phenolic compounds (Király and Farkas, 1962) and 2-0-glucosyl-2,4-dihydroxy-7-methoxyl-1,4-benzoxazin-3-one (GDMBO) (ElNagy and Shaw, 1966), but these last two differences have not been found in other comparisons of wheat cultivars resistant or susceptible to stem rust, and the relationship of high levels of these compounds to resistance appears not to be a causal one (Király and Farkas, 1962; Knott and Kumar, 1972). Correlations also have been made between resistance to *Cronartium fusiforme* and the high content of the cortical monoterpene, β-phellandrene, in slash pine (Rockwood, 1974), and the low level of this compound in loblolly pine (Rockwood, 1973); but there is no evidence that this monoterpene plays a direct role in resistance or susceptibility. Rapid increases in respiration rate have been reported in moderately and highly resistant French bean leaves after infection with *U. phaseoli* var. *typica* (= *U. appendiculatus*) (Sempio and Barbieri, 1964), and necrosis in the moderate type of resistance is accompanied by ethylene emission (Montalbini and Elstner, 1977). Increased peroxidase activity seems to accompany many [but not all (El-Gewely *et al.*, 1974)] types of necrotic-fleck resistance, irrespective of whether near-isogenic resistant and susceptible lines are compared (e.g. Daly *et al.*, 1971; Johnson and Cunningham, 1972; Montalbini, 1972).

In contrast to their intensive study in diseases involving non-biotrophic fungi, antifungal compounds which accumulate after infection [i.e. phytoalexins (Kuc, 1972)] have only rarely been looked for in rust infections. In resistant cultivars of French bean infected with *U. phaseoli* var. *typica*, phytoalexins, particularly phaseollin, accumulate at about the time the host cells become necrotic (Bailey and Ingham, 1971; ElNaghy and Heitefuss, 1976), but their effect on fungal growth is unclear (ElNaghy and Heitefuss, 1976). The presence of phytoalexins may also account for the antifungal activity of leaf diffusates from resistant coffee cultivars infected with *Hemileia vastatrix* and could be the cause of the poor growth of infection hyphae in these plants (Rodrigues *et al.*, 1975). Antifungal compounds including coniferyl alcohol and coniferyl aldehyde have been shown to accumulate more rapidly in resistant cultivars of flax infected with *M. lini* than in near-isogenic susceptible plants, and accumulation is most rapid where there is rapid restriction of fungal growth (Keen and Littlefield, 1979). Conferyl alcohol is a primary precursor in lignin biosynthesis and the observation of partially ethanol-soluble, phloroglucinol-staining, material in fresh sections of infected areas suggests that this compound is at least in the right place to potentially have an effect on fungal growth (Keen and Littlefield, 1979). Histochemical tests have also suggested an association between potentially antifungal phenolic materials in the cell wall of wheat leaf bases and resistance to *P. striiformis* (Hartley *et al.*, 1978).

IV. Possible Host Defense Mechanisms against Rust Invasion

From the histological and ultrastructural studies of rust resistance, one can make three generalizations: (1) there is a great diversity between different incompatible interactions and each one exhibits its own characteristics in terms of degree of fungal growth, type and timing of host responses, and the apparent effect of these responses on the fungus; (2) fungal development and host responses may differ at different infection sites in the same tissue; and (3) from the limited available data, there seems to be no consistent relationship between morphological features of incompatibility and the category of resistance (i.e. adult plant resistance, age-unrelated necrotic fleck resistance or slow-rusting). The first two generalizations suggest that either different features are of primary importance in different interactions (and even in different infection sites of the same tissue), or that there is a single mechanism common to all situations which is unrelated to most or all of the observed host responses. With respect to this last suggestion, it should not be ignored that biotrophic fungi, such as the rusts, induce significant metabolic changes in their *susceptible* hosts during the first few days after infection (see Chapter 5), and that once compatibility is established, haustorium-containing cells and/or surrounding tissue may be incapable of reacting in a resistant manner (Heath, 1980a; Skipp and Samborski, 1974; Tani *et al.*, 1975b). Thus, as suggested by Daly and co-workers (Daly, 1972; Daly *et al.*, 1970), the successful establishment of susceptibility may be the most important factor controlling infection type, and the observed responses associated with resistance may merely be nonspecific secondary phenomena, elicited because compatibility has not been established.

Plausible as this hypothesis may be, it certainly is more difficult [but not impossible (Daly, 1972)], to explain such "induced susceptibility" in terms of the gene-for-gene relationship which exists between many pathogens and their hosts (Ellingboe, 1976). It is conceptually much easier to assume that "induced susceptibility" is involved in establishing the "basic compatibility" (Ellingboe, 1976) between a rust fungus and its host *species*, upon which host, gene-for-gene, resistance may be superimposed. Nevertheless, it is difficult to see how even basic compatibility could be induced without involving the metabolic machinery of the host cell; yet there are at least two reports that treatments with heat and metabolic inhibitors, which may reduce both host and nonhost resistance to rust fungi, have no apparent effect on the comparable compatible interaction (Heath, 1979; Tani *et al.*, 1976). Therefore from all points of view it seems justifiable at present to assume that host resistance is due to active defense rather than an absence of induced susceptibility, and that at least some of the aforementioned responses of resistant hosts are related to the cessation of fungal growth.

On this assumption, the morphological data suggest that each host–pathogen interaction is unique, and therefore the same defense mechanisms are not necessarily involved in all (Heath, 1976; Ingram, 1978). Nevertheless, remarkably little can be indisputably deduced about the nature of these mechanisms. It does seem fairly certain that few examples of host resistance can be fully explained on the behaviour of the fungus before or during penetration. Moreover, although large reductions in penetration frequency usually have been attributed to morphological features of the plant, the involvement of phenomena other than physical barriers have by no means been ruled out and this type of resistance needs to be investigated further.

Once the fungus enters the tissue, a similar range of metabolic changes seem to accompany resistance to rust infection as found for resistance to other types of fungal pathogens, and there is similar uncertainty as to their effects on the fungus (Heath, 1980b). The role of the most prominent and universal host response, cell necrosis, has been particularly controversial in recent years (Heath, 1976; Ingram, 1978), and as Ingram (1978) has pointed out, part of the reason for this is the use of the term "hypersensitivity" to cover all forms of necrosis, thus implying that a single phenomenon is being considered. As discussed earlier, cell death probably is not a single phenomenon, and its role in resistance may be expected to vary depending on whether haustorium function is affected, or whether substances are produced (or elicited) during necrosis which affect fungal growth (Heath, 1976; 1980b; Ingram, 1978); thus the lack of a consistent relationship between necrosis and fungal growth in some interactions (Brown *et al.*, 1966; Littlefield, 1973; Mares and Cousen, 1977; Martin *et al.*, 1977; Ogle and Brown, 1971) is not a convincing argument against necrosis acting as a defense mechanism in others. Several studies *do* suggest a temporal relationship between host necrosis and an effect on fungal growth, and even where such a relationship is not apparent, host necrosis may adversely affect the fungus in a number of more subtle ways (Heath, 1976). However, it is also true that histological and ultrastructural studies suggest that cell death is not the only host response which may affect fungal growth, and reduced growth may occur in its absence.

Of the suggested roles in resistance of other host responses, some seem intuitively more probable, or are supported by better evidence, than others. All these suggestions are based on the correlation of the host response with some effect of the fungus, and some of the better examples include the increased mRNA synthesis, protein synthesis and Golgi activity which accompany reduced growth of infection hyphae of *P. coronata* f. sp.*avenae* (Tani and Yamamoto, 1979), the morphological change in the host plasmalemma which accompanies signs of starvation in haustoria of *U. phaseoli* var. *vignae* (Heath and Heath, 1971), and the appearance of phloroglucinol-staining compounds at the time of the cessation of growth of *M. lini* (Keen and Littlefield, 1979). Other correlations seem more tenuous, such as the role of microbodies in seedling resistance of safflower to *P. carthami* (Zimmer, 1970). However, in all these examples, the possibility exists that the observed correlations are coincidental.

Since resistance to rust infection is usually accompanied by more than one host response, results that indicate what host features do *not* seem to have a role in resistance or susceptibility are of value, although these too tend to be correlative and open to alternative interpretations. As mentioned earlier, resistance to stem rust does not seem to be related to pre-inoculation concentrations of GDMBO or phenolic compounds in wheat (Király and Farkas, 1962; Knott and Kumar, 1972), nor is it correlatable to levels of glycinebetaine and choline elicited after infection (Pearce and Strange, 1977) or the loss of antigenic proteins (Barna *et al.*, 1975). Wheat resistance to stem rust conditioned by the temperature sensitive Sr6 allele apparently does not involve the accumulation of 2-hydroxyputrescine amides, since such an accumualtion also takes place even when plants are grown at the temperature at which the allele is ineffective (Samborski and Rohringer, 1970). Infection type also is not related to pre-inoculation sterol levels (Nowak *et al.*, 1972), nor does peroxidase activity appear to be involved in this particular example of resistance because both resistance and ethylene-induced susceptibility are accompanied by increased activity of this enzyme (Daly *et al.*, 1970). Similarly, peroxidase and PAL activities do not seem to be responsible for the resistance of Shokan 1 oat leaves to *P. coronata* f. sp. *avenae* because similar activites are found in the same plant treated with blasticidin S to induce susceptibility (Tani and Yamamoto, 1979). However, such observations, as many others, have to be interpreted with caution, since for example, the lack of a clear understanding of the role of peroxidase in infected tissue makes it possible that this enzyme may not be performing the same function in the "artificially" susceptible tissue as it does in resistant interactions. The lack of a significant drop in peroxidase activity after established, Sr6 controlled, resistant infection types of wheat are converted to susceptible ones by a rise in temperature (Seevers and Daly, 1970b), or by ethylene treatment (Seevers *et al.*, 1971), also does not necessarily imply the unimportance of this enzyme in resistance, since peroxidase activity may reside in cells which became necrotic before susceptibility was induced. Probably the best way to demonstrate a lack of involvement in a known process in resistance is to exploit mutations affecting this process. For example, catachol oxidase seems unlikely to be involved in a necrotic fleck resistance of maize to *P. sorghi* since plants continue to be resistant when homozygous for a null mutation of this enzyme (Pryor, 1976). To continue the list of events or compounds not seemingly associated with resistance, no qualitative differences in phenolic compounds have been found in white pine trees resistant and susceptible to *Cronartium ribicola* (Hanover and Hoff, 1966). Resistance or susceptibility of French bean to *U. phaseoli* var.

typica does not seem to be based on differences in composition or content of phospholipids, phospholipid fatty acids, or sterols (Hoppe and Heitefuss, 1975a, 1975b), and the changes in host permeability which occur in incompatible interactions are found only after cell death, possibly as a consequence of phytoalexin production (ElNaghy and Heitefuss, 1976). Apart from the flax study mentioned earlier, a role for phytoalexins in host resistance has not yet been clearly demonstrated; moreover the possibility that some rust fungi are insensitive to the phytoalexins known to affect other plant pathogens is suggested by the continued normal development of *Phakaspora pachyrhizi* in susceptible soybeans in spite of the early production of such compounds (Deverall *et al.*, 1977). Encasement of haustoria in callose-containing material also does not seem to be a significant defense mechanism in resistant plants since it usually occurs after adverse effects on the haustorium can be detected (Littlefield and Heath, 1979).

Although there is much information concerning the genetic basis of rust resistance (see Chapter 3), it provides few clues as to the identity of the defense mechanisms involved. Histological studies have shown that in *M. lini*—infected flax and *P. graminis* f. sp. *tritici*—infected wheat, the morphological expression of resistance typical for the given race–cultivar interaction seems governed by the type of gene for resistance present in the cultivar (Figs 7, 8, 9) (Littlefield, 1973; Rohringer *et al.*, 1979). Since coniferyl alcohol and coniferyl aldehyde accumulate during the resistance of flax to *M. lini* conditioned by several different genes, Keen and Littlefield (1979) have suggested that the same defense mechanism is involved in all, but that the different resistance genes, or alleles, govern differences in timing of resistance expression. This may prove to be correct for resistance to *M. lini* but may not be true for other situations: the complexity of the genetic basis of some forms of rust resistance is demonstrated by the fact that the morphological expression of resistance may be altered by the genetic background in which a given gene for resistance is placed, but not always to the same extent in all tissues of the plant (Rohringer *et al.*, 1979). Similarly for apple cultivars infected with *Gymnosporangium junipera-virginianae*, it has been suggested that although two host genes control the presence or absence of fungal pycnia, modifying genes determine the extent of leaf damage (Aldwinkle *et al.*, 1977).

The observation that there are many different host responses, and various effects on the fungus, in a single host–pathogen interaction raises the question of whether more than one defense mechanism may be active in the same tissue. For example, when penetration frequency is reduced, could this be caused by a different host response or feature than the one that inhibits fungal growth in infection sites where penetration takes place? Similarly, in the Shokan 1 oat–*P. coronata* f. sp. *avenae* interaction (Tani *et al.*, 1975a), is the poor growth of the infection hypha caused by a different host–pathogen interaction than the one that causes the death of the invaded cell after a haustorium is initiated? Also, is this host necrosis involved in the subsequent cessation of fungal growth, or would growth have ceased even if the cell had not died? None of these questions has been satisfactorily answered as yet, but it has been suggested for *U. phaseoli* var. *vignae* that there are many stages during the infections process where an interaction occurs between the plant and the fungus (Heath, 1974). Intuitively it seems unlikely that all these interactions are identical, and certainly the ultrastructural evidence suggests that the plant may respond differently at each stage of infection (Heath, 1974). Susceptibility has been suggested to depend on a "correct" plant response during each interaction while an "incorrect" response at any stage may lead to resistance (Heath, 1974). A similar concept has been developed for resistance to the powdery mildew fungus, *Erysiphe graminis*, (Ellingboe, 1972), and in this case, the proportion of individuals getting over each "hurdle" is unique for each combination of gene for resistance in the host, and gene for avirulence in the pathogen. Whether a single gene for resistance exerts an effect on more than one of these potential hurdles remains to be established, and it may be that, irrespective of major gene action, the more "hostile" environment of a resistant plant makes "incorrect" host responses more likely to occur, particularly if the "correct" response depends on some metabolic activity of the fungus which may be affected by its reduced vigor. Thus, not only does this "hurdle" hypothesis allow for more than one defense mechanism to be active in a single incompatible host–pathogen interaction, but it also provides for the observed variability between infection sites in the same tissue, since fungal growth and host response at each will depend on how many hurdles each individual fungus has overcome; therefore the most common infection type found in a given host–pathogen interaction will be determined by which host response is the most effective in restricting fungal growth.

V. Differences between Host and Nonhost Defense Mechanisms

Although the defense mechanisms possessed by nonhost plants (i.e. plant *species* for which a given rust fungus is not a pathogen) are no more clearly understood than are those possessed by resistant cultivars of the host plant, there is good evidence that different mechanisms may be involved in these two types of resistance. "Incorrect" pre-penetration behaviour seems much more prevalent on nonhosts (Heath, 1974; 1977; Sempio and Barbieri, 1966;

Wynn, 1976) and is probably of greater significance in resistance, although rarely is it completely effective (Heath, 1977). Once the fungus has entered the tissue, nonhost defense mechanisms characteristically occur before, and apparently prevent, the formation of the first haustorium (Gibson, 1904; Heath, 1974, 1977; Leath and Rowell, 1966; Raggi, 1964; Sempio and Barbieri, 1966), and where it has been investigated in detail, the morphological expression of these defense mechanisms differ from the normally post-haustorial responses of the same plant species to incompatible races of the "pathogenic" rust fungus (Heath, 1972, 1974; unpublished data).

The responses seen in nonhost plants have been suggested to be part of a battery of potential defense mechanisms possessed by every plant, of which one or more is nonspecifically elicited by any plant pathogen for which the plant is not a host (Heath, 1981). Consequently, pathogenic rust fungi are those which have been able to "overcome" these nonhost defense mechanisms (Heath, 1974). Recent evidence suggests that they may be able to do so by actively preventing the expression of the plant's responses (Heath, 1980a, 1981b), and a similar suppression of its formation may explain the absence of callose in most compatible, and many incompatible, host plants even though this material should normally be formed in response to the holes in the cell wall made during haustorium initiation (Heath, 1974). If such suppression of host responses exists in compatible interactions, it adds another dimension to the theory of "induced susceptibility", where, rather than (or as well as) conditioning susceptibility *per se*, fungal activity induces the susceptible state (basic compatibility of Ellingboe, 1976) through inhibition of defense reactions. This latter type of induced susceptibility would not require any active participation by the susceptible host, and is thus in accord with the aforementioned lack of effect of metabolic inhibitors on compatible interactions.

If multiple potential defense mechanisms, different in each plant, have to be specifically overcome before basic compatibility can be established (Heath, 1974), it is not surprising that a given rust fungus can successfully attack only a small number of plant species, or that its host range does not significantly change with time. However, compared with nonhost resistance, rust fungi seem to find it relatively easy to overcome most types of gene-for-gene resistance which have been naturally or artifically re-introduced into the host (see Chapter 3). Thus it becomes important in a search for future methods of control of rust diseases to know whether the greater durability of nonhost resistance is due to the greater number of defense mechanisms involved, or whether the types of defense mechanisms characteristic of nonhosts are more difficult to overcome than the ones "put back" into host species (Heath, 1981a).

References

Aist, J. R. (1976). *A. Rev. Phytopath.* **15**, 145–163.

Aldwinckle, H. S., Lamb, R. C. and Gustafson, H. L. (1977). *Phytopathology* **67**, 259–266.

Allen, R. F. (1923a). *J. Agric. Res.* **23**, 131–152.

Allen, R. F. (1923b). *J. Agric. Res.* **26**, 571–604.

Allen, R. F. (1926). *J. Agric. Res.* **32**, 701–725.

Allen, R. F. (1927). *J. Agric. Res.* **34**, 697–714.

Antonelli, E. and Daly, J. M. (1966). *Phytopathology* **66**, 610–618.

Ashagari, D. and Rowell, J. B. (1975). *Proc. Am. Phytopath. Soc.* **2**, 82 (Abstr.)

Bailey, J. A. and Ingham, J. L. (1971). *Physiol. Pl. Pathol.* **1**, 451–456.

Barna, B., Érsek, T. and Mashaâl, S. F. (1974). *Acta Phytopath. Acad. Scie, Hungaricae* **9**, 293–300.

Barna, B., Balázs, E. and Király, Z. (1975). *Physiol. Pl. Pathol.* **6**, 137–143.

Bhattacharya, P. K. and Shaw, M. (1968). *Can. J. Bot.* **46**, 96–99.

Brown, J. F., Shipton, W. A. and White, N. H. (1966). *Ann. Appl. Biol.* **58**, 279–290.

Burrows, V. D. (1960). *Nature* **188**, 957–958

Chakravarti, B. P. (1966). *Phytopathology* **56**, 223–229.

Clifford, B. C. and Roderick, H. W. (1978). *Ann. Appl. Biol.* **89**, 295–298.

Daly, J. M. (1972). *Phytopathology* **62**, 392–400.

Daly, J. M. (1976). *In* "Encyclopedia of Plant Physiology. 4. Physiological Plant Pathology" (R. Heitefuss and P. H. Williams, eds), pp. 27–50. Springer-Verlag, Berlin, Heidelberg and New York.

Daly, J. M., Seevers, P. M. and Ludden, P. (1970). *Phytopathology* **60**, 1648–1652.

Daly, J. M., Ludden, P. and Seevers, P. (1971). *Physiol. Pl. Pathol.* **1**, 397–407.

Deverall, B. J., Keogh, R. C. and McLeod, S. (1977). *Trans. Br. Mycol. Soc.* **69**, 411–415.

El-Gewely, M. R., Smith, W. E. and Colotelo, N. (1974). *Biochem. Genetics* **11**, 103–119.

Ellingboe, A. H. (1972). *Phytopathology* **62**, 401–406.

Ellingboe, A. H. (1976). *In* "Encyclopedia of Plant Physiology. 4. Physiological Plant Pathology". (R. Heitefuss and P. H. Williams, eds), pp. 761–778. Springer-Verlag, Berlin, Heidelberg, New York.

ElNaghy, M. A. and Heitefuss, R. (1976). *Physiol. Pl. Pathol.* **8**, 269–277.

ElNaghy, M. A. and Shaw, M. (1966). *Nature* **210**, 417–418.

Frič, F. and Fuchs, W. H. (1970). *Phytopath. Z.* **57**, 161–174.

Fuchs, A., Rohringer, R. and Samborski, D. J. (1967). *Can. J. Bot.* **45**, 2137–2153.

Gavinlertvatana, S. and Wilcoxsin, R. D. (1978). *Trans. Br. Mycol. Soc.* **71**, 413–418.

Gibson, C. M. (1904). *New Phytol.* **3**, 184–191.

Hanover, J. W. and Hoff, R. J. (1966). *Physiologia Plantarium* **19**, 554–562.

Harder, D. E., Samborski, D. J., Rohringer, R., Rimmer, S. R., Kim, W. K. and Chong, J. (1980). *Can. J. Bot.* **57**, 2626–2634.

Hartley, R. D., Harris, P. J. and Russell, G. E. (1978). *Ann. Appl. Biol.* **88**, 153–158.

Heagle, A. S. and Moore, M. B. (1970). *Phytopathology* **60**, 461–466.

Heath, M. C. (1971). *Phytopathology* **61**, 383–388.

Heath, M. C. (1972). *Phytopathology* **62**, 27–38.

Heath, M. C. (1974). *Physiol. Pl. Pathol.* **4**, 403–414.
Heath, M. C. (1976). *Phytopathology* **66**, 935–936.
Heath, M. C. (1977). *Physiol. Pl. Pathol.* **10**, 73–88.
Heath, M. C. (1979). *Physiol. Pl. Pathol.* **15**, 211–218.
Heath, M. C. (1980a). *Phytopathology* **70**, 356–360.
Heath, M. C. (1980b). *Ann. Rev. Phytopathol.* **18**, 211–236.
Heath, M. C. (1981a). *In* "Plant Disease Control: Resistance and Susceptibility" (R. C. Staples and G. H. Toenniessen, eds), pp. 201–217. John Wiley, New York.
Heath, M. C. (1981b). *Physiol. Pl. Pathol.* **18**, 149–155.
Heath, M. C. and Heath, I. B. (1971). *Physiol. Pl. Pathol.* **1**, 277–287.
Hilu, H. M. (1965). *Phytopathology* **55**, 563–569.
Hoppe, H. H. and Heitefuss, R. (1975a). *Physiol. Pl. Pathol.* **5**, 263–271.
Hoppe, H. H. and Heitefuss, R. (1975b). *Physiol. Pl. Pathol.* **5**, 273–281.
Ingram, D. S. (1978). *Ann. Appl. Biol.* **89**, 291–295.
Johnson, L. B. and Cunningham, B. A. (1972). *Phytochemistry* **11**, 547–551.
Jones, D. R. and Deverall, B. J. (1977a). *Physiol. Pl. Pathol.* **10**, 275–284.
Jones, D. R. and Deverall, B. J. (1977b). *Physiol. Pl. Pathol.* **10**, 285–290.
Jones, D. R. and Deverall, B. J. (1978). *Physiol. Pl. Pathol.* **12**, 311–319.
Kaul, R. and Shaw, M. (1960). *Can. J. Bot.* **38**, 399–407.
Keen, N. T. and Littlefield, L. J. (1979). *Physiol. Pl. Pathol.* **14**, 265–280.
Kim, W. K., Rohringer, R., Samborski, D. J. and Howes, N. K. (1977). *Can. J. Bot.* **55**, 568–573.
Király, Z. and Farkas, G. L. (1962). *Phytopathology* **52**, 657–664.
Király, Z., Barna, B. and Érsek, T. (1972). *Nature* **239**, 456–458.
Kochman, J. K. and Brown, J. F. (1975). *Phytopathology* **65**, 1404–1408.
Knott, D. R. and Kumar, J. (1972). *Physiol. Pl. Pathol.* **2**, 393–399.
Kuć, J. (1972). *Ann. Rev. Phytopathol.* **10**, 207–232.
Leath, K. T. and Rowell, J. B. (1966). *Phytopathology* **56**, 1305–1309.
Littlefield, L. J. (1973). *Physiol. Pl. Pathol.* **3**, 241–247.
Littlefield, L. J. and Aronson, S. J. (1969). *Can. J. Bot.* **47**, 1713–1717.
Littlefield, L. J. and Heath, M. C. (1979). "Ultrastructure of Rust Fungi". Academic Press, New York, San Francisco and London.
Manocha, M. S. (1975). *Phytopath. Z.* **82**, 207–215.
Mares, D. J. (1979). *Physiol. Pl. Pathol.* **15**, 289–296.
Mares, D. J. and Cousen, S. (1977). *Physiol. Pl. Pathol.* **10**, 257–274.
Marryat, D. C. E. (1907). *J. Agric. Sci.* **2**, 129–138.
Marte, M. and Montalbini, P. (1972). *Phytopath. Z.* **75**, 59–73.
Martin, C. D., Littlefield, L. J. and Miller, J. D. (1977). *Trans. Br. Mycol. Soc.* **68**, 161–166.
Mendgen, K. (1977). *Naturwissenshaften* **64**, 438.
Mendgen, K. (1978). *Phytopath. Z.* **93**, 295–313.
Miller, T. and Cowling, E. B. (1977). *Phytopathology* **67**, 179–186.
Miller, T., Cowling, E. B., Powers, H. R. Jr. and Blalock, T. E. (1976). *Phytopathology* **66**, 1229–1235.
Montalbini, P. (1972). *Acta Phytopathol. Acad. Scie. Hungaricae* **7**, 41–45.
Montalbini, P. and Elstner, E. F. (1977). *Planta* **135**, 301–306.
Nowak, R., Kim, W. K. and Rohringer, R. (1972). *Can. J. Bot.* **50**, 185–190.
Ogle, H. J. and Brown, J. F. (1971). *Ann. Appl. Biol.* **67**, 309–319.
Olien, C. R. (1957). *Diss. Abstr.* **17**, 1171.
Onoe, T., Tani, T. and Naito, N. (1976). *Ann. Phytopath. Soc. Japan* **42**, 481–488.

Patton, R. F. and Spear, R. N. (1977). *Proc. Am. Phytopathol. Soc.* **4**, 85 (Abstr.)
Pearce, R. B. and Strange, R. N. (1977). *Physiol. Pl. Pathol.* **11**, 143–148.
Prusky, D., Dinoor, A. and Jacoby, B. (1980). *Physiol. Pl. Pathol.* **17**, 33–40.
Pryor, A. (1976). *Physiol. Pl. Pathol.* **8**, 307–311.
Quick, W. A. and Shaw, M. (1964). *Can. J. Bot.* **42**, 1531–1540.
Raggi, V. (1964). *Phytopath. Mediterranea* **3**, 135–155.
Rockwood, D. L. (1973). *Phytopathology* **63**, 551–553.
Rockwood, D. L. (1974). *Phytopathology* **64**, 976–979.
Rodrigues, C. J. Jr., Medeiros, E. F. and Lewis, B. G. (1975). *Physiol. Pl. Pathol.* **6**, 35–41.
Rohringer, R., Fuchs, A., Lunderstädt, J. and Samborski, D. J. (1967). *Can. J. Bot.* **45**, 863–889.
Rohringer, R., Kim, W. K. and Samborski, D. J. (1979). *Can. J. Bot.* **57**, 324–331.
Romig, R. W. and Caldwell, R. M. (1964). *Phytopathology* **54**, 214–218.
Rothman, P. G. (1960). *Phytopathology* **50**, 914–918.
Russell, G. E. (1977). *Phytopath. Z.* **88**, 1–10.
Samborski, D. J. and Rohringer, R. (1970). *Phytochemistry* **9**, 1939–1945.
Samborski, D. J. and Shaw, M. (1956). *Can. J. Bot.* **34**, 601–619.
Samborski, D. J., Kim, W. K., Rohringer, R., Howes, N. K. and Baker, R. J. (1977). *Can. J. Bot.* **55**, 1445–1452.
Seevers, P. M. and Daly, J. M. (1970a). *Phytopathology* **60**, 1322–1328.
Seevers, P. M. and Daly, J. M. (1970b). *Phytopathology* **60**, 1642–1647.
Seevers, P. M., Daly, J. M. and Catedral, F. F. (1971). *Pl. Physiol.* **48**, 353–360.
Sempio, C. and Barbieri, G. (1964). *Phytopath. Z.* **50**, 270–282.
Sempio, C. and Barbieri, G. (1966). *Phytopath. Z.* **57**, 145–158.
Sharp, E. L. and Emge, R. G. (1958). *Phytopathology* **48**, 696–697.
Shaw, M. and Hawkins, A. R. (1958). *Can. J. Bot.* **36**, 1–16.
Silverman, W. (1960). *Phytopathology* **50**, 130–136.
Skipp, R. A. and Samborski, D. J. (1974). *Can. J. Bot.* **52**, 1107–1115.
Skipp, R. A., Harder, D. E. and Samborski, D. J. (1974). *Can. J. Bot.* **52**, 2615–2620.
Sood, P. N. and Sackston, W. E. (1970). *Can. J. Bot.* **48**, 2179–2181.
Stakman, E. C. (1914). *Minn. Agricultural Exp. Stat. Bull.* 138.
Stakman, E. C. and Harrar, J. G. (1957). "Principles of Plant Pathology". The Ronald Press, New York.
Swaebly, M. A. (1956). *Diss. Abstr.* **16**, 843–844.
Tani, T. and Yamamoto, H. (1979). *In* "Recognition and Specificity in Plant Host—Parasite Interactions" (J. M. Daly and I. Uritani, eds), pp. 273–287. University Park Press, Baltimore.
Tani, T., Yoshikawa, M. and Naito, N. (1973). *Phytopathology* **63**, 491–494.
Tani, T., Yamamoto, H., Onoe, T. and Naito, N. (1975a). *Physiol. Pl. Pathol.* **7**, 231–242.
Tani, T., Ouchi, S., Onoe, T. and Naito, N. (1975b). *Phytopathology* **65**, 1190–1193.
Tani, T., Yamamoto, H., Kadota, G., Naito, N. (1976). *Technical Bull. of Fac. of Agriculture* **27**, 95–103.
Thatcher, F. S. (1943). *Can. J. Res.* **21**, 151–172.
Uritani, I. (1971). *Ann. Rev. Phytopathol.* **9**, 211–234.
Van Dyke, C. G. and Hooker, A. L. (1969). *Phytopathology* **59**, 1934–1946.
Von Broembsen, S. L. and Hadwiger, L. A. (1972). *Physiol. Pl. Pathol.* **2**, 207–215.
Wynn, W. K. (1976). *Phytopathology* **66**, 136–146.

Wynn, W. K. and Staples, R. C. (1981). *In* "Plant Disease Control: Resistance and Susceptibility" (R. C. Staples and G. H. Toenniessen, eds), pp. 45–69. John Wiley, New York.
Zimmer, D. E. (1965). *Phytopathology* **55**, 296–301.
Zimmer, D. E. (1970). *Phytopathology* **60**, 1157–1163.

7. Chemical and Biological Control of Cereal Rusts

H. BUCHENAUER

Institut für Pflanzenkrankheiten,
Universität Bonn, West Germany

CONTENTS

I. Introduction

Cereals are susceptible to numerous fungal diseases which may result in significantly reduced yields. Of the most serious and widespread diseases of cereals are those caused by fungi belonging to the group of rust fungi.

In the endeavours for controlling rusts successfully the development of

I

varieties resistant to rusts has played and will play an essential role. However, there are also shortcomings in breeding resistant varieties. For example, it is not always possible to combine resistance with other factors required and new races of diseases may appear which successfully attack previously resistant varieties. Therefore, in order to provide protection against loss of resistance by the emergence of new physiological races, the development of chemicals which effectively control cereal rusts is necessary.

Chemical control of cereal rusts had been reviewed by Dickson (1959), Rowell (1969) and Hassebrauk and Roebbelen (1975). Dickson (1959) stated that progress in the development of effective chemicals for the control of cereal rusts had been slow and the use of available fungicides appeared not to be economically feasible. Rowell (1968) reported the relationships between factors responsible for the development of rust epidemics and the effectiveness of chemicals against cereal rust. He emphasized the short duration of the activity of most fungicides relative to the long period of rust epidemics and the relationship between the minimal residues in the harvested grain and the persistence of control during the last month of crop development. Hassebrauk and Roebbelen (1975) reviewed the chemical and biological control of stripe rust.

The subject of combating rust diseases by biological control measures has received increasing interest in recent years.

II. Chemical Control

A. Sulphur

In the history of plant protection, sulphur is generally regarded as the first fungicide and it has been used to control powdery mildews of fruit trees (Forsyth, 1802; McCallan, 1956). Since sulphur possesses low mammalian toxicity and it is easily available, its effectiveness against cereal rusts was examined. As to the details of the early studies with sulphur the reader is referred to the review by Eriksson and Henning (1896). Inorganic sulphur compounds and elementary sulphur had been abundantly tested. However, because of the inadequate knowledge of physical characteristics and application techniques, these compounds rendered insufficient control of cereal rust diseases.

Because of the unsatisfactory efficacy of sulphur and other inorganic fungicides (e.g. copper compounds), great efforts were made to control rust diseases by breeding rust resistant varieties. The cognition that the emergence of physiological races of the various rust species limited the duration of resistant varieties again led, after an interruption of several decades, to intensified studies on control of cereal rusts by chemical means. Simultaneously, it was realized that the fungitoxic effectiveness of sulphur compounds largely depended on optimum particle size and correct formulation. In these studies, various preparations of sulphur were tested, for example, finely divided and colloidal sulphur preparations (Bailey and Greaney, 1925). The results showed that applications of sulphur dusts alleviated the severity of various cereal rusts and yield losses, but its effectiveness had not been regarded as sufficiently satisfying (Kightlinger, 1925; Kightlinger and Whetzel, 1926; Lambert and Stakman, 1926, 1929). Because of the exclusive protective action of sulphur the applications had to be initiated as soon as the first symptoms appeared. Because of the limited duration of the protective activity, especially under severe epidemic conditions, regularly repeated treatments with increased dosages were required to obtain satisfactory disease control. Usually under humid–warm weather conditions, these frequent treatments resulted in severe phytotoxicity problems (Gassner and Straib, 1936).

Sulphur caused lysis of germ tubes and at higher concentrations germination of uredospores was inhibited. Cereal rust species differed in their sensitivity to sulphur; for instance, *Puccinia striiformis* proved to be more sensitive than *P. graminis* f. sp. *tritici* and *P. recondita* f. sp. *tritici*.

Despite numerous studies the ultimate mode of action of sulphur in sensitive fungi has not been hitherto elucidated. First it had been suggested that either the oxidized sulphur dioxide, or the reduced form, hydrogen sulphide, is responsible for its fungitoxic activity. Eventually elemental sulphur had been regarded as the intrinsic fungitoxic agent (see Tweedy, 1969).

B. Piric Acid

In greenhouse tests Gassner and Hassebrauk (1936) examined the therapeutic effectiveness of a great number of organic chemicals against *P. recondita* f. sp. *tritici* and *P. simplex* following soil treatment of wheat and barley seedlings. Piric acid showed both protective and therapeutic activity; for instance, spray applications of wheat seedlings a few days after inoculation decreased the sporulation rate of the leaf rust fungus. The compound showed an unfavourable chemotherapeutic index, for concentrations necessary for sufficient disease control caused phytotoxicity.

C. Sulfonamides and Related Compounds

After the revolutionary chemotherapeutic effectiveness of the sulfonamides against bacterial diseases in humans and animals had been detected by Domagk (1935), the activity of these compounds against cereal rusts was tested by Hassebrauk (1938a). For instance, applications of *o*- or *p*-toluene sulfonamide metanilic acid, and 3-amino-benzenesulfonic acid (4–10 mg/ 100 cm^2 soil surface) provided complete control against the most important cereal rusts under greenhouse conditions. In field trials, however, *p*-toluene sulfonamide showed no effect on rust development, at the higher quantities applied (e.g. $3 > g/m^2$ soil surface), the compound inhibited host plant development (Hassebrauk, 1938b, 1940). Hassebrauk (1952) and Hotson (1952, 1953) showed the activity of sulfonamides against rusts was reversed by the application of *p*-aminobenzoic acid and folic acid. These investigators assumed that the sulfonamides have the same mode of action in rust fungi as in bacteria. Studies with bacteria indicated that *p*-aminobenzoic acid— representing for some bacteria an essential growth factor—and the sulfona- mides are competitive antagonists during the biosynthesis of folic acid. In further studies sodium salt of sulfanilic acid, sulfamic acid and sulfanilic acid derivatives exhibited both protective and curative properties against stem and leaf rust of wheat (Mitchell *et al.*, 1950; Hassebrauk, 1951; Livingston, 1953). These compounds also provided, under field conditions, partial con- trol of rusts when applied together with wetting agents and increased grain yield. At higher concentrations sulfamic and sulfanilic acid derivatives caused phytotoxicity, impaired baking behaviour and germination of the harvested grain (Mattern and Livingston, 1955; Forsyth and Peturson, 1958). Further shortcomings of many of these compounds were their high costs and the number of applications required for sufficient control.

D. Dithiocarbamates

After the detection of the fungitoxic activity of the dithiocarbamates (Tis- dale and Williams, 1934) and of the ethylene-bisdithiocarbamates (Dimond *et al.*, 1943), these chemicals had been thoroughly tested for their effective- ness against cereal rusts. While Zadoks (1958) was unable to demonstrate conclusive control of stripe rust with zineb (zinc ethylene-1,2-bisdi- thiocarbamate) and maneb (manganous ethylene-1,2-bisdithio- carbamate), other authors reported significantly diminished disease severity that simultaneously resulted in increase of grain yields after spray treatment with zineb (Corbaz, 1962; Nelson, 1962; Oran and Parlak, 1969). Activity against stripe rust of wheat has also been established for metiram (mixed precipitate consisting of the ammonia complex of zinc-(*N*, *N*-1,2-ethylene-bis-(dithiocarbamate)) and *N*,*N*-*poly*-1,2-ethylene-bis-

(thiocarbamoyl)-disulfide) (Singh-Verma, 1973) and the chemically related propineb [zinc- (*N*, *N*-propylene-1,2-bisdithiocarbamate,] (Mundy, 1973).

E. Dithiocarbamates and Nickel Salts

The therapeutic activity of nickel salts against rusts was first recognized by Sempio (1936) who showed that nickel added prior to inoculation to nutrient solution at a concentration of $1 \cdot 3 \times 10^{-4}$M effectively suppressed development of wheat leaf rust. However, more than 20 years elapsed before interest in nickel salts was renewed. Greenhouse experiments with numerous organic nickel compounds established that some compounds exerted both protective and eradicative properties against leaf rust of rye (Keil *et al.*, 1958a,b) and in field experiments, many of these nickel chemicals provided control against leaf and stem rust of wheat. Postinfectional treatments arrested further development of disease symptoms, e.g. three days after application a dark ring of discoloured tissue formed around the pustules (Peturson *et al.*, 1958; Forsyth and Peturson, 1959).

In field trials where dithiocarbamates and nickel salts had been applied simultaneously, the effectiveness against cereal rusts was superior to single applications of either component. Treatment with both components resulted in additive effectiveness whereby the dithiocarbamates displayed exclusively protective and the nickel salts additional eradicative activity (Forsyth and Peturson, 1960). The effectiveness of combined applications of dithiocarbamates and nickel salts against cereal rusts as well as the probable mode of action of nickel compounds in fungi and plants had been reported in detail by Rowell (1968).

Bohnen (1963) examined extensively the efficacy of "Sabithane" (combination of zineb and nickel sulfate) during the epidemic caused by stripe rusts in Switzerland in 1961. Two and three spray applications with "Sabithane" resulted in yield increases of $30 \cdot 8$ and $33 \cdot 2\%$, respectively. With regard to rust development relatively late aerial applications at GS (10.5.1–10.5.3; flowering) with "Sabithane" (3, 5 kg/ha in 70 l water containing $0 \cdot 2\%$ of the surfactant "Etaldyn") promoted yield (17%). Treatments resulted in severe nickel accumulation in grains (45 μg/kg) (Bohnen, 1968). Two aircraft applications of nickel sulfate and maneb diminished development of stem and leaf rust of wheat and increased grain yields (Rowell, 1964). Spray treatments of mancozeb (complex of zinc and maneb, containing 20% manganese and $2 \cdot 5\%$ zinc; Dithane M 45) also significantly reduced stripe leaf and stem rust incidence and enhanced yield over the control (Gupta *et al.*, 1976; Kucharek, 1977). Kucharek (1977) demonstrated that the effective use of these formulations required exact timing of applications with regard to both disease and host development. Control is directly related to the time period for which rust development is retarded (Rowell, 1964). Highest yield response was attained when spray applications were initiated before flag leaves fully emerged. This early timing suppressed disease development and potential inoculum on second and third leaf from the top and protected flag leaves from rust infection (Weihing, 1969; Kucharek, 1977). Since physiologically flag leaves and spikes are of crucial importance for grain yield, fungicidal applications against rust diseases are usually timed to protect flag leaves from disease development.

Treatments with mancozeb also resulted in a greater development of root system of wheat plants compared to plants in untreated plots. Furthermore, in treated plants a higher percentage of the second and third leaves remained green. Kucharek (1977) pointed out the significance of leaves below the flag leaf with respect to grain yield.

Nickel is readily taken up by both roots and foliage of cereals, acropetally translocated and accumulated in the tissue. The effect of nickel on germination of uredospores (e.g. germination of uredospores of *P. graminis* f. sp. *tritici*) was most marked at a nickel concentration of 4 μg/ml (Andersen, 1960). The high nickel concentrations in the tissue of treated plants suggested that the inhibition of rust development by nickel compounds is primarily due to direct fungitoxic properties of nickel and to a lesser extent to physiological alterations of the host metabolism. Various physiological effects of nickel on host plants have been observed, such as enhanced respiration (Jensen and Daly, 1960; Forsyth, 1962), chlorophyll retention, increased total *N*-content and dry weight (Bushnell, 1966).

Because of the ready accumulation of nickel in the grains, nickel salts have not been extensively used for control of rusts in cereals (Hoffman *et al.*, 1962). Because of their relatively low mammalian toxicity, the ethylene bisdithiocarbamates have been regarded as relatively harmless. Recently it was found that ethylene-thiourea, an impurity as well as metabolic and nonbiological conversion product of ethylenebisdithiocarbamate fungicides, caused toxicological hazards (Vonk and Kaars Sijpesteijn, 1970; Engst *et al.*, 1971; Fishbein and Fawkes, 1965).

F. Carboxylic Acid Anilides

1. Oxathiin-derivatives

The discovery of the 1,4-oxathiin derivatives carboxin (2, 3-dihydro-6-methyl-5-phenylcarbamoyl-1,4-oxathiin) and oxycarboxin (2,3-dihydro-5-carboxanilido-6-methyl-1,4-oxathiin-4,4-dioxid) represented a major breakthrough in the chemotherapy of plant diseases (von Schmeling and Kulka, 1966). Both chemicals possess systemic properties and display a highly selective toxic activity against Basidiomycetes.

Although in *in vitro* studies carboxin showed a higher intrinsic fungitoxicity than oxycarboxin, in *in vivo* experiments, oxycarboxin was found to be superior to carboxin in long-term effectiveness, e.g. in controlling cereal rusts (Rowell, 1967) and bean rust (Snel and Edgington, 1969). On the other hand, against diseases where usually a short-term effectiveness is required, as for example in controlling loose smut of barley and wheat, carboxin provides a higher activity than oxycarboxin.

Hardison (1971a) studied the relationship between molecular structure and chemotherapeutic activity of twelve 1,4-oxathiin derivatives against *P. striiformis* and various smut fungi in Kentucky bluegrass (*Poa pratensis*) following soil treatment. At high dosages, carboxin showed only moderate activity against rust and marked effectiveness against flag smut. Within the group of non-oxidized analogs of carboxins tested, the 3'-methyl derivative (F 306) displayed an enhanced activity but this was accompanied by increased phytotoxicity. While the monoxide analogue (F 183) proved to be better than carboxin the strongest systemic activity in controlling stripe rust was achieved by the dioxide form, oxycarboxin. Among the dioxide analogues tested, the 4'-methoxy derivative (F 837) showed chemotherapeutic activity similar to that of oxycarboxin. Soil treatment with oxycarboxin had been more effective than carboxin against stripe smut in grass plants (Hardison, 1966).

Four foliar applications with carboxin and oxycarboxin almost completely suppressed stripe rust development on three Kentucky bluegrass varieties in Hawaii and no significant differences between the two compounds were detected. On the contrary, Hardison (1966) found oxycarboxin to be more effective than carboxin against stripe rust in grass plants after foliar sprays. Spray treatments with oxycarboxin and carboxin at rates of 4·2 and 2·5 kg/ha significantly increased yields of wheat by controlling leaf and stem rust. In some trials the oxathiin fungicides proved to be superior and in others inferior to the effectiveness of nickel-zineb treatments (Hagborg, 1970). Control of *P. hordei* in barley with oxycarboxin increased grain yields by between 3·8 and 16·5% (Hagemeister and Neuhaus, 1977).

The effectiveness of oxycarboxin against leaf and stem rust of wheat was considerably improved when instead of a wettable powder (Plantvax WP) formulation an emulsifiable concentrate (Plantvax EC) of the fungicide was used (Hagborg, 1971). A comparison of a 20% emulsifiable concentrate (EC) and a 75% wettable powder (WP) of oxycarboxin indicated that the EC formulation proved to be about three times as effective as the WP formulation against stripe rust of wheat (Judge *et al.*, 1975).

Rowell (1973a) examined the long lasting effectiveness of oxycarboxin after seed treatment for controlling leaf and stem rust development on wheat. He determined the number of trapped uredospores and the disease severities in the individual plots. Oxycarboxin only strongly reduced but did not completely inhibit the initial stages of rust development after seed treatment. Depending on the progress of the rust epidemic, the effectiveness of the chemical decreased and reached 50% of disease severity of the control plots after 70 and 83 days from planting. Thereupon as the effectiveness of the fungicide declined severe epidemic disease development nullified the initial disease control. Rowell (1973a) pointed out that under these conditions of wheat growth and of rust development, the duration of effective control of rusts by seed treatment with a systemic fungicide had to exceed 80 days. Comparison of the effectiveness of oxycarboxin after seed treatment at relatively large dosages (0·16 g/kg seed, which was attained by using the sticker methyl cellulose) and after three foliar sprays (each 1·12 kg/ ha) applied at 35, 45 and 56 days following planting showed similar degree of leaf rust control during the growing period of 60 days. To obtain a significant increase in yield with either treatment, it appeared to be necessary to prevent rust development in the last 30 days of the growing period of the crops (Rowell, 1976a).

Line (1976) evaluated several systemic fungicides for control of stripe and leaf rust of wheat at several locations in north-western United States during 1974 and 1975. He reported that oxycarboxin was most effective when rains occurred after foliar applications. The author assumed that the chemical had been transported by the rain to the roots which took up and translocated the compound into the foliage. When precipitation was limited or irrigation was not used and oxycarboxin did not reach the root system the chemical was less effective.

Generally, correct timing for rust control is a critical factor to achieve the highest benefit. The numerous possible interactions between growth stages of cereal plants and disease development make it extremely difficult to define precisely the timing of fungicide application. The results suggested that both early application and protection of flag leaves are of major importance for effective control of cereal rust diseases. Mundy (1973) reported that spray treatments should be carried out before disease started to develop exponentially and the mean value of disease severity of the top two leaves did not exceed 5%.

To attain maximum yield benefit, Judge et al. (1975) evaluated different timings of oxycarboxin spray treatments using different wheat cultivars in different locations and showed differential levels of stripe rust attack. The authors found that with respect to disease development, rust severity should not exceed a level of 5% on the top two leaves. Regarding the plant development, the critical stages were GS 8–GS 10·1. Since it proved to be difficult to observe and estimate low disease incidences under practical conditions, Judge *et al.* (1975) recommended spray applications with carboxin when stripe rust appeared on flag leaf or leaf two from the top.

Since the introduction of the oxathiins in practice, development of resistance in plant pathogenic fungi is not known, although resistance to these chemicals had been obtained easily under laboratory conditions (Ragsdale and Sisler, 1970; Ben Yephet *et al.*, 1974; Van Tuyl, 1977). While Ben Yephet *et al.* (1974) found resistant strains of *Ustilago hordei* somewhat less competitive than sensitive strains, Van Tuyl (1977) could not demonstrate a correlation between increased degree of resistance and decreased pathogenicity. It has been reported in one case only that there is the development of resistance against oxycarboxin. Following continuous treatments with oxycarboxin, resistant strains of the *Chrysanthemum* rust *Puccinia horiana* emerged (Abiko *et al.*, 1975).

Generally, the higher chemotherapeutic activity of oxycarboxin compared to carboxin against cereal rusts and bean rust (Snel and Edgington, 1969) has been attributed to its pronounced stability in plant tissue. Carboxin showed a higher toxicity to germination of uredospores of rust fungi than oxycarboxin (Snel and Edgington, 1968). Carboxin was readily oxidized to the nonfungitoxic sulfoxide (90–92%) and the fungitoxic sulfone (8–10%) in two weeks old wheat (Chin *et al.*, 1970a,b) and bean plants (Snel and Edgington, 1970). Long-term studies indicated that the carboxamide linkage of carboxin was hydrolysed and the resulting aniline was bound to plant polymers, such as lignin (Snel and Edgington, 1970; Chin *et al.*, 1970a; Briggs *et al.*, 1974). Furthermore the carboxin metabolite *p*-hydroxycarboxin detected in barley plants was already bound to lignin (Briggs *et al.*, 1974).

Chin *et al.* (1970b) reported oxidation of carboxin mainly into the sulfoxide in water as well as in nonsterilized and sterilized soil. This inactivation reaction of carboxin was strongly accelerated in presence of light energy on surfaces of glass and leaves of bean plants (Buchenauer, 1975a; Wolkoff *et al.*, 1975). On the other hand, oxycarboxin proved to be very photostable (Buchenauer, 1975a). The oxidative inactivation of carboxin seemed to proceed even more rapidly on plant surfaces than in soil or within plant tissue. Thus, the weak performance of carboxin against rust diseases may be attributable to rapid inactivation both within and on the surface of plants.

Within the group of systemic fungicides the biochemical mode of action of carboxin and oxycarboxin in fungi had been most extensively elucidated. These compounds interfered, in sensitive fungi, with mitochondrial respiration (Mathre, 1971; Ulrich and Mathre, 1972; Lyr *et al.*, 1972) by interfering with the non heme-iron-sulfur protein ($FeSP_p$) of the succinodehydrogenase complex (Lyr *et al.*, 1975; Schewe *et al.*, 1979). The carboxylic acid anilide part of the molecule containing nucleophilic (amide-*N*, carboxylic group) and hydrophobic sites (phenyl ring at amide-*N*) as well as the methyl or iodide group in 2-position of e.g. the oxathiin, pyrane, benzene or thiazole ring represent the essential requirements for fungitoxic activity (Ten Haken and Dunn, 1971; Mathre, 1971; White and Thorn, 1975; Schewe *et al.*, 1979).

The effect of oxycarboxin on ultrastructural changes in *Uromyces phaseoli* during infection of leaves of French beans was studied by Pring and Richmond (1976) and Richmond and Pring (1977). On leaves of plants which previously had been treated with the chemical by root drench, uredospores germinated but germ tubes were shorter than those on control plants. While in treated plants appressoria and intercellular hyphae were formed, growth rarely proceeded beyond this infection process and no haustoria could be detected. This interference of oxycarboxin in the infection process closely corresponds to a frequent type of resistance mechanism of bean plants against *U. phaseoli* var. *vignae* (Heath, 1974).

After postinfectional treatment, oxycarboxin induced the first recognizable cytological changes in *U. phaseoli* following an exposure period of 20 hours. Mitochondria in haustoria of treated plants became swollen, their cristae disorganized and the plasmalemma surrounding the haustorium became fragmented. Two days after treatment mitochondria became disrupted, large vacuoles were present and lipids accumulated; after six days, haustoria and intercellular hyphae were dead.

In oat seedlings which had been infected with crown rust (*P. coronata* var. *avenae*) oxycarboxin induced various ultrastructural changes: Mitochondria and haustoria were significantly larger, endoplasmic reticulum was almost absent and cytoplasmic structures had disappeared. One day after fungicide exposure haustoria were difficult to fix and section and contained greatly enlarged empty mitochondria (Simons, 1975). A few hours after addition of carboxin to *Rhodotorula mucilaginosa* (Basidiomycete) mitochondria of the fungus were swollen and showed structural disintegration (Lyr *et al.*, 1972). The rapid and complete structural disintegration of mitochondria of fungi sensitive to carboxin and oxycarboxin confirmed the results on the studies of the biochemical mode of action of these fungicides in energy metabolism.

Although the present findings of the mode of action of the oxathiins indicate a direct effect on the rust fungi, there are various reports on physiological effects on plants induced by oxathiins. Foliar applications of carboxin and oxycarboxin increased protein content in wheat grain and slightly stimulated amino acid metabolism in wheat, corn, sorghum and soybean plants (Reyes *et al.*, 1969). On the other hand, oxycarboxin did not cause changes in amino acid metabolism in bean plants (Newby and Tweedy,

1973). Carboxin retarded senescence in leaves of barley (Carlson, 1970; Gross and Kenneth, 1973) and Kentucky bluegrass plants (Murdoch, 1973). Treated barley plants showed no guttation and were sensitive to drought, possibly, because stomata tended to remain open (Gross and Kenneth, 1973). The investigators assumed that *in vivo* activity of carboxin against fungal diseases resulted from hormonal imbalances of treated plants. Carboxin strongly inhibited photosynthesis and to a lesser extent respiration in leaves of barley plants immediately after spraying with high levels (e.g. 15 mM) of the fungicide. Recovery of photosynthesis started within 24 hours and returned to normal at five days after treatment (Carlson, 1970). While at high concentrations carboxin markedly impaired photosynthesis in pinto bean and corn plants, oxycarboxin and F 831 were less effective (Mathre, 1972).

Carboxin and its derivatives exhibit high selectivity. Growth and metabolism of non-target organisms (such as plants and non-sensitive fungi) are only affected when high fungicidal concentrations ($>10^{-4}$ M) are reached, whereas sensitive fungi are strongly inhibited at concentrations of 10^{-5} and 10^{-6} M (Mathre, 1972; Snel *et al.*, 1970). Structural differences within the receptor region between sensitive and less sensitive organisms (possibly different amino acid sequences) seem to be responsible for the relatively high selectivity of carboxin and its derivatives.

2. Carboxanilido derivatives

Among the numerous benzolic anilide derivatives tested benodanil (2-iodobenzanilide; BAS 3170 F) proved to be most active for rust control coupled with a great margin of crop safety. A somewhat low effectiveness and crop tolerance were exhibited by mebenil (2-methylbenzanilide; BAS 3050 F). In greenhouse tests, seed treatments with mebenil protected summer wheat plants for several weeks against stripe rust infection (Pommer and Kradel, 1969; Pommer and Osieka, 1969; Pommer, 1976).

Benodamil has been extensively examined for its efficacy against rusts on cereals and grasses. Foliar applications of the chemical reduced incidence of stem rust of wheat (Hagborg, 1972), brown rust of wheat (Frost and Hampel, 1976), stripe rust on wheat and barley (Frost *et al.*, 1973; Mundy, 1973; Frost, 1975; Frost and Hampel, 1976), brown rust of barley (Frost, 1975; Frost and Hampel, 1976; Nuttall and Mundy, 1976; Widdowson *et al.*, 1976; Heyland and Fröhling, 1977), crown rust of oats (Pommer, 1976) and stripe rust of Merion Kentucky bluegrass (Hardison, 1975a).

Results indicated that benodanil predominantly displayed protective activity and had little eradicant action (Frost and Hampel, 1976; Pommer,

1976). Benodanil protected winter wheat against severe infection of stripe rust for up to four weeks after treatment and eradicated the rust present. On the other hand, tridemorph and triforine prevented noticeable rust development for about two weeks. In these studies pyracarbolid proved to be almost as effective as benodanil (Mundy, 1973).

The curative effectiveness of benodanil could be improved by addition of tridemorph (Frost and Brown, 1973) or by using instead of a wettable powder (WP) formulation either an emulsion concentrate (EC) (containing 200 g a.i./1) or an oil formulation (25% a.i.) (Hagborg, 1972; Frost and Hampel, 1976; Pommer, 1976). In controlling stem rust of wheat a WP formulation of benodanil performed equal to an EC-formulation of oxycarboxin and was superior to oxycarboxin WP formulation (Hagborg, 1972).

Studies on transcuticular movement of benodanil through isolated cuticles revealed very slight penetration of the compound from WP formulations which was markedly increased by employing EC or oil formulations. The enhanced penetration resulted in an appreciable improvement of the systemic and curative effectiveness of benodanil (Pommer, 1976). In view of the increased uptake and efficacy of these liquid formulations, it appeared possible to reduce the application rate of benodanil.

Since benodanil predominantly displayed protective activity, for obtaining optimum and persistent yield responses correct timing of application proved to be of prime importance. Spray applications of benodanil were highly effective in controlling yellow rust of wheat when the flag leaf was protected from infection (Mundy, 1973). These findings were confirmed by results obtained by other authors. Highest yield increases resulted from applications to crop after flag leaf emergence and before flag leaf was infected (Frost et al., 1973; Frost, 1975; Frost and Hampel, 1976). In order to increase the curative effectiveness when disease is already advanced, addition of tridemorph was recommended (Frost and Brown, 1973). To ensure good brown rust control of wheat, benodanil should be applied at an early stage of disease development (Frost and Hampel, 1976). In controlling stripe rust in spring barley, the optimal timing with respect to disease development was as soon as the first symptoms appeared. Contrary to stripe rust on wheat it has not yet been possible to define the optimal application time with regard to growth stage of barley (Frost and Hampel, 1976). Correct timing for a single benodanil treatment against brown rust of barley was before, or at the beginning, of epidemic disease progress. With respect to growth stages, Heyland and Fröhling (1977) determined highest yield improvement when application was made at the beginning of shooting. At early appearance of symptoms and continuous disease development, two treatments may be required for optimal yield response (Frost et al., 1973; Frost and Hampel, 1976).

Concerning the mechanism of action of these compounds, benodanil, mebenil and pyracarbolid as well as other carboxyamide analogues interfered with the electron transport system in mitochondria of sensitive fungi (Ten Haken and Dunn, 1971; White and Thorn, 1975) and electron transport particles from bovine heart mitochondria (Schewe *et al.*, 1979). White and Thorn (1975) testing ninety-three carboxin analogues and related compounds found, with few exceptions, a close correlation between inhibition of succinic dehydrogenase activity in mitochondrial preparations from fungi (e.g. *Ustilago maydis*) and antifungal activity both *in vitro* and *in vivo*.

3. Thiazole-derivatives

In the search for new fungicides with chemotherapeutic properties Hardison (1971b) examined various thiazole compounds against smut and rust pathogens in Kentucky bluegrass. The carboxanilide moiety of these thiazole derivatives is identical to that of the 1,4-oxathiin fungicides. Among the thiazole analogues tested the parent compound, 2,2-dimethyl-5-carboxanilidothiazole (G 696) and its 1-amino substitution derivative, 2-amino-4-methyl-5-carboxanilidothiazole (F 849) exhibited systemic fungicidal activity against *P. striiformis* in *Poa pratensis* after soil treatment. Compared to oxycarboxin the thiazole compounds were less active in controlling stripe rust and more phototoxic. By analogy with the oxathiin fungicides, Hardison (1971a,b) presumed that thiazole derivatives with oxidized sulfur in the heterocycle would represent further promising candidates with therapeutic activity against smut and rust diseases.

G. Morpholine Derivatives

1. Tridemorph

Since the introduction of the systemic fungicide tridemorph (*N*-tridecyl-2, 6-dimethyl morpholine; "Calixin") the compound had been extensively evaluated for its effectiveness against various foliar diseases of cereals. The compound is highly active against powdery mildews and is widely used in controlling barley and wheat mildew (Kradel *et al.*, 1969; Jung and Bedford, 1971); its effectiveness against barley mildew is superior to that against wheat mildew.

The wider spectrum of activity including various other foliar diseases on cereals, in particular *P. striiformis, P. recondita, Septoria nodorum* and *Rhynchosporium secalis*, might contribute to additional yield benefits.

K

Tridemorph (0·52 kg a.i./ha) applied either alone or in mixture with metiram (at a rate of 0·52 kg/ha tridemorph plus 1·6 kg a.i./ha metiram) showed protective and curative activity against stripe rust of wheat and treatments resulted in yield increases (Frost and Brown, 1973). Treatment of barley plants with "Calixin M" (containing 11% tridemorph plus 36% maneb) at quantities of 4·5–5 kg/ha reduced disease severity caused by *P. striiformis* and *P. hordei* (Lartand and Lipatoff, 1976). Comparative studies showed that tridemorph was inferior to benodanil in controlling stripe rust; two spray treatments of tridemorph (each 0·52 kg/ha) were required to give a comparable degree of disease control as a single spray of benodanil (1·5 kg/ha).

A comparison of the efficacy of tridemorph between powdery mildews and rusts revealed pronounced differences. At concentrations of 0·01–10 μg/ml tridemorph suppressed mildew development, however, for control of rusts (e.g. *P. striiformis* and *P. recondita*) concentrations of 100 μg/ml were necessary. At higher concentrations of tridemorph phytotoxicity increased (Bahadur *et al.*, 1974).

The primary mode of action of tridemorph in fungi has not yet been completely elucidated. In treated barley leaves the compound inhibited haustorial formation of powdery mildews (Schlüter and Weltzien, 1971). Various biochemical effects have been observed, such as interference with respiration in the yeast *Torulopsis candida* (Bergmann *et al.*, 1975) and inhibition of incorporation of (^{14}C)-histidine in protein fraction of *Botrytis fabae* protoplasts (Fisher, 1974). Recently, Polter and Casperson (1979) demonstrated influence of tridemorph on lipid biosynthesis in *Botrytis cinerea* and Barug and Kerkenaar (1979) showed cross resistance for instance, between triadimefon-resistant strains of *Ustilago maydis* sporidia and tridemorph. These results suggests similarities in the mode of action of ergosterol biosynthesis inhibitors (e.g. triforine, triadimefon, imazalil, fenarimol and nuarimol) and tridemorph.

2. Fenpropemorph

Fenpropemorph [*cis*-4-(3-(4-*tert*-butylphenyl)-2-methylpropyl)-2,6-dimethylmorpholine] belonging also to thorpholine derivatives was recently introduced. The compound displayed both preventive and curative activity against various important foliar diseases of cereals (Bohnen and Pfinner, 1979; Pommer and Himmele, 1979; Hampel *et al.*, 1979; Bohnen *et al.*, 1979). With respect to the antifungal spectrum and the degree of activity fenpropemorph proved to be superior to tridemorph and possessed also a wider margin of plant safety. Foliar applications of fenpropemorph (at

750 g a.i./ha) exhibited excellent activity against powdery mildews on wheat and barley and protected plants under field conditions for three to four weeks. At the optimum application rate of 750 g a.i./ha the compound also protected cereals for three to four weeks against all prevalent rust species and showed effectiveness comparable to triadimefon at a dosage of 125 g a.i./ha (Bohnen *et al.*, 1979). Even when leaves showed rust infections between 5 and 10% curative applications still showed high efficiencies; infections present were either killed or further disease development was arrested (Bohnen and Pfinner, 1979).

H. Indar

Indar (4-*n*-butyl-1, 2, 4-triazole; RH-124) represents a highly selective systemic fungicide for the control of wheat leaf rust. Among several rusts studied, Indar only controlled wheat leaf rust. Even at high concentrations, this compound did not affect other rust diseases, such as bean rust, crown rust of oat and stem rust of wheat (von Meyer *et al.*, 1970; Tyagi *et al.*, 1973; Rowell, 1976b). Numerous studies indicating that seed, soil and/or foliar treatments with Indar had high potential for control of leaf rust in certain wheat growing areas, for instance, in the north-central United States and in India.

Rowell (1973b, 1976b) reported the development of leaf rust on spring wheat and its economic significance in the north-central region of the United States. Because of the limited inoculum at the beginning of the crop development the disease builds up slowly and achieves about 1% severity at the heading stage; during the following period of wheat development while inoculum density increases disease builds up rapidly and leaf rust attack results in losses as great as 45%.

Seed treatment that would control rust diseases during the growing season would provide an elegant and simple form of application. This method would be particularly suitable in areas where the growing season of cereal is short and where rusts are causing yield losses every year (Rowell, 1968).

Both seed treatment and foliar application (19 days after planting) effectively controlled leaf rust during 70 days after planting and yield was significantly increased by both application methods. With respect to the dosages applied seed treatment (at 140 g/ha) proved to be more effective than spray treatment (469–560 g/ha). In comparison with spraying, for seed treatment only 20–25% of the effective dosage were sufficient to achieve similar levels of effectiveness (Rowell, 1976a). Thus, seed treatment with Indar was more economical than foliar treatment, in regard to both the quantity of fungicide needed and the costs of application. Indar proved to be superior to oxycarboxin after seed treatment and root treatment (Rowell, 1964a; Von Meyer *et al.*, 1970).

In major wheat growing areas of India, leaf rust probably is the most important disease of wheat with frequent epidemics causing severe losses. Most of the dwarf wheat cultivars which had been introduced and extensively cultivated over large areas are susceptible to leaf rust (Joshi *et al.*, 1970; Tyagi *et al.*, 1973; Singh and Singh, 1975). In north India where wheat is sown in November and December, first symptoms of leaf rust appear in January–February and the disease reaches epidemic proportions in March and April. Because of the long vegetative period of the crop, seed treatment with Indar did not provide a corresponding long lasting protection (Tyagi *et al.*, 1973; Singh and Singh, 1975).

Application of granular formulations (0·7 and 1% granules) of Indar at a rate of 200 a.i./ha to the soil resulted in effective control of *P. recondita* in wheat within a wide margin of timing (4–63 days before appearance of disease symptoms). After treatment plots were irrigated. This method had the advantage of considerable flexibility in timing the fungicide treatment and may be of value for such regions where first rust incidence is hard to predict precisely and watering of soil is possible (Singh and Singh, 1975; Tyagi *et al.*, 1979).

By addition of Indar to the irrigation water (at a rate of 0·8–11 a.i./ha) yield losses due to leaf rust attack were diminished (Tyagi *et al.*, 1974; Singh and Singh, 1975).

Single foliar sprays of Indar at dosages higher than 0·4 l a.i./ha mitigated grain yield reductions when applied either before rust appearance or when disease intensity did not exceed 5%. However, when disease incidence reached 15%, higher quantities of the fungicide (e.g. 1·6 l a.i./ha) were necessary for effective control (Tyagi *et al.*, 1973, 1974; Rowell, 1973b). Similar results were obtained by Singh and Singh (1975). Spray treatments with Indar provided effective control within the time period of one week before until three weeks after rust symptoms appeared; effectiveness of treatments rapidly decreased at later stages of disease development. When Indar was applied during the joint stage of plant growth, the compound significantly inhibited the development of leaf rust on five hard winter cultivars and resulted in increases of grain yields (Watkins and Daupnik, 1979).

The systemic protective efficacy of Indar was significantly superior to other fungicides possessing activity against wheat leaf rust. In greenhouse tests with wheat seedlings, Indar showed higher activities after foliar, soil and seed applications than oxycarboxin, triarimol and benomyl (Rowell, 1976b). In field experiments a single foliar treatment with Indar (at 0·5–2 l/ha) was equally effective as four sprays of zineb (Singh and Singh, 1975) and one spray treatment with Indar (1·12 kg/ha) showed similar effectiveness to three sprays of oxycarboxin (each at 2·24 kg/ha) (Buchenau, 1970; Rowell, 1969; Reed and Chambers, 1973; Line et al., 1974). Indar displayed an insignificant eradicative activity after both spray and soil treatment. On the basis of ED_{50}-values as eradicant (190 μg/leaf) more than 10^5 times higher quantities of the chemical were required than when applied as protectant (0·0012 μg/leaf). The other compounds tested such as oxycarboxin, triarimol and benomyl showed similar protective and curative effectiveness in these experiments (Rowell, 1976b). The poor eradicant activity of Indar seemed to be attributable to its weak in vitro fungitoxicity against germination of uredospores of P. recondita f. sp. tritici. With other fungicides (e.g. carboxin, triarimol and benomyl), a closer relation between fungitoxicity and systemic therapeutic activity was found (Rowell, 1976b).

Seed treatment with Indar at 70 and 140 g/ha did not adversely affect plant development when seed was planted in moist soil but stand densities (85·2 and 78·5%, respectively, of the control) and seedling growth were reduced when seed was planted in relatively dry soil that delayed seed germination. Field results were confirmed by greenhouse observations indicating that when applied to seed, Indar could cause reduction in length (14%) of first leaf under unfavourable soil conditions (Rowell, 1976b).

Rowell (1976b) found a relatively low persistence for Indar in treated wheat leaves. The effectiveness of Indar decreased by 50% after 59 hours; corresponding values for oxycarboxin, triarimol and benomyl were 60, 116 and 176 hours, respectively. The long lasting effectiveness of Indar against leaf rust probably resulted from persistence in soil and on leaf surface that guaranteed a continuous uptake of the chemical throughout the crop development.

Annual applications of Indar as seed or spray treatment might therefore result in undesirable enrichment of soil residues. Under environmental conditions prevailing for instance in north-central United States, Indar showed a pronounced high soil persistence. Rowell (1976b) reported that soil residues from spray treatments at dosages of 280 and 560 g/ha delayed leaf rust development in wheat in the following year.

Studies on the effectiveness of seed treatment with Indar on four wheat cultivars differing in their susceptibility to leaf rust revealed that quantities of the fungicide required for achieving comparable levels of disease control increased as the degree of susceptibility decreased (Rowell, 1976b). The long lasting effect of Indar on rust development was more pronounced on moderately susceptible varieties than on highly sensitive ones (Rowell, 1973b).

The histological alterations during the development of P. recondita f. sp. tritici in Indar treated wheat plants was studied by Watkins et al. (1975;

1977). The chemical did not influence the initial stages of the infection process, such as germination of uredospores and substomatal penetration. The first ultrastructural differences in fungal development of untreated and treated seedlings were detectable after 12 hours and became pronounced with increasing exposure periods. After 36 hours, fungus development had ceased in treated tissue. The most conspicuous interference of Indar during infection of the leaf rust fungus was the inhibition of haustorium and the abnormal distortion and swelling of hyphae. Infection structures had greatly lost their metabolic activity. Inoculation of treated plants resulted in no visible necrotic symptoms, this response resembled natural immune reactions of wheat plants and differed from hypersensitive resistence reaction by absence of host cell necrosis.

I. Inhibitors of Ergosterol biosynthesis

Recently, various fungicides with interesting properties have been developed. These belong chemically to the group of *N*-substituted piperazine (triforine), triazole (e.g. triadimefon, triadimenol, dichlobutrazol, propiconazol, (CGA 64250), CGA 64251), pyrimidine (e.g. fenarimol, nuarimol) and imidazole derivatives (e.g. imazalil, phenapronil).

Triforine (*N, N'*-bis (1-formamido-2,2,2-trichloroethyl) piperazine) displays activity against powdery mildews, rusts and apple scab (Schicke and Veen, 1969).

Triadimefon [1-(4-chlorophenoxy)-3,3-dimethyl-1 (1,2,4-triazol-1-y1)-2-butanone] is recommended as a foliar fungicide in cereals for controlling powdery mildews, rusts, *Typhula incarnata* and *Rhynchosporium secalis* (Grewe and Büchel, 1973; Frohberger, 1973; Ebenebe and Fehrmann, 1974; Kampe, 1975; Buchenauer, 1975b, 1976a,b; Kolbe, 1976; Scheinpflug *et al.*, 1977).

Triadimenol [1-(4-chlorophenoxy)-3,3-dimethyl-1(1,2,4-triazol-1-y1)-2-butanol], chemically related to triadimefon, had been developed for seed treatment of cereals to combat numerous important diseases, such as smuts, soil-borne *Septoria nodorum*, *T. incarnata*, *R. secalis*, *Pyrenophora avenae* and *Cochlibolus sativus*. Beyond that, seed treatment with triadimenol protected cereal seedlings against mildew and rust infections (Frohberger, 1977, 1978).

Dichlobutrazol [1-(2,4-dichlorophenyl)-4,4-dimethyl-2-(1,2,4-triazol-1-y1) pentan-3-o1] (Bent and Skidmore, 1979; von Zitzewitz and Heckele, 1979), propiconazol [1-(2-(2,4-dichlorophenyl)-4-propyl-1,3-dioxolan-2-yl-methyl)-1,2,4-triazole] and CGA 64251 [1-(2-(2,4-dichlorophenyl)-4-ethyl-1,3-dioxolan-2-y1-methyl)-1H-1,2,4-triazole] (Urech *et al.*, 1979; Staub *et al.*, 1979a,b) exert a broad spectrum activity like triadimefon and triadimenol.

Fenarimol [α-(2-chlorophenyl)-α-(4-chlorophenyl)-5-pyrimidine-methanol] and nuarimol [α-(2-chlorophenyl-α-(4-fluorophenyl)-5-pyrimidinemethanol] show structural similarities with triarimol.

While the application range of fenarimol primarily includes orchards, vegetables and ornamentals the use of nuarimol is confined to cereals. The antifungal spectrum of fenarimol and nuarimol largely resembled that of triarimol and triadimefon (Brown *et al.*, 1975; Döhler and Merz, 1979).

Imazalil [1-(2-(2,4-dichlorophenyl)-2-propenyloxy) ethyl-1H-imidazole] is especially useful for seed treatment to control *Pyrenophora graminea* and *Cochliobolus sativus* in barley and *Pyrenophora avenae* in oats (Bartlett and Ballard, 1975). Phenapronil (α-butyl-α-phenyl-1H-imidazole-1-propanenitrile) exhibits activity against powdery mildews, rusts, apple scab and *Septoria* spp. (Carley, 1979).

The broad spectrum activity of this group of fungicides including fen-propemorph ensures effective control of economically important cereal diseases.

In growth chamber tests triforine displayed both protective and eradicative activity against wheat leaf rust. Foliar treatments (500 μg/ml) completely protected plants for at least nine days and prevented pustules development within three days after application. This compound retarded germination of uredospores on wheat leaves following leaf or soil application (Ebenebe *et al.*, 1971).

In field tests, triforine protected barley plants against stripe rust and eradicated disease present. Reinfection occurring in treated plants within two weeks was insignificant, but in comparison to benodanil (providing protection for four weeks) triforine was inferior in its anti-rust activity.

Numerous investigators have reported excellent effectiveness of triadimefon in control of cereal rusts following foliar treatments (Maykuhs and Hoppe, 1976; Scheinpflug *et al.*, 1977; Rowley *et al.*, 1977; Reschke, 1978). Triadimefon proved to be superior to all other fungicides against rust fungi at present registered in its duration of persistance. Under conditions in Western Germany protection of triadimefon against stripe rust normally lasted for four weeks and yields were improved between 11 and 27%. By comparison, oxycarboxin suppressed disease for a period of two weeks and the efficacy of pyracarbolid, triforine and the mixture of tridemorph plus metiram was only short-lasting. Propineb was ineffective (Maykuhs and Hoppe, 1976). In the Netherlands, a single spray treatment at GS 6·0–8·0 with 500 g a.i./ha of triadimefon against leaf rust of wheat resulted in significant increases of grain yields (26%). A second application with the same rate at GS 9·0–10·1. further enhanced yields to 44–66%. There was a correlation between disease incidence and yield increases. Optimum timings for spray treatments against leaf rust on winter wheat and brown rust on winter barley proved to be as soon as visible disease incidence appeared and transferred to crops in the Netherlands. This implied that the chemical should be applied against leaf rust of wheat at GS 9·0–10·5·1. and against brown rust of barley at GS 8·0–9·0 (Wäckers et al., 1978).

Martin and Morris (1979) reported optimal timings for control of rusts in winter and summer barley with triadimefon (at 500 g/ha) in Great Britain. Barley varieties susceptible to brown rust should be treated with the appearance of the first visible symptoms, generally at GS 6·0–9·0. If the treatments carried out later, the effects on disease suppression and yield improvements were less pronounced. Under environmental conditions favouring early and severe disease development, 3–4 weeks after first treatment a second application might be necessary.

Of the fungicides evaluated against stripe and leaf rust of wheat at different locations in North-Western United States in 1974 and 1975 triadimefon was most effective and improved yield most (Line, 1976).

Single sprays with triadimefon (at 0·56–2·32g/ha) ensured control of stripe rust on Kentucky bluegrass for 2–3 months (Hardison, 1975a).

In the case of slow stripe rust development in wheat, a single treatment of triadimefon at a concentration of 62·5 g a.i./ha showed activity, however, under severe epidemic disease progression higher fungicidal concentrations (125–250 g a.i./ha) were required (Siebert, 1976).

In recent studies several new fungicides have been evaluated for control of cereal rusts after foliar applications. Diclobutrazol (at 125 g a.i./ha) showed high protective and curative properties for control of stripe and leaf rust of wheat and stripe and brown rust of spring barley. Persistent action against rusts lasted for about six weeks (Bent and Skidmore, 1979; Zitzewitz and Heckele, 1979). Propiconazol (at 125 g a.i./ha) exhibited marked protective and curative activity against cereal rusts (Urech et al., 1979; Staub et al., 1979; Watkins and Doupnik, 1979). While propiconazol and triadimefon (each 125 g a.i./ha) showed a similar effectiveness against P. hordei, propiconazol exerted a somewhat longer persistence against P. striiformis (Urech et al., 1979).

To determine the duration of protective activity of triadimefon in greenhouse tests wheat seedlings (1–2 leaf stage) were sprayed with triadimefon (250 and 500 μg/ml) and inoculated at certain times after treatment. The treated leaves as well as new ones remained completely protected against mildew infections for 15 days. After this period, treated leaves showed little leaf rust infection but disease incidence on the new leaves did not differ from that of control plants (Scheinpflug et al., 1977; 1978). In order to evaluate the curative and eradicative activity of triadimefon, wheat seedlings infected with leaf rust and barley seedlings infected with powdery mildew were sprayed at different times (5, 10 and 15 days) after inoculation. Results showed that all treatments, independent of application time, inhibited further disease development. For example, treatments five days after inoculation prevented formation of mildew and leaf rust spores, application after ten and 15 days arrested further development of rust disease and a dark ring around the pustules was formed (Scheinflug et al., 1977; 1978).

Comparative studies of the field experiments on the persistence of triadimefon against mildew and the various cereal rusts indicated that the compound protected cereals against mildew between four and six weeks whereas the effectiveness against rusts began to decrease following three weeks (Frohberger, 1976; Scheinpfiug et al., 1977).

Foliar applications revealed that development of cereal mildews was completely prevented at $2·5\ \mu g/ml$ triadimefon. For controlling wheat leaf rust, a concentration of $50\ \mu g/ml$ was required (Frohberger, 1976).

Field observations have repeatedly shown different sensitivity of the various cereal rusts against triadimefon irrespective of the differential environmental demands required for development of each rust fungus. The sensitivity of the different rust fungi against triadimefon increased in the order: P. graminis ≥ P. recondita > P. striiformis (Frohberger, 1980).

Experiments on the movement of triadimefon after treatment of a leaf section of wheat or barley plants with (^{14}C)-triadimefon confirmed that the major portion of the fungicide penetrated (28–51%) was translocated acropetally and merely a small percentage (0·5–2%) moved basipetally (Führ et al., 1978; Buchenauer, 1979b).

Various representatives of this group of fungicides (e.g. triforine, triadimefon, nuarimol, phenapronil, diclobutrazol, propiconazol) and fenpropemorph exhibit broad spectrum activity and high potentiality for controlling various economically important foliar diseases of cereals. In cereal growing areas where various foliar diseases of varying intensities occur, year to year spray treatments with one of these fungicides can result in optimal control of diseases. Hagemeister and Neuhaus (1977) demonstrated that simultaneous control of leaf rust and mildew in barley with triforine was superior in its effectiveness to single applications of either carboxin or chinomethionate but somewhat inferior to the combination of the two latter fungicides. Furthermore, the marked curative action and the long-lasting effectiveness of these broad spectrum fungicides allows a somewhat wider margin in application timing and a reduction in number of treatments. On the other hand, the specific control of individual diseases requires careful observations of the development of each single disease and occasionally, several applications.

Triadimefon, fenarimol, nuarimol, phenapronil, diclobutrazol and propiconazol are rapidly taken up by roots and leaves of cereal plants and acropetally translocated with the transpiration stream. The fungitoxic agent in shoot extracts of barley and wheat plants which had been previously treated with the various fungicides were detected by bioautographic and autoradiographic methods. While the toxic agents in extracts of fenarimol, nuarimol, phenapronil, diclobutrazol and CGA 64250 treated plants showed chromatographic behaviour identical to the corresponding original compound triadimefon had been transformed into another compound with enhanced fungitoxic properties than triadimefon. Further studies with barley plants revealed that triadimefon was increasingly converted with extended exposure periods into the two diasteromers of triadimenol. After an exposure period of four to five days about 50% of triadimefon had been predominantly converted into the diastereomers of triadimenol. Of the two isomers, the threo-form proved to be superior in its fungitoxic and systemic activity against *Cladosporium cucumerinum* and *Erysiphe graminis* f. sp. *hordei* on barley plants. Triadimenol is highly persistent in plant tissues (Buchenauer, 1979a,b).

Because of the high mammalian toxicity of mercury fungicides, it is necessary to substitute the mercury derivatives by less toxic organic substances. Compounds with systemic properties are available which may substitute the mercury compounds.

Triadimenol, apart from controlling all smuts and *Pyrenophora avenae* showed marked effectiveness against mildews, *Typhula incarnata, Rhynchosporium secalis* and early stages of rust diseases. While all special forms of cereal powdery mildews proved to be highly sensitive to triadimefon, there are differences in the sensitivity within the various cereal rusts towards triadimenol; the order of sensitivity resembled that of triadimefon (Frohberger, 1978; 1980).

Experimental results indicated that the primary mode of action of triforine, triadimefon, triadimenol and nuarimol in sensitive fungi is through the interference with ergosterol biosynthesis. While the chemicals in *Ustilago maydis* and *U. avenae* sporidia severely inhibited ergosterol biosynthesis, C-4-methyl, C-4,4-dimethyl and C-14-methyl sterols accumulated. These results suggested that the chemicals selectively inhibited oxidative C-14-demethylation reaction during ergosterol biosynthesis (Sherald *et al.*, 1973; Buchenauer, 1975b, 1976b, 1977, 1979b). Preliminary results suggest that phenapromil, diclobutrazol, CGA 64250 and CGA 64251 also inhibit ergosterol biosynthesis in *U. avenae* sporidia. Ergosterol represents the major sterol component in many fungal species and is essential for the maintenance of structure and function of fungal membranes. Inhibition of ergosterol biosynthesis may result in changed structure of newly synthesized membranes and consequently in the inhibition of fungal development.

Moreover, with extended exposure periods, free fatty acids increasingly accumulated in treated sporidia. These high concentrations of free fatty acids that are toxic to fungal cells may sustain the fungitoxic action of this group of fungicides. Unfortunately, so far no informations are available on the biochemical mode of action of these fungicides in rust fungi.

No information is available yet regarding whether or not this group of rust active chemicals interfere with oxidative demethylation processes during sterol biosynthesis in rust fungi. It is known that the 4-desmethylsterol fraction of uredospores of rust species such as *P. graminis* f. sp. *tritici* and *P. striiformis* differ markedly from spores of fungi belonging to other groups. The major difference between rust spores and those of other fungi tested was the absence of ergosterol and the presence of C_{29}-sterols. The sterol compositon of uredospores of both *P. graminis* f. sp. *tritici* and *P. striiformis* were qualitatively and quantitatively similar. Of the 4-desmethyl sterol fraction the predominant component was identified as a Δ^7-C_{2g} sterol (stigmast-Δ^7-enol), the minor component was assumed to be its unsaturated isomer (Weete and Laseter, 1974).

Greenhouse and growth chamber studies showed significant side effects of triadimefon, triadimenol, fenarimol, nuarimol, propiconazol and diclobutrazol of various plant species. For instance, seed and soil treatments inhibited growth of roots, shoots and coleoptiles of barley and wheat seedlings (Buchenauer, 1975b; Buchenauer and Grossmann, 1977; Buchenauer and Röhner, 1977; Förster, 1979). These compounds also severely inhibited shoot growth of tomato and cotton plants (Buchenauer and Grossmann, 1977). Growth inhibition of plants induced by the fungicides was either completely or partly reversed by exogeneously applied gibberellic acid (Buchenauer and Grossmann, 1977; Buchenauer and Röhner, 1977). Growth retardation of the chemicals probably resulted from an interference by the fungicides with gibberellic acid biosynthesis. At higher concentrations of triadimefon, triadimenol and nuarimol phytosterole metabolism was also affected (Buchenauer and Röhner, 1977; Buchenauer, 1979b).

Fully grown primary leaves of triadimefon and triadimenol treated barley seedlings were shorter but wider and contained higher fresh and dry weights. Furthermore, in shoots of treated seedlings the quantities of free amino acids were increased and transpiration was reduced. While respiration was not affected, photosynthesis was diminished. Treatment also resulted in significant retardation of degradation of pigments and nucleic acids. In field studies, seed treatments with triadimenol (25–100 g/100 kg seed) did not affect leaf emergence and primary leaves of treated seedlings were somewhat larger than those of control plants. Further plant development was not influenced. Spray treatments with triadimefon (125–250 g a.i./ha) at GS 9 did not appreciably affect metabolic process of the flag leaves (Förster, 1979).

III. Biological Control

A. Hyperparasitism

Numerous fungi associated with spores in sori of the rust fungi affect the development of rust spores by varying degrees (Schröder and Hassebrauk, 1957). These fungi may be either hyperparasites or commensales of the rust fungi (Hassebrauk and Roebbelen, 1975). The elucidation of the role of secondary organisms and parasites has stimulated a great deal of interest in research.

Tuberculina maxima has been shown to be a common mycoparasite of various rust fungi on pines (Hubert, 1935). Hassebrauk (1936) reported experiments with parasitic fungi isolated from sori of cereal rusts, such as *P. graminis* f. sp. *tritici, P. striiformis, P. recondita* and *P. graminis* f. sp. *avenae*. The following parasitic organisms had been included (Hassebrauk, 1937): *Verticillium albo minimum* (A. et R. Sartory et Meyer) Westerdijk, *Verticillium compactiusculum* Sacc., *Verticillium malthousei* Ware, *Cephalosporium lefroyi* Horne.

Darluca filum (Biv. Bern. ex Fr.) Cast the imperfect stage of *Eudarluca caricis* is a ubiquitous hyperparasite associated with many species of rust fungi of cereals, e.g. *P. graminis, P. recondita, P. striiformis, P. coronata* and *P. sorghi* (Schroeder and Hassebrauk, 1957; Nicolas and Villanueva, 1965; Bean, 1968; Kranz, 1969) as well as with *Cronartium fusiforme* Hedgc. and Hunt ex Cumm. and *Cronartium strobilium* (Arthur) Hedge and Halm which cause important diseases on pines (Kuhlman and Matthews, 1976).

Cunningham (1967) reported that of the many hyperparasites of rust fungi in the tropics, *D. filum* was most commonly observed. Surveys on natural distribution of *D. filum* in Kenya and Guinea indicated that between 0–99% of uredosori of all rust species studied were infested by the hyperparasite and few pycnidia seemed to inhibit the development of teliospores (Kranz, 1969).

Among the spores of uredial sori, *D. filum* forms clumps of shiny black sphaerial pycnidia producing two celled conidia abundantly. The fungus may also parasitize pycnial, aecial and telial spores.

Light microscopic studies on hyperparasitism showed direct penetration of uredospores of *P. recondita* by the hyphae of *D. filum* without formation of adsorption structures (von Schroeder and Hassebrauk, 1957). Carling *et al.* (1976) reported ultrastructural studies of the parasitism of *P. graminis* by *D. filum*. They also confirmed that the mycoparasite directly penetrated walls of uredospores by possibly both mechanical and enzymic processes without forming any specialized penetration structures, such as appressoria or penetration pegs. After penetration cytoplasmic cell content was disorganized. Cytoplasmic constituents and cell wall material may serve as nutrients for the invading fungal hyperparasite.

Factors affecting growth and survival of the mycoparasites are not completely understood. *D. filum* grows on several natural substrates (Nicolas and Villanueva, 1965). Bean (1968) showed differences in nutrient requirements between mycelial and conidial isolates of *D. filum*. Since conidial isolates were predominantly associated with rust growth, conidial, rather than mycelial, isolates should be included in studies affecting growth, survival and pathogenicity.

The optimum temperature for conidia production was 30°C and conidia were still viable after five months of dry storage under outdoor temperature conditions (Bean and Rambo, 1968). Germination rate of conidia of *D. filum* was significantly stimulated in the presence of uredospores of *P. recondita* and *Uromyces phaseoli*, whereas spores of non-rust fungi did not affect germination. This enhanced germination and longevity of *D. filum* is probably caused by chemical compounds from the rust uredospores (Swendsrud and Calpouzos, 1970).

In artificial inoculation experiments Fedorintchik (1955) found that under favourable environmental conditions for *D. filum*, 98% of the leaf rust sori were damaged or destroyed. Besides leaf rust *D. filum* also parasitized, after artificial inoculation, *P. dispersa* on rye, *P. hordei* on barley, *P. graminis* and *P. coronata* on oats (Fedorintchik, 1939).

In greenhouse experiments for studying resistance against rusts, *D. filum* may often cause severe damages (Hassebrauk, 1936), but under field conditions epidemic rust developments frequently occur even in the presence of the hyperparasite.

Biological control of rust diseases by spraying cereals with *D. filum* spore suspensions before rust infection seems unlikely. The inoculation of *D. filum* spores three days prior to inoculation with rust spores resulted in even more rust infection than when both organisms were applied simultaneously, or when the rust fungus was applied first followed by *D. filum*. The percentage of uredia containing pycnidia of *D. filum* increased with prolonging mist period from three to 15 days (Swendsrud and Calpouzos, 1972). Similar results had been reported by von Schroeder and Hassebrauk (1957). Field experiments indicated that biological control of *P. recondita* by using *D. filum* was unsuitable under normal weather conditions in Northern Germany.

McKenzie and Hudson (1976) isolated, from pustules of teliospores of *P. graminis* f. sp. *tritici*, *Gonatobotrys simplex* (Corda) and *Verticillium lecanii* (Zimm.) Viegas. It was observed that *V. lecanii* grew almost exclusively on rust pustules.

Hassebrauk (1936) made the interesting observation that simultaneous inoculation of barley plants with spore suspensions of *P. hordei* and *Cephalosporium lefroyi* or oat plants with *P. coronata* and *C. lefroyi* resulted in no production of uredospores but of teliospores.

Biali *et al.* (1972) reported a similar phenomenon: teliospore production of various cereal rusts (*P. coronata, P. hordei, P. graminis* f. sp. *avenae* and *P. recondita* f. sp. *tritici*) was stimulated by the rust parasite *Aphanocladium album*, a fungus closely related to *C. lefroyi*. Since the epidemic development of cereal rusts is caused by uredospores, the authors considered *A. album* suitable for preventing rust epidemics by early induction of teliospore production.

The studies of Forrer (1976, 1977a,b) indicated that the induction of teliospores of cereal rusts could be due to metabolic products of *A. album*. The products were present in the culture filtrate after an incubation of the fungus for five days. After purification, the metabolic products induced precocious teliospore development in *P. graminis* f. sp. *tritici, P. dispersa* and *P. sorghi*. The fungi, however, differed in their sensitivity. To increase formation of teliospores after leaf application 0·24 μg/cm^2 of the purified extract were sufficient for *P. graminis* f. sp *tritici* while for *P. dispersa* and *P. sorghi* 0·94 μg/cm^2 leaf area were required. The active agents were also taken up by roots and translocated acropetally (Forrer, 1977b).

B. Induced Resistance

Plants can also be protected against diseases by applying procedures resembling those of immunization of vertebrates which forms the basis to prevent infectious diseases in animals (Kuc, 1966; Kuc and Caruso, 1977). Plants inoculated or treated with nonpathogens, non virulent races of pathogens, heat inactivated pathogens and high molecular substances of virulent agents, often show resistance to diseases caused by fungal pathogens.

In general, all plants have effective mechanisms for disease resistance and the degree of resistance may depend on the speed and extent of defense reactions. Yarwood (1956) demonstrated cross protection or induced resistance between rust fungi of different genera. When sunflower plants first inoculated with uredospores of the nonpathogenic fungus *Uromyces phaseoli* (= *U. appendiculatus*) were subsequently inoculated with *Puccinia helianthii* the number of pustules that subsequently developed was markedly reduced as compared with the number that developed on control plants. Similarly, inoculations of bean plants with the nonpathogenic fungus *P. helianthii* protected bean leaves from infection by *U. phaseoli*.

In greenhouse experiments Johnston and Huffman (1958) established a similar mechanism of induced resistance between the leaf rust fungus (*P. recondita* f. sp. *tritici*) and the oat crown rust organism (*P. coronata* var. *avenae*). Inoculation of wheat seedlings with uredospores of the crown rust fungus prior to inoculation with the pathogenic leaf rust resulted in a reduced number of pustules and in a different infection type. For instance, pustules were generally much smaller than those on plants exclusively inoculated with leaf rust. The induced resistance reaction was only localized in the tissue. The investigators assumed that the mechanisms was partly due to killing and plugging of stomata.

Littlefield (1969) reported induced resistance of flax to virulent races of *Melampsora lini* by prior inoculation of the plants with avirulent races of the rust fungus.

Preinoculation (three to six days) of wheat seedlings with avirulent races of *P. graminis* f. sp. *tritici* activated resistance mechanisms of the plant tissue to virulent stem rust races and number of pustules per leaf area was reduced by 80%.

Histological investigations revealed that the avirulent race was able to penetrate the leaves and one to two days after inoculation necrotic cells around the infection site were detected. As defense mechanism Cheung and Barber (1972) proposed that enzymes liberated during germination of the avirulent uredospores (Cheung and Barber, 1971) possibly would induce the synthesis of antifungal compounds responsible for resistance against the virulent rust race.

Induction of resistance in wheat seedlings was obtained when non-virulent and virulent strains of *P. striiformis* were applied almost simul-taneously (Johnson and Allen, 1975). Therefore, the authors postulated that by use of multiline varieties—consisting of a mixture of lines having similar phenotypic features but differing in resistance behaviour to rust races—would provide an environment to test the effect of induced resistance under field conditions. From the results of subsequent field trials Johnson and Taylor (1976) concluded that in some multiline cultivars induced resist-ance could be detected.

The mechanisms explaining the phenomenon of induced resistance have not been established yet. In general they can be based on activation of single or multiple defense reactions (Kuc and Caruso, 1977).

References

Abiko, K., Kishi, K. and Yoshioka, A. (1975). *Ann. Phytopathol. Soc. Japan* **41**, 100.
Andersen, A. S. (1960). The duration of effectiveness of activity of systemic fungi-

cides in wheat seedlings against stem rust. M.Sc. thesis, University of Minnesota, Minneapolis.

Bahadur, P., Sinha, V. C. and Upadhyaya, Y. M. (1974). *Indian Phytopathol.* **27**, 410–412.

Bailey, D. L. and Greaney, F. J. (1925). *Sci. Agr.* **6**, 113–117.

Bartlett, D. H. and Ballard, N. E. (1975). *Proc. 8th Br. Insectic. Fungic. Conf.* **1**, 205–211.

Barug, D. and Kerkenaar, A. (1979). *Meded. Fac. Landbouww. Rijksuniv. Gent* **44(1)**, 421–428.

Bean, G. A. (1968). *Phytopathology* **58**, 252–253.

Bean, G. A. and Rambo, G. W. (1968). *Phytopathology* **58**, 883.

Bent, K. J. and Skidmore, A. M. (1979). *Proc. 1979 Br. Crop Protect. Conf.—Pests and Diseases* **2**, 477–484.

Ben-Yephet, Y., Henis, Y. and Dinoor, A. (1974). *Phytopathology* **64**, 51–56.

Bergmann, H., Lyr, H., Kluge, E. and Ritter, G. (1975). *In* "Systemfungizide" (H. Lyr and C. Polter, eds). Akademie-Verlag, Berlin.

Biali, M., Dinoor, A., Eshed, N. and Kenneth, R. (1972). *Ann. Appl. Biol.* **72**, 37–42.

Bohnen, K. (1963). *Mitt. Biol. Bundesanst. Land-Forstw. (Berlin-Dahlem)* **108**, 125–129.

Bohnen, K. (1968). *Agric. Aviat.* **10**, 12–13.

Bohnen, K. and Pfinner, A. (1979). *Meded. Fac. Landbouww. Rijksuniv. Gent* **44(2)**, 487–499.

Bohnen, K., Siegle, H. and Löcher, F. (1979). *Proc. 1979 Br. Crop. Protect. Conf.—Pests and Diseases,* 541–548.

Briggs, P. W., Wright, R. H. and Hackett, A. M. (1974). *Pestic. Sci.* **5**, 599.

Brown, I. F. Jr., Taylor, H. M. and Hall, H. R. (1975). *Phytopathology* (Proc.) **2**, 31.

Buchenau, G. W. (1970). *Fungic. Nematicide Tests* **26**, 104.

Buchenauer, H. (1975a). *Pestic. Sci.* **6**, 525–535.

Buchenauer, H. (1975b). *Mitt. Biol. Bundesanst. Land-und Forstw. (Berlin-Dahlem)* **165**, 154–155.

Buchenauer, H. (1976a). *Pflanzenschutz-Nachrichten (Bayer)* **29**, 267–280.

Buchenauer, H. (1976b). *Pflanzenschutz-Nachrichten (Bayer)* **29**, 281–302.

Buchenauer, H. (1977). *Proc. 1977 Br. Crop. Protect. Conf.—Pests and Diseases,* Vol. III, 699–711.

Buchenauer, H. (1979a). *IX Intern. Congr. Plant Protect.,* Washington, DC, USA, August 5–11, Abstr. No. 939.

Buchenauer, H. (1979b). Untersuchungen zur Wirkungsweise und zum Verhalten verschiedener Fungizide in Pilzen und Kulturpflanzen. Inaugural dissertation, University of Bonn.

Buchenauer, H. and Grossmann, F. (1977). *Neth. J. Pl. Pathol.* **83**, 93–103.

Buchenauer, H. and Röhner, E. (1977). *Mitt. Biol. Bundesanst. Land-Forstw., Berlin-Dahlem* **178**, 158.

Bushnell, W. R. (1966). *Can. J. Bot.* **44**, 1485–1493.

Carley, H. E. (1979). *IX. Intern. Congr. Plant Protect.,* Washington, DC, USA, August 5–11, Abstr. No. 306.

Carling, D. E., Brown, M. F. and Millikan, D. F. (1976). *Phytopathology* **66**, 419–422.

Carlson, L. W. (1970). *Can. J. Pl. Sci.* **50**, 627–630.

Cheung, D. S. M. and Barber, H. N. (1971). *Arch. Microbiol.* **77**, 139–146.

Cheung, D. S. M. and Barber, H. N. (1972). *Trans. Br. Mycol. Soc.* **58**, 333–336.

Chin, W. T., Stone, G. M. and Smith, A. E. (1970a). *J. Agric. Food Chem.* **18**, 709.

Chin, W. T., Stone, G. M. and Smith, A. E. (1970b). *J. Agric. Food Chem.* **18**, 731.

Corbaz, R. (1962). *Agric. Romande, ser. A,* **1**, 44–45.

Cunningham, J. L. (1967). *Phytopathology* **57**, 645.

Dickson, J. G. (1959). *Bot. Rev.* **25**, 486–513.

Dimond, A. E., Heuberger, J. W. and Horsfall, J. G. (1943). *Phytopathology* **33**, 1095.

Döhler, R. and Merz, M. V. (1979). *Mitt. Biol. Bundesanst. Forstlandw., Berlin-Dahlem,* **191**, 185.

Domagk, G. (1935). *Angew. Chemie* **48**, 657.

Ebenebe, Ch., Fehrmann, H. and Grossmann, F. (1971). *Plant Dis. Repr.* **55**, 691–697.

Ebenebe, Ch., Fehrmann, H. (1974). *Z. PflKrankh. PflSchutz* **81**, 711–716.

Engst, R., Schnaak, W. and Lewerenz, H. J. (1971). *Z. Lebensm. Unters. Forsch.* **146**, 91.

Eriksson, J. and Henning, E. (1896). "Die Getreideroste, ihre Geschichte und Natur sowie Maßregeln gegen dieselben". Stockholm.

Fedorintchik, N. S. (1939). *Rev. Appl. Mycol.* **18**, 580–581.

Fedorintchik, N. S. (1955). *Rev. Appl. Mycol.* **34**, 141.

Fishbein, L. and Fawkes, J. (1965). *J. Chromatog.* **19**, 364.

Fisher, D. J. (1974). *Pestic. Sci.* **5**, 219.

Forrer, H. R. (1976). *Fourth European and Mediterranean Cereal Rusts Conference,* Interlaken (Switzerland), 5–10th September, p. 98.

Forrer, H. R. (1977a). *Phytopathol. Z.* **88**, 306–311.

Forrer, H. R. (1977b). Der Einfluß von Stoffwechselprodukten des Mycoparasiten *Aphanocladium album* auf die Sporenbildung von Getreiderostpilzen. Dissertation ETH Zürich.

Förster, H. (1979). Nebenwirkungen der systemischen Fungizide Triadimefon und Triadimenol auf den Stoffwechsel von Getreidepflanzen, insbesondere von Gerste. Dissertation, University Stuttgart-Hohenheim.

Forsyth, F. R. (1962). *Can. J. Bot.* **38**, 415–423.

Forsyth, F. R. and Peturson, B. (1958). *Can. J. Pl. Sci.* **38**, 173–180.

Forsyth, F. R. and Peturson, B. (1959). *Phytopathology* **49**, 1–3.

Forsyth, F. R. and Peturson, B. (1960). *Plant Dis. Repr.* **44**, 208–211.

Forsyth, W. (1802). *In* "A Treatise on the Culture and Management of Fruit Trees". Nichols, London.

Frohberger, P. E. (1973). *Mitt. Biol. Bundesanst. Land-Forstw., Berlin-Dahlem* **151**, 61–74.

Frohberger, P. E. (1976). Private communication.

Frohberger, P. E. (1977). *Mitt. Biol. Bundesanst. Land-Forstw., Berlin-Dahlem* **178**, 150–151.

Frohberger, P. E. (1978). *Pflanzenschutz-Nachrichten (Bayer)* **31**, 11–24.

Frohberger, P. E. (1980). Private communication.

Frost, A. J. P. (1975). *Cereal Rust Bull.* **3**.

Frost, A. J. P. and Brown, E. A. (1973). *Proc. 7th Br. Insectic. Fungic. Conf.* **1**, 105–110.

Frost, A. J. P., Jung, K. U. and Bedford, J. L. (1973). *Proc. 7th Br. Insectic. Fungic. Conf.* **1**, 111–118.

Frost, A. J. P. and Hampel, M. (1976). *Fourth European and Mediterranean Cereal Rusts Conference*, Interlaken (Switzerland), 5–10th September, 99–101.

Führ, F., Paul, V., Steffens, W. and Scheinpflug, H. (1978). *Pflanzenschutz-Nachrichten (Bayer)* **31**, 116–131.

Gassner, G. and Hassebrauk, K. (1936). *Phytopathol. Z.* **9**, 427–454.

Gassner, G. and Straib, W. (1936). *Phytopathol. Z.* **9**, 479–505.

Grewe, F. and Büchel, K. H. (1973). *Mitt. Biol. Bundesanst. Land-Forstw., Berlin-Dahlem* **151**, 208–209.

Gross, Y. and Kenneth, R. (1973). *Ann. Appl. Biol.* **73**, 307–318.

Gupta, I. L., Sharma, B. S. and Dalela, G. G. (1976). *Ind. J. Mycol. Pl. Pathol.* **5**, 115.

Hagborg, W. A. F. (1970). *Can. J. Pl. Sci.* **50**, 631–641.

Hagborg, W. A. F. (1971). *Can. J. Pl. Sci.* **51**, 239–241.

Hagborg, W. A. F. (1972). *Can. J. Pl. Sci.* **52**, 665–667.

Hagemeister, U. and Neuhaus, W. (1977). *Arch. Phytopath. PflSchutz, Berlin* **13**, 391–398.

Hampel, M., Löcher, F. and Saur, R. (1979). *Meded. Fac. Landbouww. Rijksuniv. Gent* **44(2)**, 511–518.

Hardison, J. R. (1966). *Plant Dis. Repr.* **50**, 624.

Hardison, J. R. (1971a). *Phytopathology* **61**, 731–735.

Hardison, J. R. (1971b). *Phytopathology* **61**, 1396–1399.

Hardison, J. R. (1975). *Plant Dis. Repr.* **59**, 652–655.

Hassebrauk, K. (1936). *Phytopathol. Z.* **9**, 513–516.

Hassebrauk, K. (1937). *Phytopathol. Z.* **10**, 465.

Hassebrauk, K. (1938a). *Phytopathol. Z.* **11**, 14–36.

Hassebrauk, K. (1938b). *Angew. Bot.* **20**, 366–373.

Hassebrauk, K. (1940). *Phytopathol. Z.* **12**, 509–510.

Hassebrauk, K. (1951). *Phytopathol. Z.* **17**, 384–400.

Hassebrauk, K. (1952). *Phytopathol. Z.* **19**, 56–78.

Hassebrauk, K. and Roebbelen, G. (1975). *Mitt. Biol. Bundesanst., Berlin-Dahlen* **164**, 183.

Heath, M. C. (1974). *Physiol. Pl. Pathol.* **4**, 403–414.

Heyland, K.-U. and Fröhling, J. (1977). *Z. PflKrankh. PflSchutz* **84**, 451–467.

Hoffman, I., Carson, R. B. and Forsyth, F. R. (1962). *J. Sci. Food Agr.* **13**, 423–425.

Hotson, H. H. (1952). *Phytopathology* **42**, 11.

Hotson, H. H. (1953). *Phytopathology* **43**, 659–662.

Hubert, H. (1935). *Phytopathology* **25**, 253–261.

Jensen, S. G. and Daly, J. M. (1960). *Phytopathology* **50**, 640–641.

Johnson, R. and Allen, D. J. (1975). *Ann. Appl. Biol.* **80**, 359–363.

Johnson, R. and Taylor, A. J. (1976). *Fourth European and Mediterranean Cereal Rusts Conference*, Interlaken (Switzerland), 5–10th September, 49–51.

Johnston, C. O. and Huffmann, M. D. (1958). *Phytopathology* **48**, 69–70.

Joshi, L. M., Renfro, B. L., Saari, E. E., Wilcoxson, R. D. and Raychaudhuri, S. P. (1970). *Plant Dis. Repr.* **54**, 391–394.

Judge, F. D., Hall, D. W. and Jackson, D. (1975). *Proc. 8th Br. Insectic. Fungic. Conf.* **3**, 967–974.

Jung, K. U. and Bedford, J. L. (1971). *Proc. 6th Br. Insectic. Fungic. Conf.* **1**, 75.

Kampe, W. (1975). *Mitt. Biol. Bundesanst. Land-Forstw., Berlin-Dahlem* **165**, 153.

Keil, H. L., Frohlich, H. P. and Van Hook, J. O. (1958a). *Phytopathology* **48**, 652–655.

Keil, H. L., Frohlich, H. P. and Glassick, C. E. (1958b). *Phytopathology* **48**, 690–694.
Kightlinger, C. V. (1925). *Phytopathology* **25**, 611–613.
Kightlinger, C. V. and Whetzel, H. H. (1926). *Phytopathology* **16**, 64.
Kolbe, W. (1976). *Pflanzenschutznachrichten (Bayer)* **29**, 310–335.
Kradel, J., Effland, H. and Pommer, E. H. (1969). *Gesunde Pflanze* **21**, 121.
Kranz, J. (1969). *Phytopathol. Z.* **65**, 43–53.
Kuc, J. (1966). *Ann. Rev. Microbiol.* **20**, 337–364.
Kuc, J. and Caruso, F. L. (1977). *In* "Host Plant Resistance to Pests", pp. 78–89. American Chemical Society, Washington.
Kucharek, T. A. (1977). *Plant Dis. Repr.* **61**, 71–75.
Kuhlmann, E. G. and Matthews, F. R. (1976). *Phytopathology* **66**, 1195–1197.
Lambert, E. B. and Stakman, E. C. (1926). *Phytopathology* **16**, 64.
Lambert, E. B. and Stakman, E. C. (1929). *Phytopathology* **19**, 631–643.
Lartand, G. and Lipatoff, V. (1976). *Meded. Fac. Landbouww. Rijksuniv. Gent* **41**, 677–685.
Line, R. F. (1976). *Fourth European and Mediterranean Cereal Rusts Conference,* Interlaken (Switzerland) 5–10th September, 105–108.
Line, R. F., Walder, J. T. and Hewitt, B. V. (1974). *Fungic. Nenaticide Tests* **30**, 105–106.
Littlefield, L. J. (1969). *Phytopathology* **59**, 1323–1328.
Livingston, J. E. (1953). *Phytopathology* **43**, 496–499.
Lyr, H., Ritter, G. and Casperson, G. (1972). *Z. Allgem. Mikrobiol.* **12**, 275.
Lyr, H., Schewe, T., Zanke, D. and Müller, W. (1975). *In* "Systemfungizide" (H. Lyr and C. Polter, eds), p. 153. Akademie-Verlag, Berlin.
Martin, T. J. and Morris, D. B. (1979). *Pflanzenschutz-Nachrichten (Bayer)* **32**, 31–82.
Mathre, D. E. (1971). *J. Agric. Food Chem.* **19**, 872.
Mathre, D. E. (1972). *Bull. Environ. Contam. Toxicol.* **8**, 311–316.
Mattern, P. J. and Livingston, J. E. (1955). *Cereal Chem.* **32**, 208–211.
Maykuhs, F. and Hoppe, H. (1976). *Nachrichtenbl. Deut. Pflanzenschutzd. (Braunschweig)* **28**, 40–42.
McCallan, S. E. A. (1956). *Phytopathology* **46**, 582.
McKenzie, E. H. C. and Hudson, H. J. (1976). *Trans. Br. Mycol. Soc.* **66**, 223–238.
Meyer, Von, W. C., Greenfield, S. A. and Seidel, M. C. (1970). *Science* **169**, 997–998.
Mitchell, J. W., Hotson, H. H. and Bell, F. H. (1950). *Phytopathology* **40**, 873.
Mundy, E. J. (1973). *Pl. Pathol.* **22**, 171–176.
Murdoch, C. L., Laemmlen, F. F. and Parvin, P. E. (1973). *Plant Dis. Repr.* **57**, 217–219.
Nelson, W. L. (1962). *Phytopathology* **52**, 746.
Newby, L. C. and Tweedy, B. G. (1973). *Phytopathology* **63**, 686–688.
Nicolas, G. and Villanueva, J. R. (1965). *Mycologia* **57**, 782–788.
Nuttall, M. and Mundy, E. J. (1976). *Exp. Husbandry* **30**, 95–112.
Oran, Y. K. and Parlak, Y. (1969). *Bitki Koruna Bült.* **9**, 87–98.
Peturson, B., Forsyth, F. R. and Lyon, C. B. (1958). *Phytopathology* **48**, 655–657.
Polter, C. and Casperson, G. (1975). *In* "Systemfungizide", pp. 225–240. Akademie-Verlag, Berlin.
Pommer, E. H. (1976). *Fourth European and Mediterranean Cereal Rusts Conference,* Interlaken (Switzerland) 5–10th September, 109–111.

Pommer, E. H. and Himmele, W. (1979). *Meded. Fac. Landbouww. Rijksuniv. Gent* **44(2)**, 499–510.

Pommer, E. H. and Kradel, J. (1969). *Proc. 5th Br. Insectic. Fungic. Conf.* **2**, 563.

Pommer, E. H. and Osieka, H. (1969). *Z. PflKrankh. PflSchutz* **76**, 33–35.

Pring, R. J. and Richmond, D. V. (1976). *Physiol. Pl. Pathol.* **8**, 155–162.

Ragsdale, N. N. and Sisler, H. D. (1970). *Phytopathology* **60**, 1422–1427.

Reed, H. E. and Chambers, A. Y. (1973). *Fungic. Nematicide Tests* **29**, 99.

Reschke, M. (1978). *DLG-Mitt.* **93**, 384–385.

Reyes, J. C., Moyer, J. L., Hausing, E. D. and Paulson, G. M. (1969). *Phytopathology* **59**, 1046.

Richmond, D. V. and Pring, R. J. (1977). *Neth. J. Pl. Path.* **83**, 403–410.

Rowell, J. B. (1964). *Phytopathology* **54**, 999–1008.

Rowell, J. B. (1967). *Plant Dis. Repr.* **51**, 336–339.

Rowell, J. B. (1968). *Ann. Rev. Phytopathol.* **6**, 243–262.

Rowell, J. B. (1969). *Fungic. Nematicide Tests* **25**, 98.

Rowell, J. B. (1973a). *Plant Dis. Repr.* **57**, 567–571.

Rowell, J. B. (1973b). *Plant Dis. Repr.* **57**, 653–657.

Rowell, J. B. (1976a). *Fourth European and Mediterranean Cereal Rusts Conference,* Interlaken (Switzerland), 5–10th September, 112–113.

Rowell, J. B. (1976b). *Phytopathology* **66**, 1129–1134.

Rowley, N. K., Wainwright, A. and Chipper, M. E. (1977). *Proc. 1977 Br. Crop. Protec. Conf.—Pests and Diseases* **1**, 17–24.

Scheinpflug, H., Paul, V. and Kraus, P. (1977). *Mitt. Biol. Bundesanstalt (Berlin-Dahlem)* **177**, 178–179.

Scheinpflug, H., Paul, V. and Kraus, P. (1978). *Pflanzenschutz-Nachrichten (Bayer)* **31**, 101–115.

Schewe, T., Müller, W., Lyr, H. and Zanke, D. (1979). *In* "Systemfungizide" (H. Lyr and C. Polter, eds), pp. 241–252. Akademie-Verlag, Berlin.

Schicke, P. and Veen, K. H. (1969). *Proc. 5th Br. Insectic. Fungic. Conf.* **2**, 569.

Schlüter, K. and Weltzien, H. C. (1971). *Meded. Fac. Landbouww. Rijksuniv. Gent* **36**, 1159.

Schmeling, von, B. and Kulka, M. (1966). *Science* **152**, 659–660.

Schroeder, von, H. and Hassebrauk, K. (1957). *Ztbl. Bakt. II* **110**, 675–696.

Sempio, C. (1936). *Riv. Patol. Vegetale* **21**, 201–278.

Sherald, J. L., Ragsdale, N. N. and Sisler, H. D. (1973). *Pestic. Sci.* **4**, 719–727.

Siebert, R. (1976). *Pflanzenschutz-Nachrichten (Bayer)* **29**, 303–309.

Simons, M. D. (1975). *Phytopathology* **65**, 388–392.

Singh, A. and Singh, S. L. (1975). *Plant Dis. Repr.* **59**, 743–747.

Singh-Verma, S. B. (1973). *Mitt. f.d. Landbau (BASF)* **6**, 1–45.

Snel, M. and Edgington, L. V. (1968). *Phytopathology* **58**, 1068.

Snel, M. and Edgington, L. V. (1969). *Phytopathology* **59**, 1050.

Snel, M. and Edgington, L. V. (1970). *Phytopathology* **60**, 1708.

Snel, M., Schmeling, von, B. and Edgington, L. V. (1970). *Phytopathology* **60**, 1164–1169.

Staub, Th., Schwinn, F. and Urech, P. (1979a). *IX. Intern. Congr. Plant Protection,* Washington, DC, USA, August 5–11, Abstr. 306.

Staub, Th., Schwinn, F. and Urech, P. (1979b). *Phytopathology* **69**, 1046.

Swendsrud, D. P. and Calponzos, L. (1970). *Phytopathology* **60**, 1445–1447.

Swendsrud, D. P. and Calponzos, L. (1972). *Phytopathology* **62**, 931–932.

Ten Haken, P. and Dunn, C. L. (1971). *Proc. 6th Br. Insectic. Fungic. Conf.* **2**, 453.

Tisdale, W. H. and Williams, I. (1934). *Chem. Abstr.* **28**, 6984.

Tweedy, B. G. (1969). *In* "Fungicides" (D. C. Torgeson, ed.), Vol. 2. Academic Press, New York and London.

Tyagi, P. D., Grover, R. K. and Ahuja, S. C. (1973). *Ann. Appl. Biol.* **75**, 387–391.

Tyagi, P. D., Grover, R. K. and Singh, M. (1974). *Z. PflKrankh. PflSchutz* **81**, 265–268.

Tyagi, P. D., Singh, M. and Parkash, V. (1979). *Z. PflKrankh. PflSchutz* **86**, 751–754.

Ulrich, J. T. and Mathre, D. E. (1972). *J. Bacteriol.* **110**, 628.

Urech, P. A., Schwinn, F. J., Speich, J. and Staub, Th. (1979). *Proc. 1979 Br. Crop. Protect. Conf.—Pests and Diseases* **2**, 508–515.

Van Tuyl, J. M. (1977). *Meded. Landbouwhogeschool Wageningen* **77(2)**, 136.

Vonk, J. W. and Sijpesteijn, A. K. (1970). *Ann. Appl. Biol.* **65**, 489.

Wäckers, R. W., van den Berge, C. and Spek, J. (1978). *Pflanzenschutz-Nachrichten* **31**, 151–162.

Watkins, J. E. and Doupnik, B. L. (1979). *Phytopathology* **69**, 544.

Watkins, J. E., Statler, G. D. and Littlefield, L. J. (1975). *Am. Phytopathol. Soc.* **2**, 61–62.

Watkins, J. E., Littlefield, L. J. and Statler, G. D. (1977). *Phytopathology* **67**, 985–989.

Weete, J. D. and Laseter, J. L. (1974). *Lipids* 575–581.

Weihing, J. L. (1969). *Am. Phytopathol. Soc.* **24**, 83.

White, G. A . and Thorn, G. D. (1975). *Pestic. Biochem. Physiol* **5**, 380–395.

Widdowson, F. V., Jenkyn, J. F. and Penny, A. (1976). *J. Agric. Sci.* **86**, 271–280.

Wolkoff, A. W., Onuska, F. I., Comba, M. E. and Larose, R. H. (1975). *Anal. Chem.* **47**, 754.

Yarwood, C. E. (1956). *Phytopathology* **46**, 540–544.

Zadoks, J. C. (1958). Gele-Roestber. Nr. 8, Nederl. Graan Centr.

Zitzewitz, von, W. and Heckele, K. H. (1979). *Mitt. Biol. Bundesanstalt Land-Forstw., Berlin-Dahlem,* **191**, 184.

Subject Index

A

Acrolein, *see* 2-Propenal
Actinomycin D, 166
Aecia, 8–10
Aeciospores, 8–12, 163
  definition, 3–4, 8
Agropyron scrabrum, 140
α-Amanitin, 169, 195
Amino acid
  excretion, 98–101
  leakage, 80, 101–102
  metabolism, 96–104
Amphispores, 14
Amylase, 64
Appressorium, 157, 224
Arabitol, 159
Aspergillus nidulans, 74
Aurintricarboxylic acid, 186
Autoecious life cycle, 2, 122
Autofluorescence, 227
Autoradiography
  of *in vitro* labelled proteins, 188–189
Axenic culture, 37–80
  definition, 38
  from mycelium in infected plants,
    52–57
  from spores, 39–52
  metabolism, 85–117
  variability in growth initiation,
    79–80
Axenic culture media, *see also* specific
  components
  for established cultures, 61–66
  for *Puccinia graminis*, 39–46, 61–66
  supplements to the basic medium,
    41–42

B

Barberry, *see Berberis vulgaris*
Barley, *see Hordeum vulgare*

Basidia, 19
Basidiospores, 19
  definition, 4
Berberis vulgaris, 137–138
BIM, *see* 3,3′-bis-Indolylmethane
Biogenic radiation, 28
Biological control of rust infection,
  270–273
Blackberry rust, *see Phragmidium
  violaceum*
Blasticidin S, 233, 238
Blister rust, *see Cronartium ribicola*
Breeding programs, 142–143
Brown leaf rust of wheat, *see Puccinia
  recondita*

C

Callose, 226, 229, 239, 241
Canadian strain, 68–72, 78
Carbohydrate levels
  in rust-infected plants, 105–106
Carbohydrate metabolism
  of axenic cultures, 105–116
  in infected plant cells, 206–217
  of spore germination, 158–161
Carbohydrate sources
  in culture of *Puccinia graminis*,
    40–41, 62–64
Carbon source
  for established cultures, 62
Carboxin, 252–257
Carboxylic acid anilides,
  carboxanilido derivatives of,
    257–259
  fungitoxic activity of, 252–259
  mechanism of action of, 255–257
  oxathiin derivatives of, 252–257
  thiazole derivatives of, 259
Carnation rust, *see Uromyces dianthi*

281